A Green Vision Towards a Renewable Energy Future

Suphi S. Oncel
Editor

A Green Vision Towards a Renewable Energy Future

Volume I: Challenges and Potentials

Springer

Editor
Suphi S. Oncel
Bioengineering Department
Ege University
Bornova, Izmir, Türkiye

ISBN 978-3-031-93759-0 ISBN 978-3-031-93760-6 (eBook)
https://doi.org/10.1007/978-3-031-93760-6

This Springer imprint is published by the registered company Springer Nature Switzerland AG
The registered company address is: Gewerbestrasse 11, 6330 Cham, Switzerland

To my dear family who makes this life valuable and E.A.-S.Y.

Preface

Today, facing the facts related to the climate change, the urge of our energy need poses a basic challenge to the survival on Earth. As we stand in the eye of an environmental crisis, it is a necessity to question our dependence on fossil fuels. The increase in greenhouse gases because of these fossil fuels pushes our planet's compensation capabilities resulting in severe weather oscillations, extreme pollution, and lost biodiversity, leading to destruction of health and economy.

As mentioned in recent studies, a strong response is needed to eliminate the anthropogenic effect on the environment. This should be started with the transition from fossil to renewable energy. The potential of clean and sustainable energy sources derived from the sun, wind, water, waste, and geothermal offers a promising pathway to not only mitigate climate change but also redefine our civilization.

The first volume of this book focuses on the renewable energy with a special emphasis on the realization without underestimating the pros and cons. Comprising the key points like economy and education, the volume aims to highlight the potential of renewable energy perspective to lead to a greener future.

Renewable energy can hold the key to transform the world, and having a high opinion of this topic will have a support for the change not only for today but also for tomorrow. The journey towards this ultimate goal leading to a greener future is not easy, but this book is an attempt to have a word in this challenge.

As the editor of this book, I must thank for the valuable contributions of the collaborating authors. I am glad to have their support to increase the impact of our words to reach a better future for our children.

Bornova, Izmir, Türkey Suphi S. Oncel

Contents

Renewable Energy Through Achievements and Challenges

Emek Aslan, Çağla Yarkent, Sait Sargın, and Suphi S. Oncel

1 Introduction

As energy is the main driver of our daily lives, it is considered indispensable for the sustainability of life (Li et al., 2022; Nassar et al., 2022; Sen & Ganguly, 2017). Energy promotes human development as well as increases economic growth and production (Amjith & Bavanish, 2022; Gasparatos et al., 2017; Pugalendhi et al., 2024). Economic development includes industrialization and urbanization, and population growth supports the increase in these developments (Fang et al., 2022). According to reports, the global population is anticipated to expand to 8 billion by 2030 and reach 9.31 billion in 2050, while the human population is expected to increase to 12.3 billion by 2100 (Kant Bhatia et al., 2021; Li et al., 2022; Worku et al., 2024). The collective effect rapidly raises the demand for energy as the population and economy continue to accelerate (Guo et al., 2022; Kim & Alameri, 2020). In 2014, the worldwide energy consumption of primary energy was almost 160.31 million MWh, and this amount will climb to 240.32 million MWh by 2040 with population growth, economic expansion, and technological advancements (Mentel et al., 2023). According to the US Energy Information Administration (EIA), the world's energy consumption would rise by nearly 50% between 2018 and 2050 (Guo et al., 2022). So, there is a critical call to find energy suppliers to sustain life on Earth.

E. Aslan
Agricultural Biotechnology Department, Ondokuz Mayıs University, Samsun, Turkiye

Ç. Yarkent · S. S. Oncel (✉)
Bioengineering Department, Ege University, İzmir, Turkiye
e-mail: suphi.oncel@ege.edu.tr

S. Sargın
Bioengineering Department, Bursa Technical University, Bursa, Turkiye

S. S. Oncel (ed.), *A Green Vision Towards a Renewable Energy Future*,
https://doi.org/10.1007/978-3-031-93760-6_1

Today, a huge part of energy demand is supplied by fossil fuels. Fossil fuels, including coal, natural gas, and oil, have been extensively used to produce energy and are expected to continue through at least 2050 (Worku et al., 2024; Yang et al., 2021). These nonrenewable energy sources enhance the ecological footprint (Li et al., 2022). The main terrifying consequences of using fossil fuels consist of resource depletion, climate change, and global warming (Chao et al., 2023). The existing fossil fuels are not actually limitless (Amjith & Bavanish, 2022; Betz et al., 2023; Maheshwari et al., 2022). It is expected that oil and gas may persist for 50 more years, and coal and uranium for around 100 more years (Mentel et al., 2023; Pugalendhi et al., 2024). The continued use of fossil fuels not only leads to the depletion of natural resources but also to the loss of biodiversity (Fang et al., 2022). Both have important effects on ecological balance as they boost climate change and carbon dioxide emissions (Aghabalayev & Ahmad, 2023). Therefore, there must be a balance between energy production and environmental sustainability (Pugalendhi et al., 2024).

The extensive and widespread use of fossil fuel derivatives has resulted in environmental pollution, and many health issues for humans, such as cardiovascular diseases, infectious diseases, and cancer (Esmaeili, 2022; Kim & Alameri, 2020; Romanello et al., 2023; Tomarov & Shipkov, 2023). For example, rising carbon emissions create serious air pollution, leading to tragic events. According to the World Health Organization, this situation results in nearly 4.2 million premature deaths per year (Magazzino et al., 2022). Furthermore, the WHO mentioned that climate change is currently responsible for over 150,000 deaths per year, with the figure anticipated to rise to almost 250,000 per year between 2030 and 2050 (Cheng et al., 2022). In 2019, the World Meteorological Organization warned that the continual rise of greenhouse gases (GHGs) has established a long-term trend and will cause greater harm to the climate in the future (Yang et al., 2022).

As a result of the accelerated intensity of global warming and climate change, brutal and frequent changes happen in the natural environment, posing problems for humans as well as other living creatures on Earth (Fang et al., 2022; Magazzino et al., 2022). Fossil fuels release GHGs into the atmosphere, including volatile compounds, carbon dioxide, nitrogen oxides, and solid particles, leading to global climate change. Furthermore, carbon-based fuels are estimated to meet 85% of the world's energy needs. The annual emission of carbon dioxide into the atmosphere to meet global energy demand is estimated at 36 billion tonnes. In the coming decades, approximately 90% of these emissions will be from fossil fuels (Alem et al., 2023; Worku et al., 2024; Yue et al., 2021). GHG emissions, particularly carbon dioxide emissions, are the main players in climate change, and their continual release causes global warming (Ampese et al., 2022; Cheng et al., 2022). The extensive GHG emissions have brought an increase in global temperature (Chen et al., 2022). Over the last decades, the Earth has warmed by more than 1 °C (Aghabalayev & Ahmad, 2023; Lorente et al., 2023). The United Nations estimates temperatures will increase by up to 1.5 °C by 2040 compared to the levels of 1900. As a result, global warming has a threatening effect on the natural balance of the Earth. For instance, it prompts an increase in the frequency of extreme weather events, unparalleled in history, such as torrential rainfall in certain regions of the world, and rising droughts and deserts in others (Lorente et al., 2023). These scary events can end up with the extinction of species, loss of biodiversity, floods, forest fires, ocean acidification, melting of glaciers, and sea level rise (Chen et al., 2022; Zhang et al., 2022). For example, global warming increases the level of seawater acidification, causing coral bleaching and the extinction of marine life (Magazzino et al., 2022). On the other hand, the increase in global temperature primarily affects individuals over the age of 65 and newborns under the age of one year. Today, they encounter twice the number of hot days that they would have experienced between 1986 and 2005. The number of heat-related deaths of adults (older than 65 years) increased by 85% compared with 1990–2000 (Romanello et al., 2023). As global GHG emissions are predicted to rise by 50% by 2050, it is expected that there will be an increase in temperature (Chen et al., 2022).

The future of the global energy system is profoundly unclear in all scenarios, and action must be taken for the survival of climate, human civilization, and other living creatures (Li et al., 2022; Raimi et al., 2024). According to the World Environment Protection Agency, it is important to develop alternative fuels with qualities similar to those of fossil fuels (Esmaeili, 2022). In this regard, the term "renewable energy sources" is considered as clean energy sources with significant potential to compete with fossil fuels, which can be called polluting energy sources (Lorente et al., 2023). In 2019, the EIA announced that renewable energy will be the leading source of primary energy consumption by 2050 (Capurso et al., 2022; Guo et al., 2022; Liang et al., 2022).

The interest in renewable energy was discussed based on the data obtained from the Web of Science and Espacenet by searching the keyword "renewable energy" between 2001 and 2024 (Fig. 1). A total of 297,179 scientific papers were published in the journals, and approximately 97% (287,889) of them were published between those years. This shows that the interest in this subject was increasing rapidly between these years. As Fig. 1a illustrates, the great momentum began in 2009 and has continued unabated. In the patent search, it was seen that a total of 288,562 studies were patented between 2001 and 2024. The distribution of these patents is shown in Fig. 1b. It is apparent that patented research on "renewable energy" technology/application has risen since 2009 and continues. One of the most noteworthy features is that the number of publications and patents follow similar trends. There is a link between both, and each stimulates interest in the other.

In order to underline the trend in studies on "renewable energy," the literature was analyzed bibliographically using VosViewer Software (Version 1.6.20). The keyword "renewable energy" was searched on the Web of Science, and the studies between 2020 and 2024 were listed based on their relevance. The bibliographic database of the most relevant 25,000 studies was obtained, and the co-occurrence of other keywords with "renewable energy" was investigated. The keywords were categorized as three clusters with 1456 items and 437,491 links. Figure 2a shows the common themes according to the interconnected relationship between the keywords. Each circle represents a keyword, and its size indicates the number of keywords involved in the studies. The links show the co-occurrence of the terms, with

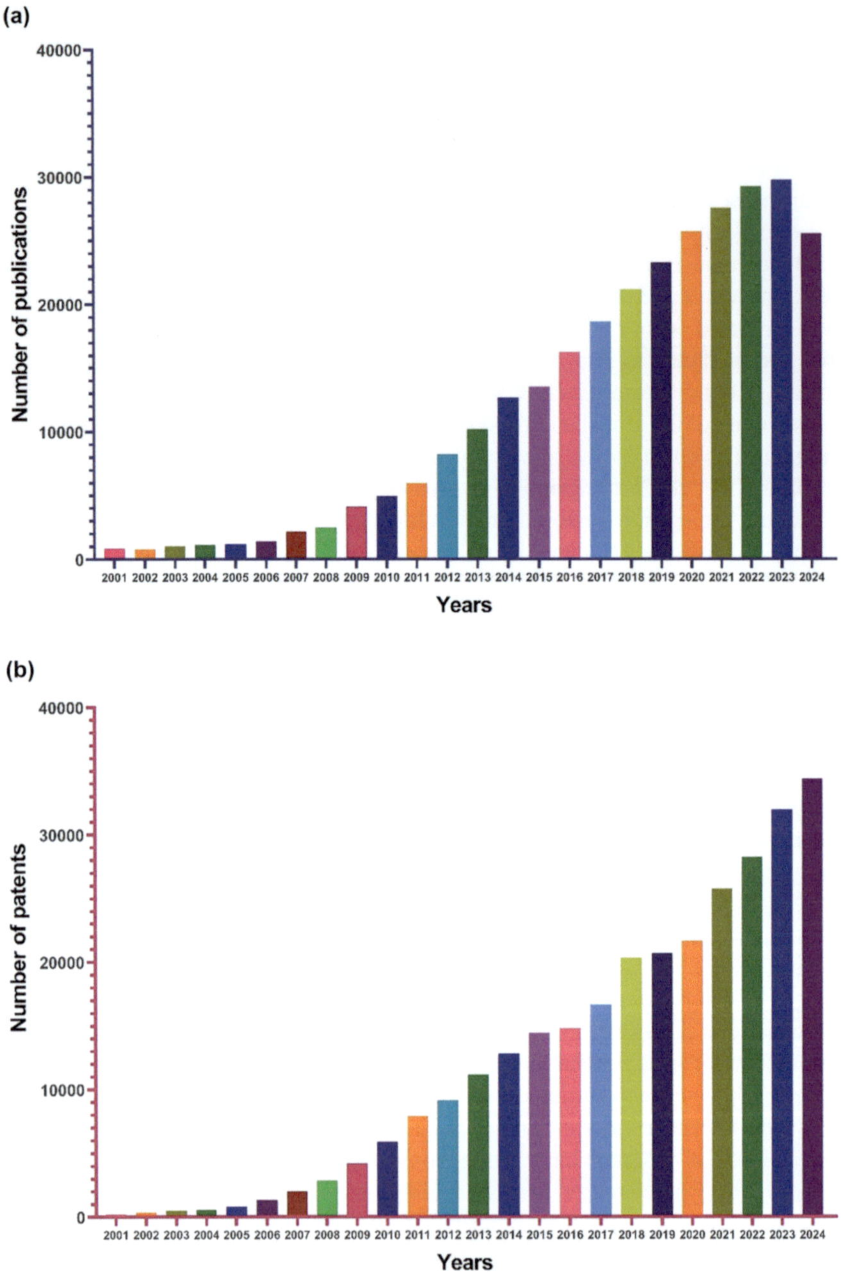

Fig. 1 The interest in "renewable energy" in scientific papers and patents. (**a**) According to the Web of Science database, and (**b**) Espacenet database, the number of studies per year (between 2001 and 2024) regarding the "renewable energy" keyword

thicker lines indicating a higher co-occurrence. This visualization technique divides the keywords into three clusters based on their interconnections. The first cluster is red and indicates the strong relationship between the studies on "renewable energy consumption," "country," "emission," "impact," "policy," "effect," and "economic growth." This pattern shows that renewable energy consumption has gained

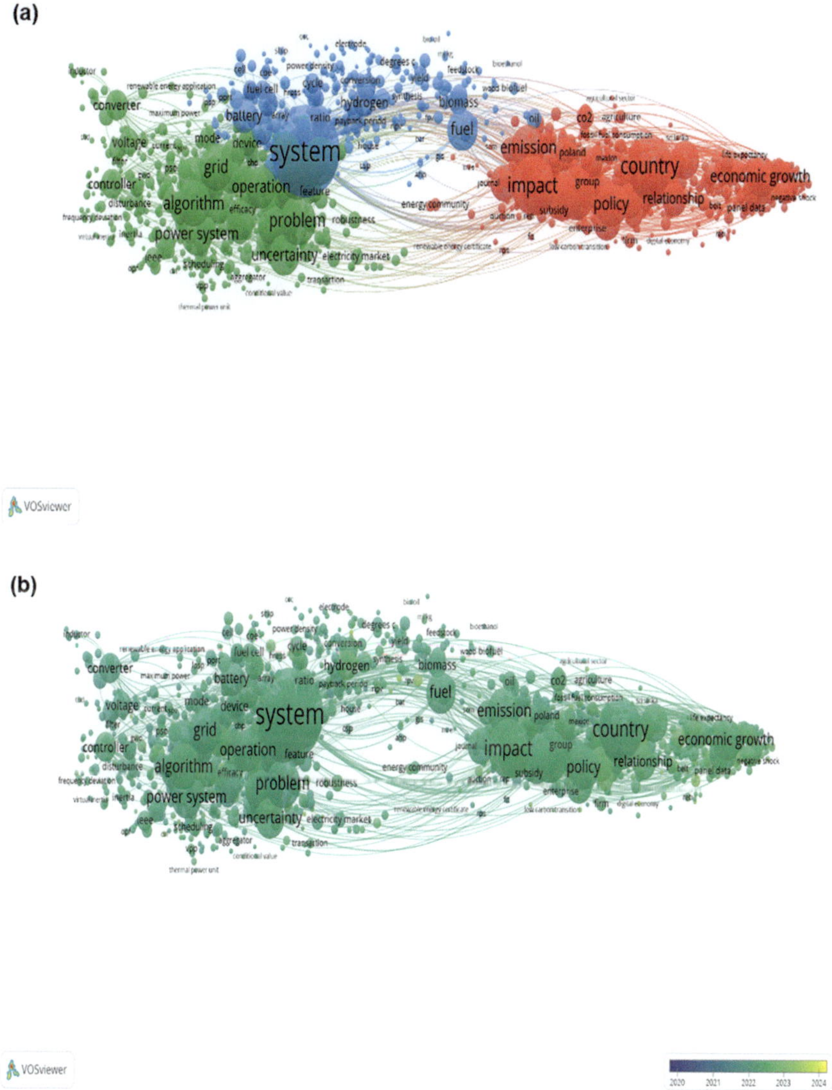

Fig. 2 (a) Network, (b) overlay, and (c) density visualizations of keywords co-occurrences and connections obtained via the VOSviewer software

(c)

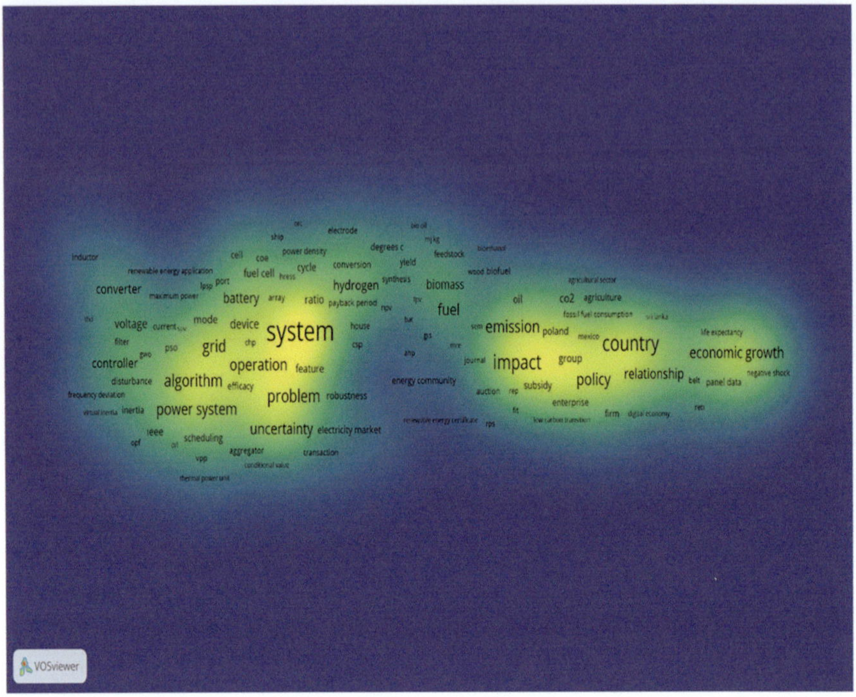

Fig. 1.2 (continued)

attention due to issues regarding GHG emissions, and their effects and impacts on countries and economic growth. In addition, policies introduced by countries or global organizations boost the studies on renewable energy consumption. The second cluster, green, demonstrates that the studies on "power systems" and strategies to increase the efficiency of these systems by investigating "algorithms", "grid" models, and "operation" technologies solve the "problems" and "uncertainty." This situation increases awareness of the problems in the energy field and results in the establishment of a strong connection with the studies on "renewable energy consumption" (in the red cluster). The third cluster is shown in blue color and behaves like a bridge between the green cluster and red clusters. As seen, the awareness of the problems and the focus on renewable energy consumption enhance interest in studies on "systems" and renewable energy sources, such as "hydrogen" and "biomass," to obtain "fuel" from them. Figure 2b shows the database evaluated to emphasize the trend in this topic from 2020 (dark blue) to 2024 (yellow). The overlay analysis demonstrated that the studies with the mentioned keywords have had their place since 2020, but the important point is that some novel applications, such as "green hydrogen," "green finance," and "digital economy," have gained more attention recently. Figure 2c shows that the studies on "renewable energy" have focused on topics such as "system", "power systems", "algorithm", "grid," "operation," "uncertainty" and "renewable energy consumption," "emission," "impact," "policy," "effect," "country," and "economic growth."

2 Renewable Energy Concept

Many global initiatives have identified the necessity for developing practical, cost-effective, and dependable sustainable energy alternatives with the potential to supplant most fossil-based sources. Climate Action, one of the United Nations' 17 sustainable development goals, comprises seven targets focused on reducing climate change and its effects on ecosystems. In addition to Green Chemistry, chemists have been provided with guiding principles for over 30 years to improve sustainability and reduce dependence on fossil fuels through renewable raw materials. Even with the focus on more sustainable alternatives, high cost, land use, and limited production scale impede the development of practical options for non-fossil-based sources (Anastas & Warner, 2000; Burrington, 2024). Implementing such practical renewable options should also consider the suitability of renewable fuels for the end use for which it replaces a fossil-based fuel. Failure to match the fuel to the application risks significant infrastructure changes and associated costs, introduction delays, and limited market penetration. In this context, electrification offers the advantage of providing a widespread and established energy platform. It has inherent efficiency advantages over combustion-based alternatives in transportation and other applications where highly efficient electric motors offer an alternative (Burrington, 2024; Ganesh et al., 2021; Takach et al., 2022).

The most significant challenges are food waste, agrochemicals, mixed plastics, pharmaceuticals, municipal solid waste, and animal waste. Lignocellulosic feedstocks are recognized as the most suitable source for renewable energy production due to their abundance, affordability, and accessibility (Singh et al., 2022; Osman et al., 2023; Worku et al., 2024).

The energy industry must assume a significant role in reducing emissions. Nevertheless, utilizing low-carbon fuels like hydrogen is also a significant consideration. To achieve this, clean energy technology must be rapidly adopted. A significant research program is dedicated to developing innovative technologies for renewable energy sources to replace fossil fuels (Khaligh et al., 2023; Pareek et al., 2019).

An increasing number of countries have enacted legislation encouraging investment in hydrogen technologies. Hydrogen is the most abundant and simplest element in the Earth's crust. It is a chemical element that readily forms compounds with other elements, making it ubiquitous in various substances, including water, hydrocarbons, and alcohol. Also, hydrogen can be obtained from geothermal, hydroelectric, solar, and wind sources (Peng et al., 2023; Verma et al., 2024). As a viable energy carrier and a vital component of energy security, hydrogen's versatility in domestic energy sources is a significant advantage. Producing hydrogen with different materials and technological methods is a highly favorable approach. Many techniques can produce hydrogen, including electrolytic, thermal, and photolytic processes. However, the long-term sustainability of hydrogen depends on the energy efficiency of its production (Jeje et al., 2024; Le et al., 2024; Worku et al., 2024).

In light of the information mentioned above, there is an apparent necessity to transform the concept of renewable energy into a more accessible and straightforward one. Upon analysis of recent research, books, and compilation publications, it becomes evident that there are specific sectoral categories within the field of renewable energy. Especially, the energy demand of today's world is met by six energy production systems: hydropower, wind energy, solar energy, bioenergy, geothermal energy, and marine energy (Fig. 3). Since 2018, a huge part of renewable energy has been obtained from hydropower plants. The amount of energy produced by wind power plants increased in 2022, and it is predicted that it will continue to grow at an accelerated rate. Solar energy ranks third in terms of renewable energy output. While it provided a similar contribution to renewable energy production as bioenergy in 2018, its contribution increased in 2022 and is expected to continue this way. The importance of bioenergy in renewable energy production has been evident since

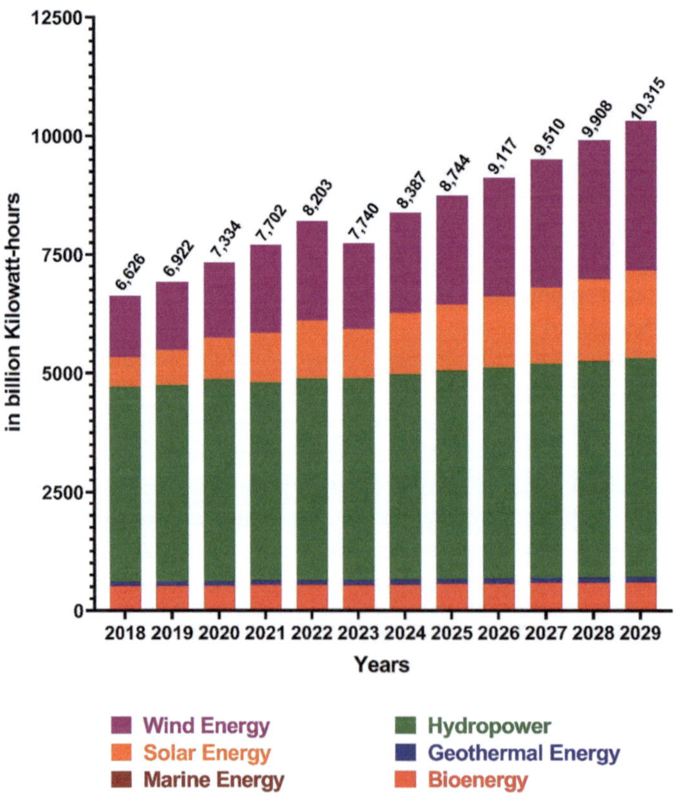

Fig. 3 The worldwide renewable energy production in billion kilowatt-hours (The data is obtained from Statista Market Insights.)

2018. However, it is also noteworthy that this situation has not changed effectively over the years. Geothermal energy and marine energy production need new developments to increase their efficiency to provide a greater contribution to renewable energy applications. Overall, it is supposed that the total amount of renewable energy output will rise from 6,626 billion kWh to 10,315 billion kWh between 2018 and 2029. This indicates that renewable energy production will increase by approximately 56% in 2029 compared to 2018.

Titles such as Solar, Hydropower, Geothermal, Wind, Biomass, and Marine Energy encapsulate the core concept of renewable energy (Fig. 4). A conceptual approach to renewable energy can make the subject more comprehensible for those interested in the area and those less familiar with it. Considering the point that reached globally regarding renewable energy, it is necessary to bring the concepts under the roof of a concept for local governments, people working in this field in companies, and the public. In this regard, the "Renewable Energy Concept" definition is considerable in this respect.

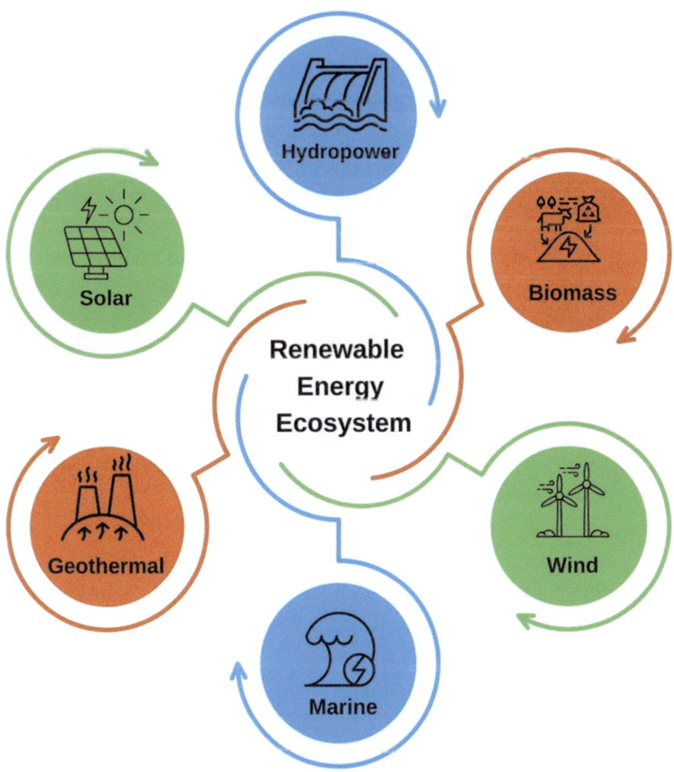

Fig. 4 Renewable energy ecosystem

2.1 Hydropower Energy

Hydropower plays a significant role in global electricity generation landscape due to its lack of associated GHG emissions and its ability to leverage natural water for energy storage. Hydropower is the largest renewable energy source, accounting for over 18% of global energy consumption. Hydropower uses a dam or other means to capture and harness the kinetic energy of water flow, typically from a river or stream, to generate electricity. The evolution of hydropower has been a gradual process throughout history. The earliest known applications of this technology can be traced back to ancient Rome, China, and Afghanistan (Wasti et al., 2022; Xu et al., 2023). Large-scale developments in hydropower emerged during the sixteenth century, and by the end of the twentieth century, it had become a widely adopted energy source. In recent decades, hydropower has again become a subject of considerable interest as countries and energy producers seek sustainable solutions to ensure energy security and address concerns about climate change. As a clean alternative to fossil fuels and a significant source of GHG emissions, hydropower is also a field of interest and investment for green banks and carbon funds (Gyamfi et al., 2020; Tefera & Kasiviswanathan, 2022).

Production, transmission, and energy consumption invariably require water. Conversely, energy is a prerequisite for irrigation, refinement, and water movement. The interrelationship between water and energy represents a fundamental aspect of our modern world, with the specific dynamics of this relationship varying based on geographical location or technological foundation (Zakariazadeh et al., 2024). The relationship between water and energy can be broadly categorized into two distinct areas: the energy required for producing water and the water needed for generating energy. These interrelationships and linkages are collectively called the "energy–water nexus" (Fig. 5). Regarding the water requirements of the power generation sector, it is notable that a considerable proportion of water is utilized for cooling towers of coal- or gas-fired thermal power plants. This is particularly evident in the need for further research that addresses the multifactorial influence of supplying a water system with renewable energy sources. In order to achieve a more comprehensive assessment, it would be beneficial to consider several factors, including

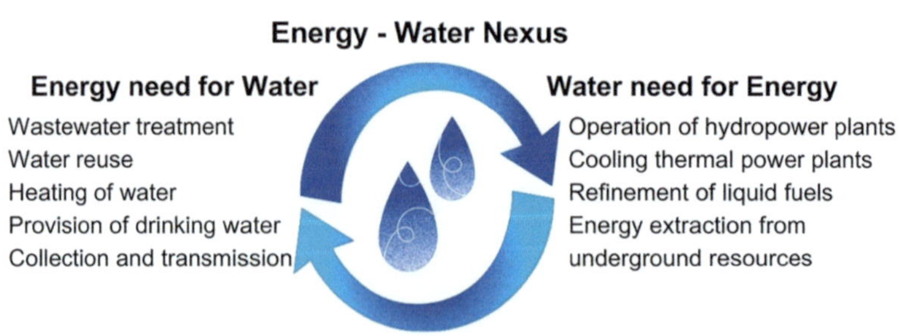

Fig. 5 Energy–water nexus (Modified from Zakariazadeh et al., 2024)

climate, region, type of renewable energy, and demand patterns (Fayiah et al., 2020; Hamiche et al., 2016; Jin et al., 2019; Zakariazadeh et al., 2024).

2.2 *Wind Energy*

Wind energy is one of the fundamental sources of renewable energies. It can be pivotal in achieving energy independence, particularly in regions with abundant and consistent wind energy sources. The contemporary concept of wind energy is converting kinetic energy from moving air, which is used to drive the wind turbine rotor. The conversion process results in the generation of mechanical power, which is subsequently transformed into electrical energy. The energy from the wind resource can be utilized in various ways, including pumping water, wind-assisted ignition, or as an air mover (Roga et al., 2022; Vanegas Cantarero, 2020). Wind is a free and inexhaustible energy source, generating electricity at the point of use, obviating the need for fuel transportation. Even under intermittent conditions, the wind is rarely still; sunlight is typically available when the wind is not blowing. Furthermore, wind energy has a relatively low environmental impact compared to other energy sources. It requires only a tiny land area, and the local community can share the benefits (Baxter et al., 2020; Amjith & Bavanish, 2022).

Wind energy utilization has constituted a fundamental aspect of renewable energy sources. Its disadvantages are susceptibility to high fluctuations and geographical dependence. Economical wind energy storage technologies must be available for this energy source to be applied to distributed generation needs. It is reported that wind has been used for a considerable length of time, dating back thousands of years, as a source of energy for ships, for driving pumps in simple windmills, and for irrigation, thus leading to the invention of water pumps (Baxter et al., 2020; Breyer et al., 2022). The most significant application of wind energy is electricity generation, as the world strives to generate more electricity from renewable sources to reduce carbon and GHG emissions (Letcher, 2023). Large intermittency and geographical dependency are among the disadvantages of wind energy. To use this energy source for distributed generation, economical technologies must be made available to store wind energy. The wind was used to power boats, to drive pumps in simple windmills, and to irrigate the land with water pumps. Today, as the world seeks to generate more electricity from renewable sources to reduce carbon and GHG emissions, its most significant use is in power generation (Letcher, 2023).

Significant advancements in wind turbine technology have made over the past decade. Modern wind turbines can be divided into two main categories: horizontal-axis turbines, which are typically employed in large-scale commercial settings, and vertical turbines, which are utilized in smaller-scale and urban applications. Wind energy is the kinetic energy of moving air. When moving air hits a turbine, it slows down, and its kinetic energy is converted into the mechanical energy of the rotating turbine blade. This energy is then potentially converted into electrical energy through a generator (Li et al., 2020; Wang et al., 2020; Pan et al., 2021). The speed of the moving fluid is a determining factor in the force and power generated. In the most straightforward fluid power theory, the moving fluid slows to a standstill upon impact and delivers 100% of the energy to the turbine. The advancements of turbine design, manufacturing, materials, aerodynamics, and control have enabled wind energy to become one of the principal sources of clean, renewable, cost-effective, and sustainable energy (Vermillion et al., 2021; Tan et al., 2022).

Advances in wind turbine manufacturing that seek to optimize wind capture include using newer composite materials and more aerodynamically efficient designs. Other relevant technologies include connecting power generation to the smart grid through a direct-connected drive system and developing energy storage systems such as batteries. Upon completion of construction, these systems will facilitate the transformation of wind into a competitive baseload power source, ensuring the uninterrupted provision of power regardless of wind conditions (Santhosh et al., 2020; Zhang et al., 2020; Jaiswal et al., 2022).

The development and utilization of wind energy offer numerous benefits. The environmental advantages are most frequently cited as the primary rationale for this development. Wind energy generation is a nonpolluting process that does not result in the emission of GHGs, commonly associated with the operation of fossil fuel-based power generation facilities. Furthermore, wind farms have minimal water consumption (Varma et al., 2020). Consequently, wind energy can reduce atmospheric pollution. Furthermore, wind energy has the potential to stimulate economic development in enterprise zones, where numerous companies are engaged in the manufacturing, installation, and maintenance of wind turbines (Deng et al., 2020; Sharma et al., 2022).

In recent years, the cost of wind energy has fallen significantly. Further cost reductions are anticipated as technology advances and the number of turbines installed increases. This has made wind energy a more affordable option and has led to the development of wind farms and other incentives. Additionally, generating more wind energy in regions with high winds will reduce the cost of electricity generated from these regions. Wind farms are increasing, and technology is improving, making it possible to utilize less windy areas as well (Hoicka et al., 2021; Roga et al., 2022).

2.3 Solar Energy

Solar energy represents one of the most abundant forms of renewable energy, prompting many to speculate that it could serve as a potential solution to the global energy crisis. The global trend is toward reducing the use of fossil fuels as the detrimental impact on the environment becomes increasingly apparent. Solar energy has the potential to play a significant role in this transition, particularly with the advent of innovative techniques for harvesting this energy in more efficient and diverse ways. As the most accessible and practical form of renewable energy, solar energy can make a substantial contribution to alleviating the energy crisis and addressing environmental concerns such as global warming (Obaideen et al., 2021; Ang et al., 2022).

Solar energy is used on a large scale worldwide because it can be converted into other forms of energy, such as electricity, heat, and hydrogen. Centralized power plants using concentrating solar collectors or smaller distributed systems using panel collectors are used on a large scale. Photocatalysis, using the UV portion of the solar spectrum, is an efficient and economical way of reducing and eliminating elevated levels of contaminants in water, primarily by breaking down organic contaminants in the presence of a catalytic converter (Pfeifroth et al., 2024). Small-scale solar water heaters are often used in developing countries, particularly in areas with high solar radiation levels (Reichle, 2023). In addition to heating water, it is possible for residential and commercial buildings to harvest solar energy for space cooling (Maka & Alabid, 2022).

Solar energy offers a promising avenue for agricultural applications, including irrigation and electricity generation. Additionally, solar drying technologies provide a viable means for producing alternative, safe, and environmentally conscious commodities (Kumar et al., 2023). The potential applications of solar drying extend to the dehydration of a diverse range of materials, including fruits, spices, vegetables, medicinal plants, aromatic flowers, prunes, milk, and various food products. Furthermore, solar drying can be employed to sterilize milk and reduce dye through solar heating. Additionally, solar energy can be harnessed to disinfect or stabilize foodstuffs (Elahi et al., 2022; Majeed et al., 2023; Rahman et al., 2022).

Solar energy offers several advantages, the most significant of which are its beneficial environmental effects. These include a reduction in the emissions of GHGs and the pollutants produced by the combustion of fossil fuels. Furthermore, solar energy plays a significant role in mitigating climate change. Concerning energy independence and economic stability, solar photovoltaics has created several million new employment opportunities in renewable energy (Balali & Stegen, 2021; Ziegler & Trancik, 2021).

The high initial capital costs involved in solar installations, the intermittency of sunlight, and the general lack of awareness and experience in the field of solar energy are among the main challenges currently facing the solar energy industry. Additional challenges pertain to the required area for solar panels, aesthetic considerations, recycling and waste collection, and resource availability. Therefore, there is a continuous and urgent need to stimulate research and investment in solar technology systems, especially photovoltaic systems. Scalable photovoltaic technologies enable distributed energy generation solutions (Kumar et al., 2023; Nastasi et al., 2022; Verduci et al., 2022).

2.4 *Biomass Energy*

Biomass energy represents a renewable energy source that can reduce the use of fossil fuels and the environmental impact of energy. The cultivation of biomass has the potential to enhance the land's structural value and protect the environment. The effective use of biomass energy could have significant effects as a renewable energy resource. The major advantages of this approach include reducing greenhouse gas emissions, increasing energy security, creating new job opportunities in several sectors, and increasing revenues in the community (Alper et al., 2020). Biomass is derived from various organic materials, including plant matter, animal waste, municipal waste, manure, field crops, trees, forest plants, and residue from the agricultural and forestry industries (Kalak, 2023; Siwal et al., 2021). The sustainable use of biomass energy may be feasible on marginal lands. However, accurately identifying the available biomass resources is a fundamental requirement for developing a suitable biomass production system that can serve as an economically and ecologically viable alternative. The characteristics of biomass energy are also critical, including its availability, potential contribution to the global carbon dioxide balance, and the environmental impacts associated with its use, considering the benefits and negative aspects that its deployment can generate (Priyadharsini et al., 2023; Saleem, 2022).

While the definition of biomass encompasses many different types of materials, the subject focuses on scrap, waste, and biodegradable materials regarding energy. When these materials are used to create usable energy, the term "biomass" is applied. The source of the biomass can be crucial to fuel quality. The structure of the plant and the amount of material available can benefit the feedstock. In addition, technologies can be developed to convert these short, complex structures into new critical industrial yeast and animal feed storage formula products (Siwal et al., 2021; Szyba & Mikulik, 2022).

The biomass types utilized for energy production encompass urban waste, including organic matter accumulated primarily in municipalities and organic waste derived from residential, industrial, agricultural, and commercial solid waste. Additionally, medical waste may be included (Kalak, 2023). The term "wood" encompasses a range of materials, including wood chips, sawdust, shells, pulp and paper, wood residues and scraps, coal, and other residues derived from cellulose. It also includes the direct burning of bark and the residuals from chemical pulping. The diverse range of vegetable and animal sources is also crucial in renewable energy, as they represent the primary sources of the generated goods. These goods can provide renewable resources that account for more than 60% of global electricity output based on biomass material (Abdulyekeen et al., 2021; Esmaeili, 2022).

Bioethanol is emerging as a prominent biofuel alternative, ranking alongside biodiesel and biohydrogen in terms of significance. Its manufacturing process is relatively uncomplicated, and the associated raw materials are widely accessible, contributing to its growing prominence (Alfonsín et al., 2019; Packiyadhas & Shanmuganantham Selvanantham, 2020). The significant economic incentive for using bioethanol as a substitute for gasoline has prompted the development of biofuels in various countries, including Brazil, Japan, United States, China, and some European countries (Tazikeh et al., 2022). The production of biofuels from sustainable natural resources and green waste presents the dual benefit of reducing carbon emissions and reducing dependence on oil. India provides a notable example, with more than 80% of rural energy consumption derived from biomass, including firewood, manure, agricultural waste, and various organic sources (Uría-Martínez et al., 2018; Neto & Gallo, 2021; Priyadharsini et al., 2023).

As another bioenergy source, algae have great potential for sustainable energy production. The potential of macroalgae biomass as a clean, sustainable, and renewable biofuel source on a global scale has prompted researchers to focus their investigations on the development of algae-based bioethanol technology to compete with or even replace fossil fuels (P. Priyadharsini et al., 2023). The success of genetically modified algae hinges on their ability to meet these conditions and demonstrate efficient translation of heterologous genes from multiple other algae species' nuclear genomes. Genetically modified algae used for fuel production should demonstrate the capacity to withstand temperature, light, and salt fluctuations and tolerate diverse geographical locations and circumstances (Surendhiran & Sirajunnisa, 2018). The latest innovations in genetic engineering and molecular biology have enabled the identification and analysis of novel biochemical pathways in various model algae species. Furthermore, genome editing methods have been developed, which can be used to produce new strains with enhanced biofuel production capabilities (Ambaye et al., 2021). Furthermore, integrating metabolic engineering and computational methodologies can facilitate the development of improved strains with enhanced lipid production and affordable downstream processing, which may ultimately lead to the creation of algal biofuels (Godbole et al., 2021; Priyadharsini et al., 2022).

2.5 *Geothermal Energy*

Geothermal energy is a renewable option with relatively low emissions compared to fossil fuels, a smaller environmental footprint, favorable resource availability, and the potential to generate base-load electricity compared to other renewable energy sources such as solar, wind, and biomass (Ball, 2021). Compared to existing energy sources, renewable energy offers essential advantages, such as maximum fuel lifetime, low environmental impact, and marginal operating costs (Ang et al., 2022; Umar et al., 2024). Geothermal energy is one of the most promising and innovative renewable sources with financial and environmental benefits. Geothermal energy plays a unique role in the energy portfolio. It is unique in terms of capability, technology, and resources. It provides clean and reliable energy as a sustainable, clean source that can provide electricity and heat (Baba & Chandrasekharam, 2022; Benti et al., 2023).

Geothermal energy, as the name suggests, refers to the heat extracted from the Earth's core. It uses the naturally occurring heat inside the Earth to provide a reliable and abundant renewable energy source. This energy source has been used for centuries, with modern usage beginning in the nineteenth century. Taking advantage of the Earth's vast underground and underwater reservoirs makes tapping into this resource possible, which replenishes itself relatively quickly. The geothermal approach opens up more sustainable and environmentally friendly energy production opportunities (Igwe, 2021; Nikolaevich et al., 2021; Dhaidan et al., 2024).

Geothermal resources are classified according to their temperature and can be categorized as high- or low-temperature. Shallow ground can extract geothermal heat for efficient heating systems. This abundant resource can be used in various fields such as power generation, heating, drying process, agricultural activities, and industrial processing (Lund & Toth, 2021; Jiansheng et al., 2022; Sharmin et al., 2023).

The resource base for geothermal power plants depends on the availability of hot fluids or rocks heated from below and, in principle, covers almost the entire Earth. In practice, geographical constraints, such as the absence of tectonic plate boundaries in most parts of the world, further limit the areas with favorable geothermal energy production (Ball, 2021). Low- to medium-temperature thermal waters are the most abundant geothermal resources but are less productive compared to high-temperature sources. Low-temperature geothermal resources can be a relevant application for space heating, generally of lower value than direct electricity sales (Igwe, 2021; Islam et al., 2022).

Geothermal power plants recover heat from geothermal reservoirs at depths of 1–10 km and generally at temperatures above 250 °C. There are different technological configurations to convert geothermal energy into electricity and to ensure efficient utilization of the extracted fluid to minimize the environmental impacts of using this renewable energy source. While the initial power generation efficiency of these geothermal power plants was extremely low (only 3–5%), various technological advances have been made to improve the efficiency of the cycle, often with considerations to reduce the emitted pollutants such as sulfuric acid and particulate matter (Hackstein & Madlener, 2021). Geothermal power plants can be divided into dry steam, flash steam, and binary cycle, depending on the mechanism used to generate electricity. Flash steam and binary cycle plants have a minimal environmental impact and are now being built in large numbers. Along with these, some challenges exist in developing and implementing geothermal power plants. Geothermal resources are often located in remote and pristine areas, and the long-term effects of microseismic events need to be assessed. In addition, new cooling technologies and zero-emission solutions using lightweight, low-cost materials are needed to support power generation (Barasa Kabeyi & Olanrewaju, 2022; Zuffi et al., 2022).

As geothermal technology continues to evolve, many innovative processes are being used to exploit geothermal resources. Technological advances have been made in drilling, particularly in drilling fluids that can withstand high temperatures. Other areas include innovative heat exchange systems, use of nanoparticles in thermal fluids and turbines, and modular designs for smaller geothermal systems

(Mahmoud et al., 2021). Research and development continue in the above areas and various countries worldwide to make the process more efficient and reduce costs (Ma et al., 2023; Puppala et al., 2022; Soltani et al., 2021).

The use of binary cycle systems is recommended for source temperatures below 150 °C to minimize the loss of volatile components and address the issues of scaling and corrosion (Yilmaz et al., 2023). In the United States, advanced geothermal systems are being developed using hydraulic fracturing and hot dry rock methods, which can produce hot fluids above 200 °C. Furthermore, research is being conducted on supercritical water and hot dry rock systems for potential incorporation into future designs (Yang et al., 2021). Geothermal electricity offers several advantages, such as addressing greenhouse issues, providing continuous energy production, and replacing conventional sources of electricity, vehicle fuels, natural gas, and coal (Ziółkowski et al., 2021). These parameters may influence the future development of geothermal electricity initiatives.

"Geothermal power generation" describes a renewable energy source with minimal GHG emissions. In all respects, and especially regarding GHG emissions, the environmental benefits of geothermal energy are evident. Nevertheless, several environmental concerns, including GHG emissions, remain unresolved. In particular, the high water consumption associated with geothermal power generation could contribute to the scarcity of water resources in regions where such resources are already scarce. Similarly, the discharge of geothermal fluids rich in metals and other substances can cause environmental harm if geothermal water is not discharged into specific water bodies or if proper wastewater treatment is not carried out. If the plant is situated within a caldera system, the underground fluids may contain elevated radionuclides or elements highly enriched with radionuclides (Alsaleh et al., 2023). The construction of a modern power plant may require thousands of meters of boreholes, which can consume tens of square kilometers of surface land or both. Furthermore, the use of geothermal resources can disrupt local ecosystems. For example, when mercury concentration in surface groundwater is relatively high, land plants and surface water discharge can be degraded by intense mercury upwelling (Baba & Chandrasekharam, 2022; Ball, 2021; Soltani et al., 2021).

There is a notable emphasis on nonengineering advances when considering innovative prospects. The long-term viability of projects, societal acceptance, the accumulation of prior knowledge, the expansion of resources, and the attraction of new participants to the field are significant factors that will contribute to the continued growth of the geothermal industry. It is evident that a pivotal nonengineering innovation, namely in finance and policy, will influence the prospective trajectory of the geothermal industry. Significant investment must be made in experimental and innovative research (Cousse et al., 2021; Spijkerboer et al., 2022). The lack of development in new knowledge will render policy and finance unable to follow a demand they are unaware of. Through workshops and scientific events, extensive collaboration occurs between government, industry, and academia (Cook et al., 2022). In addition to implementing the strategies mentioned above, efforts are being made to raise public awareness, engage leaders, and promote education at all levels.

2.6 Marine Energy

The seas cover 70% of the Earth's surface and are one of the most efficient, effective, and environmentally friendly sources of energy in the world (Aghabalayev & Ahmad, 2023). The idea to use oceans for energy production dates back to 1799, and it is estimated that their energy could power the entire globe twice over (Aghabalayev & Ahmad, 2023; Guo et al., 2022; He et al., 2023). Hence, ocean energy plays a key role in mitigating climate change and reducing carbon dioxide emissions (Aghabalayev & Ahmad, 2023; He et al., 2023). In addition to these beneficial properties, ocean energy offers significant advantages, such as high power density, high availability (up to 90%), and high predictability (Guo et al., 2022; Kuang et al., 2016; Shadmani et al., 2022; Xu et al., 2022). Therefore, this renewable energy source is considered one of the most cost-effective way to obtain power, especially in rural regions as it is easy to use, clean, and sustainable (Aghabalayev & Ahmad, 2023).

Ocean energy systems utilize the motions of the waves, the currents created by the tides, and the variation in seawater salinity and temperature (Aghabalayev & Ahmad, 2023; Chávez et al., 2023; Gasparatos et al., 2017; Kuang et al., 2016). The utilization of wave energy is regarded as an environmentally friendly approach to harvesting energy due to its little effect on the oceanic environment (Guo et al., 2022). Water waves are generated by the wind moving on the surface of the water (He et al., 2023; Stutzmann & Csoklich, 2022). The main force in this phenomenon is the difference in streaming speeds between the water and the wind, resulting in shear forces on the water. Wave energy has high potential for generating electricity and reducing carbon emissions (Aghabalayev & Ahmad, 2023; Guo et al., 2022). Although wave energy is regarded as an unexplored resource, its worldwide reserve is estimated to be 1–10 TW. The predicted exploitable wave power source was approximately 29,500 TWh/year, which overcame the worldwide electricity consumption of around 22,300 TWh/year in 2018 (Guo et al., 2022; He et al., 2023).

There are several techniques for harvesting energy from water waves. Basically, they have articulated and cylindrical parts, which move with the incoming waves and pass that action to hydraulic rams, which charge the hydraulic accumulators. Then, induction motors located inside the structure generate electricity. While Pelamis with a 750 kW capacity was the first large-scale wave energy converter developed using this technology in 2004, the first power plant was established on the coast of Portugal in 2008, but was taken offline the same year (Chávez et al., 2023). The majority of current systems rely on point absorber buoys (Xu et al., 2022). They work by collecting wave motion and transmitting it to a shaft attached to a station on the seabed, where it is converted into energy. There are various installations around the coasts of Australia (240 kW per buoy) and the United States (20 kW per buoy) (Chávez et al., 2023). Another technique is based on the oscillating water column, which is designed to follow the change in water level caused by incoming waves and tides, forcing air through a wind turbine and producing power (He et al., 2023). Unfortunately, no systems have been built beyond the prototype stage (Chávez et al., 2023). Although the energy power in coastal areas is easier to collect, these units have very low energy density per unit length (Stutzmann & Csoklich, 2022). So, despite their enormous potential, none of the wave energy plants have reached commercial viability (He et al., 2023).

Wave energy converters come with some disadvantages. A generator can be optimized to operate at a high speed and can work efficiently around this level. In case of extreme waves, they can damage this mechanism and its reliability (Guo et al., 2022). Therefore, operation of these tools in such offshore and harsh environments (intense rainfall, storms, hurricanes, typhoons, etc.) incurs high costs during their installation, operation, and maintenance phases (Aghabalayev & Ahmad, 2023; Guo et al., 2022). Hence, survivability in extreme situations is still considered challenging (Guo et al., 2022; Shadmani et al., 2022). Besides, wave power varies with the seasons; it often decreases in the summer and increases in the winter. As a result, it is difficult to maintain nearly consistent energy output levels throughout the year (Østergaard et al., 2022).

A tidal wave is caused by asymmetries in the gravity and centrifugal acceleration of water on various sides of the Earth's surface in relation to the Sun and Moon. Tides can be divided into two groups: solar and lunar tides. Solar tides are created on the two sides of the Earth's surface nearest to and farthest from the Sun and are caused due to variations in the gravitational attraction of the Sun and the centrifugal forces that occur from the Earth that orbits around the Sun. At the center of the Earth, both forces perfectly cancel each other out (Stutzmann & Csoklich, 2022). Even though the history of the use of tidal waves for energy production dates back to the 1920s, a large-scale tidal power plant was successfully established and started production in France in 1966 (Frey, 2021). Today, the investigations on this topic show that it is not clear whether tidal power plants are economically viable or not (Gasparatos et al., 2017). As another option, osmosis power plants evaluate the salinity difference between waters, for instance, salty ocean water and sweet river water. Their main characteristic is a semipermeable membrane that is permeable to the solvent (in this case, it is H_2O) but impermeable to the solute (NaCl). A basic osmotic cell is comprised of two compartments separated by a membrane with this ability (Stutzmann & Csoklich, 2022). The difference between the upper layers of the ocean and the lower layers creates thermal gradients, which can be utilized for energy production in oceans (Chávez et al., 2023).

3 The Role of Digital Economy, Artificial Intelligence, and Deep Learning in the Renewable Energy Field

The rise of the digital economy and the advent of artificial intelligence (AI) and deep learning have opened a spectrum of possibilities for applying these technologies across many fields. It will become feasible to diminish energy consumption to a certain degree due to the continually increasing global population and the detrimental utilization of existing resources (Godil et al., 2021). Meeting the world's current energy requirements presents a significant challenge. To assist with these core objectives, ongoing enhancement and incorporation of up-to-date technologies and AI are crucial for achieving solutions. This approach can assist in evaluating the efficacy of previous strategy implementation across various domains, including agriculture, communications, environmental informatics, and economics (Le et al., 2024; Vanegas Cantarero, 2020)

The concept of digital twinning has the potential to facilitate the achievement of sustainable development goals across a range of fields through the application of digitalization, sustainability, efficiency, and the dissemination of appropriate data and advanced technologies (Al-Othman et al., 2022).

The main objective of this approach can be summarized as the recognition of the link between the transformation of digitalization in renewable industries or renewable energy sources and AI in many areas, including electric vehicles, smart grids, cybersecurity, energy storage systems, net metering systems, bidirectional systems, the Internet of things (IoT), and drones (Chen et al., 2021). The results of numerous studies teaching AI to work intelligently in artificial systems or to manage the human–machine interface are being tested in the laboratory or through pilot applications in innovative network systems or building models (Ahmad et al., 2022).

Implementing digital solutions in energy systems can offer many advantages, including the potential for cost reductions of up to 30%. Such solutions can empower energy producers, consumers, and asset managers, creating a decentralized energy ecosystem (Singh et al., 2022). Digital solutions with advanced analytics and machine learning are employed in energy systems to optimize generation, distribution, and energy use by elucidating energy production and supply curves. Furthermore, digital solutions facilitate real-time monitoring across expansive and distributed energy facilities. Illustrative examples include real-time tracking of numerous solar power plants dispersed across multiple regions, predictive models for real-time fault detection in wind farm equipment, and applications for monitoring battery usage at customers' premises (Ahmad & Zhang, 2021). Notably, digital applications integrate all stakeholders, including end users, energy service companies, and original equipment manufacturers (OEM) (Ahmad et al., 2022; Breyer et al., 2022).

A case study in the renewable energy industry demonstrates that holistic predictive maintenance with a combination of digital solutions at the asset and system levels yielded a return on investment that was more than twice the projected amount within the anticipated timeframe (Polyanska et al., 2022). This suggests that successful operations can be achieved in the hydropower and solar industries and provides an example of how the digital economy and digital solutions are poised to benefit the renewable energy sector and create significant change for a better and sustainable future (Hortovanyi et al., 2023; Moura & Kohl, 2020).

AI has been implemented in renewable energy in many ways. A primary objective of the AI community's activities in the field of renewable energy has been the application of AI to predictive maintenance and optimization of renewable assets (Çinar et al., 2020;Qureshi et al., 2024).

Machine learning has proven effective in addressing various challenges within the energy sector. These include predictive maintenance, asset degradation prediction, failure prediction, energy forecasting, and optimization problems. Among the most prevalent and extensively utilized machine learning techniques in renewable energy are artificial neural networks (ANNs) (Antonopoulos et al., 2020; Sircar et al., 2021).

Numerous researchers have incorporated AI into conventional numerical weather forecasting tools, data, and models frequently utilized in power generation forecasts. Nevertheless, an increasing number of meteorology and energy systems experts are questioning the value and methodology of this approach. Wind farms, power plants, and other energy-related applications have forecasting capabilities powered by neural networks and AI (Boza & Evgeniou, 2021; Shi et al., 2020).

AI could be a superior optimization technology regarding grid battery capacity. Consumers can participate in energy trading in the digital economy beyond the accuracy of distributed blockchains and adopt several of the most influential AI conservation retrofit mechanisms (Al-Othman et al., 2022).

AI forecasting techniques, including fuzzy logic, artificial neural networks, and deep learning, are utilized to predict the availability of renewable energy sources based on many environmental variables. These include temperature, solar radiation, wind speed and direction, elevation, season, humidity, and cloud cover (Rusilowati et al., 2024). One of the primary advantages of these algorithms is their capacity to provide highly accurate forecasts, a significant improvement over traditional methods (Hannan et al., 2021).

Applying deep learning in various fields has led to identifying patterns in energy consumption data, thereby enhancing energy efficiency. This is achieved through the optimization and real-time management of the energy system, which can also be utilized to manage demand response in the energy sector. Deep learning algorithms optimize energy systems in real time, as exemplified by their use in chiller plants. Furthermore, deep learning has been utilized to assess energy consumption in buildings, thereby offering a promising avenue for enhancing building energy optimization in the future (Himeur et al., 2021; Kristian et al., 2024; Olu-Ajayi et al., 2022).

The impact of deep learning on energy efficiency is to evaluate the average cost efficiency of energy waste in power plants. The advent of deep learning has given rise to various energy-related issues and challenges for energy researchers and engineers, offering opportunities to enhance operational efficiency and reduce energy costs and environmental pollution (Olu-Ajayi et al., 2022; Pham et al., 2020). Several issues need to be resolved to allow commercial and user acceptance of this approach to reduce energy waste using deep learning in renewable energy. These issues include scalability and scaling of results, system interoperability and standardization, future intelligent energy system flexibility, and increasing return on investment (Elsisi et al., 2021; Olu-Ajayi et al., 2022; Yang et al., 2020).

The application of deep learning techniques has the potential to enhance the performance of renewable energy technologies, thereby accelerating the planning process. This, in turn, could contribute to a cleaner and greener energy benefit for the unserved energy sector in the future (Abualigah et al., 2022; Rangel-Martinez et al., 2021).

The integration of AI and deep learning into a variety of technologies serves to mitigate technical challenges. Nevertheless, some significant technical issues must be resolved before AI and deep learning can be effectively applied to renewable energy (Aldoseri et al., 2023; Soori et al., 2023). These include maintenance barriers, pollution, aging, dust, and cooling combinations. The data quality required by AI and deep learning presents a substantial challenge, as poor-quality data can lead to poor results. Furthermore, not every location is suitable for implementing AI and deep learning. Model interpretability and bias from the algorithms used for training data present additional challenges (Kristian et al., 2024; Sarker, 2021; Zhang et al., 2022).

To apply AI effectively in the energy sector, it is essential to have a comprehensive and reliable preliminary database. This phase is referred to as AI training, which employs a variety of machine learning algorithms to predict energy sector datasets based on visual image data analysis (Abualigah et al., 2022; Alabi et al., 2022). Additional nontechnical challenges include the establishment of standard practices among suppliers. Training for the workforce will be a significant constraint, as AI integration will necessitate the workforce to possess AI skills. Balancing regulation and innovation is also crucial due to public safety concerns, and AI developers need to collaborate with stakeholders for ethical AI (Abualigah et al., 2022; Alabi et al., 2022; Anushalini & Sri Revathi, 2024; Benti et al., 2023).

It is anticipated that ecosystem partnerships will play a key role in shaping the future of the classical and AI-embedded sector ecosystem. New technology integrators, data scientists, and AI programmers with industry-specific knowledge will perceive an expanding opportunity to develop innovative technical solutions that anticipate industry demand for sustainable operations (Song et al., 2022; van Dyck et al., 2023). Finally, there is a need for the affordability of renewable energy solutions to complement technology-enhanced efficiency. As the regulatory framework evolves, new technologies could increase (Aboy et al., 2022; Moro-Visconti et al., 2023).

4 The Importance of Renewable Energy Production in Industry 5.0 Concept

During the 1860s, England experienced a transformative surge in industrial development, marked by the induction of steam-powered manufacturing equipment that initiated the era of “industry age.” This phenomenon known as Industry 1.0 signified a major shift in the narrative of human progress. Then, in the early 1900s, the advent of Industry 2.0 introduced mass production techniques and the assembly line, initiating a new era known as the “electric age,” which marked a turning point in human history. In the decade of the 1970s, Industry 3.0 marked the introduction of information technology, including automation systems and computers, into the manufacturing process. As a further stage in industrialization, the ideas of AI, machine learning, and IoT were integrated into industrial production, kicking off the Industry 4.0 period. Since the first revolution, advances in information technology have been applied to industry, enabling the creation of complicated production processes, more controlled production based on automation, and the long-term, effective, and permanent functioning of machines. With Industry 4.0, machine efficiency has grown due to AI and informative technologies tailored to the industry (Gomathi et al., 2023; Leng et al., 2022).

Although Industry 4.0 has enabled the adoption of novel technologies that have significantly enhanced industrial efficiency, it has also introduced new bottlenecks in areas like human well-being, resilience, and sustainability (Bouaziz et al., 2024). This situation has prompted a new industrial revolution, and many nations, particularly highly industrialized ones, have hastened the transition to Industry 5.0 (Plata, 2022). Today, with Industry 5.0, the human factor is being considered alongside industrial efficiency, and the industry is striving for a more humanistic approach. With Industry 5.0, society's desire for customization has been met, and a symbiotic ecology between machines and humans in production has emerged (Gomathi et al., 2023; Leng et al., 2022). Therefore, it promotes human–machine interaction and, most significantly, focuses on economic development balanced with the environment, sustainability, and renewability (Plata, 2022).

Industry 5.0 emphasizes the importance of people and machines working together to achieve a shared objective rather than competing. Thus, human–machine interactions based on AI will be promoted, and quicker and better automation, paired with the power of the human brain, will enable more advanced manufacturing. The most significant step in the transformation from Industry 4.0 to Industry 5.0 will be to gain substantially more power in production by merging the best of the human and machine worlds (Grabowska et al., 2022; Sharma et al., 2022). As the growing population and industrialization have increased the global demands for energy, it is crucial to adapt power plant to the Industry 5.0 concept in order to meet the demand. Furthermore, the move to Industry 5.0 has ushered in a time of industrial production that stresses environmental stewardship alongside economic progress. Instead of relying solely on automation and data analysis, businesses prioritize human intelligence, creativity, and sustainable practices, thereby supporting the reconciliation of economic growth and environmental sustainability, which have previously struggled to meet on common ground. Industry 5.0 is a more inclusive approach that encourages the development of sustainable development approaches that will also bring answers to environmental concerns such as natural resource depletion and biodiversity loss. As a result, a balanced development that integrates industry with sustainability principles, concentrating not only on economic development but also on people and the world, becomes conceivable (Rame et al., 2024).

Industry 5.0 focuses on three pillars: human-centricity, sustainability, and resilience (Chivilò & Meneghetti, 2023; Leng et al., 2022). Previously, employees learned the operating principles of machines and adjusted their job appropriately. However, Industry 5.0 inverted this notion, emphasizing the importance of humans in the workplace and building technologies that are human-compatible. As a result, manufacturers shifted their focus to developing new technologies that satisfy the needs of human workers rather than the other way around. For example, no matter how advanced the technology used is, people's health is affected by working in areas with high carbon emissions due to fossil fuel consumption. Therefore, it is essential to use renewable energy source and green technology. On the other hand, in the Industry 5.0 concept, the goal of machines is to save employees time and effort (Adel, 2022). Industry 5.0 emphasizes that machines are responsible for monotonous and risky tasks, whereas humans choose imaginative and proficient solutions (Chivilò & Meneghetti, 2023). Here, humans have a huge effect on the production process through their ingenuity, as technological advances are insufficient to meet customer expectations. The function of the human person brings fresh ideas for using technology to meet expectations through customization (Adel, 2022). Industry 5.0 will result in a more efficient manufacturing process with additional value, better dependable autonomy, and lower waste and costs due to effective collaboration between humans and machines (Leng et al., 2022).

From a human-centric standpoint, one of the most critical aspects for businesses to be able to build production using the Industry 5.0 strategy is to educate personnel about the machines. Human workers need to learn how to utilize and manage robots (Adel, 2022). Virtual training allows for cost-effective and safe training without disrupting production, making it easier to integrate new technologies into current production lines (Chivilò & Meneghetti, 2023). Employee engagement and interest are enhanced when their understanding of the production chain system grows (Adel, 2022). As a result, full worker acceptance of machines will lead to a safer workplace (Chivilò & Meneghetti, 2023).

From a sustainability standpoint, a production platform has to run with a minimum environmental effect (Chivilò & Meneghetti, 2023). In this regard, the essential step is the transition from fossil fuels to renewable energy sources by exploiting more techniques to obtain energy from renewable energy resources. One economic strategy is to locate a manufacturing plant near raw materials and in locations with low production prices. Furthermore, cloud manufacturing enables management of equipment and operations throughout the production lifecycle (Adel, 2022).

From a resilience standpoint, manufacturers must strengthen the resilience of their production lines to improve and protect their products against disorders and disasters such as COVID-19 (Adel, 2022). In this regard, the utilization of renewable energy sources instead of fossil fuels is an essential strategy to be not dependent on expirable energy sources. For the manufacturing phase, continuous monitoring and preventative maintenance can address production line resilience issues (Chivilò & Meneghetti, 2023). Furthermore, artificial IoT-based systems can improve organizational resilience by identifying and backing up redundant items Chivilò & Meneghetti, 2023; Leng et al., 2022).

5 Conclusion

Renewable energy production is an undebatable way to sustain life on Earth. Today, renewable energy is produced from different sources by utilizing various natural resources (Table 1). Although each renewable energy source has some disadvantages, it is known that they are more beneficial and indispensable for the future than nonrenewable energy sources.

There are technical and social suggestions to overcome the disadvantages observed in these systems. Technical considerations need to be taken into account in the construction of renewable energy generating platforms: (1) expanding the geography of the use of energy sources; (2) conducting research to improve energy production technologies in order to increase their efficiency, reliability, and environmental performance; (3) studying and thoroughly analyzing risks during various stages of energy project implementation for the construction of energy production plants; (4) identifying chances for utilizing various systems in energy production technologies; and (5) developing and constructing environmentally friendly energy production complexes.

Social priorities should be taken into consideration: (1) identifying partners and donors from universities and international research centers to contribute to the establishment of renewable power plants; (2) encouraging and supporting local and foreign investors to invest in renewable energy projects, including large-scale installation and production; (3) establishing research institutes and centers specializing in

Table 1 Renewable energy sources comparison (Modified from Rezk et al., 2021, Global Renewable Energy Overview 2020 www.irena.org, The Role of Renewable Energy in a Sustainable Future 2019 www.worldenergy.org, Biomass: A Renewable Energy Source 2018 www.nrel.gov)

Renewable energy source	Resources	Advantages	Disadvantages
Solar energy	Solar panels, photovoltaic cells	Renewable, low operating costs	Weather-dependent, high initial costs
Wind energy	Wind turbines	Clean energy, low operating costs	Noise and aesthetic concerns, variability in wind
Hydropower	Dams, river flows	High efficiency, energy storage capability	Ecosystem impacts, site selection challenges
Biomass energy	Organic waste, agricultural products	Waste management, carbon-neutral potential	Emissions, need for agricultural land
Geothermal energy	Underground hot water and steam	Continuous energy source, low emissions	Limited geographical availability, high initial costs
Marine energy	Ocean waves and tidal movements	Continuous and predictable energy source	Technology costs, environmental impacts

renewable energies, providing software and technologies, and holding seminars and training courses to raise awareness among residents about the importance of switching to renewable energies; and (4) incorporating the environmental factor into economic calculations to improve the competitiveness of renewable energy sources. If the efforts on these considerations are increased, it can boost the utilization of existing renewable energy resources with higher capacity and lower environmental impact.

References

Abdulyekeen, K. A., Umar, A. A., Patah, M. F. A., & Daud, W. M. A. W. (2021). Torrefaction of biomass: Production of enhanced solid biofuel from municipal solid waste and other types of biomass. *Renewable and Sustainable Energy Reviews, 150*, 111436. https://doi.org/10.1016/j.rser.2021.111436

Aboy, M., Minssen, T., & Kop, M. (2022). Mapping the patent landscape of quantum technologies: Patenting trends, innovation and policy implications. *IIC International Review of Intellectual Property and Competition Law, 53*(6), 853–882. https://doi.org/10.1007/s40319-022-01209-3

Abualigah, L., Zitar, R. A., Almotairi, K. H., Hussein, A. M., Elaziz, M. A., Nikoo, M. R., & Gandomi, A. H. (2022). Wind, solar, and photovoltaic renewable energy systems with and without energy storage optimization: A survey of advanced machine learning and deep learning techniques. *Energies, 15*(2), 578. https://doi.org/10.3390/en15020578

Adel, A. (2022). Future of industry 5.0 in society: human-centric solutions, challenges and prospective research areas. *Journal of Cloud Computing, 11*(1), 40. https://doi.org/10.1186/s13677-022-00314-5

Aghabalayev, F., & Ahmad, M. (2023). Does innovation in ocean energy generations-related technologies in G7 countries reduce carbon dioxide emissions? Role of international collaboration in green technology development and commercial and monetary policies. *Environmental Science and Pollution Research, 30*(6), 14545–14564. https://doi.org/10.1007/s11356-022-23081-x

Ahmad, T., & Zhang, D. (2021). Using the internet of things in smart energy systems and networks. *Sustainable Cities and Society, 68*, 102783. https://doi.org/10.1016/j.scs.2021.102783

Ahmad, T., Zhu, H., Zhang, D., Tariq, R., Bassam, A., Ullah, F., AlGhamdi, A. S., & Alshamrani, S. S. (2022). Energetics systems and artificial intelligence: Applications of industry 4.0. *Energy Reports, 8*, 334–361. https://doi.org/10.1016/j.egyr.2021.11.256

Alabi, T. M., Aghimien, E. I., Agbajor, F. D., Yang, Z., Lu, L., Adeoye, A. R., & Gopaluni, B. (2022). A review on the integrated optimization techniques and machine learning approaches for modeling, prediction, and decision making on integrated energy systems. *Renewable Energy, 194*, 822–849. https://doi.org/10.1016/j.renene.2022.05.123

Aldoseri, A., Al-Khalifa, K. N., & Hamouda, A. M. (2023). Re-thinking data strategy and integration for artificial intelligence: Concepts, opportunities, and challenges. *Applied Sciences (Switzerland), 13*(12), 7082. https://doi.org/10.3390/app13127082

Alem, A. F., Worku, A. K., Ayele, D. W., Wubieneh, T. A., Teshager, A. A., Kndie, T. M., Admasu, B. T., Teshager, M. A., Asege, A. A., Ambaw, M. D., Zeleke, M. A., Shibesh, A. K., & Yemata, T. A. (2023). Ag doped Co3O4 nanoparticles for high-performance supercapacitor application. *Heliyon, 9*(2), e13286. https://doi.org/10.1016/j.heliyon.2023.e13286

Alfonsín, V., Maceiras, R., & Gutiérrez, C. (2019). Bioethanol production from industrial algae waste. *Waste Management, 87*, 791–797. https://doi.org/10.1016/j.wasman.2019.03.019

Al-Othman, A., Tawalbeh, M., Martis, R., Dhou, S., Orhan, M., Qasim, M., & Ghani Olabi, A. (2022). Artificial intelligence and numerical models in hybrid renewable energy systems with fuel cells: Advances and prospects. *Energy Conversion and Management, 253*, 115154. https://doi.org/10.1016/j.enconman.2021.115154

Alper, K., Tekin, K., Karagöz, S., & Ragauskas, A. J. (2020). Sustainable energy and fuels from biomass: A review focusing on hydrothermal biomass processing. *Sustainable Energy and Fuels, 4*(9), 4390–4414. https://doi.org/10.1039/d0se00784f

Alsaleh, M., Yang, Z., Chen, T., Wang, X., Abdul-Rahim, A. S., & Mahmood, H. (2023). Moving toward environmental sustainability: Assessing the influence of geothermal power on carbon dioxide emissions. *Renewable Energy, 202*, 880–893. https://doi.org/10.1016/j.renene.2022.11.060

Ambaye, T. G., Vaccari, M., Bonilla-Petriciolet, A., Prasad, S., van Hullebusch, E. D., & Rtimi, S. (2021). Emerging technologies for biofuel production: A critical review on recent progress, challenges and perspectives. *Journal of Environmental Management, 290*, 112627. https://doi.org/10.1016/j.jenvman.2021.112627

Amjith, L. R., & Bavanish, B. (2022). A review on biomass and wind as renewable energy for sustainable environment. *Chemosphere, 293*, 133579. https://doi.org/10.1016/j.chemosphere.2022.133579

Ampese, L. C., Sganzerla, W. G., Di Domenico Ziero, H., Mudhoo, A., Martins, G., & Forster-Carneiro, T. (2022). Research progress, trends, and updates on anaerobic digestion technology: A bibliometric analysis. *Journal of Cleaner Production, 331*, 130004. https://doi.org/10.1016/j.jclepro.2021.130004

Anastas, P. T., & Warner, J. C. (2000). *Green chemistry: theory and practice*. Oxford University Press.

Ang, T. Z., Salem, M., Kamarol, M., Das, H. S., Nazari, M. A., & Prabaharan, N. (2022). A comprehensive study of renewable energy sources: Classifications, challenges and suggestions. *Energy Strategy Reviews, 43*, 100939. https://doi.org/10.1016/j.esr.2022.100939

Antonopoulos, I., Robu, V., Couraud, B., Kirli, D., Norbu, S., Kiprakis, A., Flynn, D., Elizondo-Gonzalez, S., & Wattam, S. (2020). Artificial intelligence and machine learning approaches to energy demand-side response: A systematic review. *Renewable and Sustainable Energy Reviews, 130*, 109899. https://doi.org/10.1016/j.rser.2020.109899

Anushalini, T., & Sri Revathi, B. (2024). Role of machine learning algorithms for wind power generation prediction in renewable energy management. *IETE Journal of Research, 70*(4), 4319–4332. https://doi.org/10.1080/03772063.2023.2205838

Baba, A., & Chandrasekharam, D. (2022). Geothermal resources for sustainable development: A case study. *International Journal of Energy Research, 46*(14), 20501–20518. https://doi.org/10.1002/er.7778

Balali, Y., & Stegen, S. (2021). Review of energy storage systems for vehicles based on technology, environmental impacts, and costs. *Renewable and Sustainable Energy Reviews, 135*, 110185. https://doi.org/10.1016/j.rser.2020.110185

Ball, P. J. (2021). A review of geothermal technologies and their role in reducing greenhouse gas emissions in the USA. *Journal of Energy Resources Technology, Transactions of the ASME, 143*(1), 010903. https://doi.org/10.1115/1.4048187

Barasa Kabeyi, M. J., & Olanrewaju, O. A. (2022). Geothermal wellhead technology power plants in grid electricity generation: A review. *Energy Strategy Reviews, 39*, 100735. https://doi.org/10.1016/j.esr.2021.100735

Baxter, J., Walker, C., Ellis, G., Devine-Wright, P., Adams, M., & Smith Fullerton, R. (2020). Scale, history and justice in community wind energy: An empirical review. *Energy Research & Social Science, 68*, 101532.

Benti, N. E., Chaka, M. D., & Semie, A. G. (2023). Forecasting renewable energy generation with machine learning and deep learning: Current advances and future prospects. *Sustainability (Switzerland), 15*(9), 7087. https://doi.org/10.3390/su15097087

Betz, U. A. K., Arora, L., Assal, R. A., Azevedo, H., Baldwin, J., Becker, M. S., Bostock, S., Cheng, V., Egle, T., Ferrari, N., Schneider-Futschik, E. K., Gerhardy, S., Hammes, A., Harzheim, A., Herget, T., Jauset, C., Kretschmer, S., Lammie, C., Kloss, N., et al. (2023). Game changers in science and technology—now and beyond. *Technological Forecasting and Social Change, 193*, 122588. https://doi.org/10.1016/j.techfore.2023.122588

Bouaziz, N., Bettayeb, B., Sahnoun, M., & Yassine, A. (2024). Incorporating uncertain human behavior in production scheduling for enhanced productivity in Industry 5.0 context. *International Journal of Production Economics, 274*, 109311. https://doi.org/10.1016/j.ijpe.2024.109311

Boza, P., & Evgeniou, T. (2021). Artificial intelligence to support the integration of variable renewable energy sources to the power system. *Applied Energy, 290*, 116754. https://doi.org/10.1016/j.apenergy.2021.116754

Breyer, C., Khalili, S., Bogdanov, D., Ram, M., Oyewo, A. S., Aghahosseini, A., Gulagi, A., Solomon, A. A., Keiner, D., Lopez, G., Ostergaard, P. A., Lund, H., Mathiesen, B. V., Jacobson, M. Z., Victoria, M., Teske, S., Pregger, T., Fthenakis, V., Raugei, M., et al. (2022). On the history and future of 100% renewable energy systems research. *IEEE Access, 10*, 78176–78218. https://doi.org/10.1109/ACCESS.2022.3193402

Burrington, J. D. (2024). Renewable energy technical potential performance for zero carbon emissions. *ACS Omega, 9*(24), 25841–25858. https://doi.org/10.1021/acsomega.4c00273

Capurso, T., Stefanizzi, M., Torresi, M., & Camporeale, S. M. (2022). Perspective of the role of hydrogen in the 21st century energy transition. *Energy Conversion and Management, 251*, 114898. https://doi.org/10.1016/j.enconman.2021.114898

Chao, T., Yunbao, X., Chengbo, D., Bo, L., & Ullah, S. (2023). Financial integration and renewable energy consumption in China: Do education and digital economy development matter? *Environmental Science and Pollution Research, 30*(5), 12944–12952. https://doi.org/10.1007/s11356-022-22852-w

Chávez, V., Bárcenas, J. F., Martínez, M. L., Mateos, E., Zúñiga-Ríos, A., Guimarais, M., Wojtarowski, A., Landgrave, R., Ceballos Canché, C. H., & Silva, R. (2023). Potential sites for the use of ocean energy in the Mexican Caribbean. *Energy Sources, Part B: Economics, Planning and Policy, 18*(1), 2160524. https://doi.org/10.1080/15567249.2022.2160524

Chen, C., Hu, Y., Marimuthu, K., & Kumar, P. M. (2021). Artificial intelligence on economic evaluation of energy efficiency and renewable energy technologies. *Sustainable Energy Technologies and Assessments, 47*, 101358. https://doi.org/10.1016/j.seta.2021.101358

Chen, L., Msigwa, G., Yang, M., Osman, A. I., Fawzy, S., Rooney, D. W., & Yap, P. S. (2022). Strategies to achieve a carbon neutral society: a review. *Environmental Chemistry Letters, 20*(4), 2277–2310. https://doi.org/10.1007/s10311-022-01435-8

Cheng, D., Ngo, H. H., Guo, W., Chang, S. W., Nguyen, D. D., Bui, X. T., Wei, W., Ni, B., Varjani, S., & Hoang, N. B. (2022). Enhanced photo-fermentative biohydrogen production from biowastes: An overview. *Bioresource Technology, 357*, 127341. https://doi.org/10.1016/j.biortech.2022.127341

Chivilò, M., & Meneghetti, A. (2023). An Industry 5.0 perspective on feeding production lines. *Sustainability (Switzerland), 15*(22), 16088. https://doi.org/10.3390/su152216088

Çinar, Z. M., Nuhu, A. A., Zeeshan, Q., Korhan, O., Asmael, M., & Safaei, B. (2020). Machine learning in predictive maintenance towards sustainable smart manufacturing in industry 4.0. *Sustainability (Switzerland), 12*(19), 8211. https://doi.org/10.3390/su12198211

Cook, D., Karlsdóttir, I., & Minelgaite, I. (2022). Enjoying the Heat? Co-Creation of Stakeholder Benefits and Sustainable Energy Development within Projects in the Geothermal Sector. *Energies, 15*(3), 1029. https://doi.org/10.3390/en15031029

Cousse, J., Trutnevyte, E., & Hahnel, U. J. J. (2021). Tell me how you feel about geothermal energy: Affect as a revealing factor of the role of seismic risk on public acceptance. *Energy Policy, 158*, 112547. https://doi.org/10.1016/j.enpol.2021.112547

Deng, Q., Alvarado, R., Toledo, E., & Caraguay, L. (2020). Greenhouse gas emissions, non-renewable energy consumption, and output in South America: the role of the productive structure. *Environmental Science and Pollution Research, 27*(13), 14477–14491. https://doi.org/10.1007/s11356-020-07693-9

Dhaidan, N. S., Al-Shohani, W. A. M., Abbas, H. H., Rashid, F. L., Ameen, A., Al-Mousawi, F. N., & Homod, R. Z. (2024). Enhancing the thermal performance of an agricultural solar greenhouse by geothermal energy using an earth-air heat exchanger system: A review. *Geothermics, 123*, 103115. https://doi.org/10.1016/j.geothermics.2024.103115

Elahi, E., Khalid, Z., & Zhang, Z. (2022). Understanding farmers' intention and willingness to install renewable energy technology: A solution to reduce the environmental emissions of agriculture. *Applied Energy, 309*, 118459. https://doi.org/10.1016/j.apenergy.2021.118459

Elsisi, M., Tran, M. Q., Mahmoud, K., Lehtonen, M., & Darwish, M. M. F. (2021). Deep learning-based industry 4.0 and internet of things towards effective energy management for smart buildings. *Sensors (Switzerland), 21*(4), 1–19. https://doi.org/10.3390/s21041038

Esmaeili, H. (2022). A critical review on the economic aspects and life cycle assessment of biodiesel production using heterogeneous nanocatalysts. *Fuel Processing Technology, 230*, 107224. https://doi.org/10.1016/j.fuproc.2022.107224

Fang, J., Gozgor, G., Mahalik, M. K., Mallick, H., & Padhan, H. (2022). Does urbanisation induce renewable energy consumption in emerging economies? The role of education in energy switching policies. *Energy Economics, 111*, 106081. https://doi.org/10.1016/j.eneco.2022.106081

Fayiah, M., Dong, S. K., Singh, S., & Kwaku, E. A. (2020). A review of water–energy nexus trend, methods, challenges and future prospects. *International Journal of Energy and Water Resources, 4*(1), 91–107. https://doi.org/10.1007/s42108-020-00057-6

Frey, F. (2021). Putting oceans to work: tidal energy in the USA and the USSR, 1930–1970. *History and Technology, 37*(4), 487–504. https://doi.org/10.1080/07341512.2022.2038982

Ganesh, K. N., Zhang, D., Miller, S. J., Rossen, K., Chirik, P. J., Kozlowski, M. C., Zimmerman, J. B., Brooks, B. W., Savage, P. E., Allen, D. T., & Voutchkova-Kostal, A. M. (2021). Green chemistry: A framework for a sustainable future. *Organic Process Research and Development, 25*(8), 1455–1459. https://doi.org/10.1021/acs.oprd.1c00216

Gasparatos, A., Doll, C. N. H., Esteban, M., Ahmed, A., & Olang, T. A. (2017). Renewable energy and biodiversity: Implications for transitioning to a green economy. *Renewable and Sustainable Energy Reviews, 70*, 161–184. https://doi.org/10.1016/j.rser.2016.08.030

Godbole, V., Pal, M. K., & Gautam, P. (2021). A critical perspective on the scope of interdisciplinary approaches used in fourth-generation biofuel production. *Algal Research, 58*, 102436. https://doi.org/10.1016/j.algal.2021.102436

Godil, D. I., Yu, Z., Sharif, A., Usman, R., & Khan, S. A. R. (2021). Investigate the role of technology innovation and renewable energy in reducing transport sector CO2 emission in China: A path toward sustainable development. *Sustainable Development, 29*(4), 694–707. https://doi.org/10.1002/sd.2167

Gomathi, L., Mishra, A. K., & Tyagi, A. K. (2023). Industry 5.0 for healthcare 5.0: Opportunities, challenges and future research possibilities. In *7th international conference on trends in electronics and informatics, ICOEI 2023—Proceedings* (pp. 204–213). https://doi.org/10.1109/ICOEI56765.2023.10125660

Grabowska, S., Saniuk, S., & Gajdzik, B. (2022). Industry 5.0: improving humanization and sustainability of Industry 4.0. *Scientometrics, 127*(6), 3117–3144. https://doi.org/10.1007/s11192-022-04370-1

Guo, B., Wang, T., Jin, S., Duan, S., Yang, K., & Zhao, Y. (2022). A review of point absorber wave energy converters. *Journal of Marine Science and Engineering, 10*(10), 1534. https://doi.org/10.3390/jmse10101534

Gyamfi, B. A., Bein, M. A., & Bekun, F. V. (2020). Investigating the nexus between hydroelectricity energy, renewable energy, nonrenewable energy consumption on output: evidence from E7 countries. *Environmental Science and Pollution Research, 27*(20), 25327–25339. https://doi.org/10.1007/s11356-020-08909-8

Hackstein, F. V., & Madlener, R. (2021). Sustainable operation of geothermal power plants: why economics matters. *Geothermal Energy, 9*(1), 10. https://doi.org/10.1186/s40517-021-00183-2

Hamiche, A. M., Stambouli, A. B., & Flazi, S. (2016). A review of the water-energy nexus. *Renewable and Sustainable Energy Reviews, 65*, 319–331. https://doi.org/10.1016/j.rser.2016.07.020

Hannan, M. A., Al-Shetwi, A. Q., Ker, P. J., Begum, R. A., Mansor, M., Rahman, S. A., Dong, Z. Y., Tiong, S. K., Mahlia, T. M. I., & Muttaqi, K. M. (2021). Impact of renewable energy utilization and artificial intelligence in achieving sustainable development goals. *Energy Reports, 7*, 5359–5373. https://doi.org/10.1016/j.egyr.2021.08.172

He, F., Liu, Y., Pan, J., Ye, X., & Jiao, P. (2023). Advanced ocean wave energy harvesting: current progress and future trends. *Journal of Zhejiang University: Science A, 24*(2), 91–108. https://doi.org/10.1631/jzus.A2200598

Himeur, Y., Ghanem, K., Alsalemi, A., Bensaali, F., & Amira, A. (2021). Artificial intelligence based anomaly detection of energy consumption in buildings: A review, current trends and new perspectives. *Applied Energy, 287*, 116601. https://doi.org/10.1016/j.apenergy.2021.116601

Hoicka, C. E., Savic, K., & Campney, A. (2021). Reconciliation through renewable energy? A survey of Indigenous communities, involvement, and peoples in Canada. *Energy Research and Social Science, 74*, 101897. https://doi.org/10.1016/j.erss.2020.101897

Hortovanyi, L., Morgan, R. E., Herceg, I. V., Djuricin, D., Hanak, R., Horvath, D., Mocan, M. L., Romanova, A., & Szabo, R. Z. (2023). Assessment of digital maturity: the role of resources and capabilities in digital transformation in B2B firms. *International Journal of Production Research, 61*(23), 8043–8061. https://doi.org/10.1080/00207543.2022.2164087

Igwe, C. I. (2021). Geothermal energy: A review. *International Journal of Engineering Research & Technology, 10*(3), 655–661. Retrieved from www.ijert.org

Islam, M. T., Nabi, M. N., Arefin, M. A., Mostakim, K., Rashid, F., Hassan, N. M. S., Rahman, S. M. A., McIntosh, S., Mullins, B. J., & Muyeen, S. M. (2022). Trends and prospects of geothermal energy as an alternative source of power: A comprehensive review. *Heliyon, 8*(12), 101897. https://doi.org/10.1016/j.heliyon.2022.e11836

Jaiswal, K. K., Chowdhury, C. R., Yadav, D., Verma, R., Dutta, S., Jaiswal, K. S., & Karuppasamy, K. S. K. (2022). Renewable and sustainable clean energy development and impact on social, economic, and environmental health. *Energy Nexus, 7*, 100118.

Jeje, S. O., Marazani, T., Obiko, J. O., & Shongwe, M. B. (2024). Advancing the hydrogen production economy: A comprehensive review of technologies, sustainability, and future prospects. *International Journal of Hydrogen Energy, 78*, 642–661. https://doi.org/10.1016/j.ijhydene.2024.06.344

Jiansheng, W., Lide, S., Qiang, Z., & Jintao, N. (2022). Numerical investigation on power generation performance of enhanced geothermal system with horizontal well. *Applied Energy, 325*, 119865. https://doi.org/10.1016/j.apenergy.2022.119865

Jin, P., Fang, D., & Chen, B. (2019). Water-energy nexus based on modified multiregional input-output model within China. *Energy Procedia, 158*, 4092–4098. https://doi.org/10.1016/j.egypro.2019.01.826

Kalak, T. (2023). Potential use of industrial biomass waste as a sustainable energy source in the future. *Energies, 16*(4), 1783. https://doi.org/10.3390/en16041783

Kant Bhatia, S., Palai, A. K., Kumar, A., Kant Bhatia, R., Kumar Patel, A., Kumar Thakur, V., & Yang, Y. H. (2021). Trends in renewable energy production employing biomass-based biochar. *Bioresource Technology, 340*, 125644. https://doi.org/10.1016/j.biortech.2021.125644

Khaligh, V., Ghezelbash, A., Akhtar, M. S., Zarei, M., Liu, J., & Won, W. (2023). Optimal integration of a low-carbon energy system—A circular hydrogen economy perspective. *Energy Conversion and Management, 292*, 117354. https://doi.org/10.1016/j.enconman.2023.117354

Kim, J. H., & Alameri, S. A. (2020). Harmonizing nuclear and renewable energy: Case studies. *International Journal of Energy Research, 44*(10), 8053–8061. https://doi.org/10.1002/er.4987

Kristian, A., Sumarsan Goh, T., Ramadan, A., Erica, A., & Visiana Sihotang, S. (2024). Application of AI in optimizing energy and resource management: Effectiveness of deep learning models. *International Transactions on Artificial Intelligence (ITALIC), 2*(2), 99–105. https://doi.org/10.33050/italic.v2i2.530

Kuang, Y., Zhang, Y., Zhou, B., Li, C., Cao, Y., Li, L., & Zeng, L. (2016). A review of renewable energy utilization in islands. *Renewable and Sustainable Energy Reviews, 59*, 504–513. https://doi.org/10.1016/j.rser.2016.01.014

Kumar, C. M. S., Singh, S., Gupta, M. K., Nimdeo, Y. M., Raushan, R., Deorankar, A. V., Kumar, T. M. A., Rout, P. K., Chanotiya, C. S., Pakhale, V. D., & Nannaware, A. D. (2023). Solar energy: A promising renewable source for meeting energy demand in Indian agriculture applications. *Sustainable Energy Technologies and Assessments, 55*, 102905. https://doi.org/10.1016/j.seta.2022.102905

Le, T. T., Sharma, P., Bora, B. J., Tran, V. D., Truong, T. H., Le, H. C., & Nguyen, P. Q. P. (2024). Fueling the future: A comprehensive review of hydrogen energy systems and their challenges. *International Journal of Hydrogen Energy, 54*, 791–816. https://doi.org/10.1016/j.ijhydene.2023.08.044

Leng, J., Sha, W., Wang, B., Zheng, P., Zhuang, C., Liu, Q., Wuest, T., Mourtzis, D., & Wang, L. (2022). Industry 5.0: Prospect and retrospect. *Journal of Manufacturing Systems, 65*, 279–295. https://doi.org/10.1016/j.jmsy.2022.09.017

Letcher, T. (Ed.). (2023). *Living with climate change*. Elsevier.

Li, J., Wang, G., Li, Z., Yang, S., Chong, W. T., & Xiang, X. (2020). A review on development of offshore wind energy conversion system. *International Journal of Energy Research, 44*(12), 9283–9297. https://doi.org/10.1002/er.5751

Li, R., Wang, X., & Wang, Q. (2022). Does renewable energy reduce ecological footprint at the expense of economic growth? An empirical analysis of 120 countries. *Journal of Cleaner Production, 346*, 131207. https://doi.org/10.1016/j.jclepro.2022.131207

Li, R., Zhang, F., & Wang, Q. (2022). How does the EU's COVID-19 economic recession impact the renewable energy of other countries? The spillover effect. *Energy Strategy Reviews, 40*, 100825. https://doi.org/10.1016/j.esr.2022.100825

Liang, C., Umar, M., Ma, F., & Huynh, T. L. D. (2022). Climate policy uncertainty and world renewable energy index volatility forecasting. *Technological Forecasting and Social Change, 182*, 121810. https://doi.org/10.1016/j.techfore.2022.121810

Lorente, D. B., Mohammed, K. S., Cifuentes-Faura, J., & Shahzad, U. (2023). Dynamic connectedness among climate change index, green financial assets and renewable energy markets: Novel evidence from sustainable development perspective. *Renewable Energy, 204*, 94–105. https://doi.org/10.1016/j.renene.2022.12.085

Lund, J. W., & Toth, A. N. (2021). Direct utilization of geothermal energy 2020 worldwide review. *Geothermics, 90*, 101915. https://doi.org/10.1016/j.geothermics.2020.101915

Ma, Z. D., Zhang, Y. P., Saw, L. H., Cui, X., Jia, G. S., & Jin, L. W. (2023). Investigation on local geothermal energy attenuation after long-term operation of ground heat exchanger with considering aquifer effect. *Geothermics, 107*, 102608. https://doi.org/10.1016/j.geothermics.2022.102608

Magazzino, C., Toma, P., Fusco, G., Valente, D., & Petrosillo, I. (2022). Renewable energy consumption, environmental degradation and economic growth: the greener the richer? *Ecological Indicators, 139*, 108912. https://doi.org/10.1016/j.ecolind.2022.108912

Maheshwari, P., Haider, M. B., Yusuf, M., Klemeš, J. J., Bokhari, A., Beg, M., Al-Othman, A., Kumar, R., & Jaiswal, A. K. (2022). A review on latest trends in cleaner biodiesel production: Role of feedstock, production methods, and catalysts. *Journal of Cleaner Production, 355*, 131588. https://doi.org/10.1016/j.jclepro.2022.131588

Mahmoud, M., Ramadan, M., Pullen, K., Abdelkareem, M. A., Wilberforce, T., Olabi, A. G., & Naher, S. (2021). A review of grout materials in geothermal energy applications. *International Journal of Thermofluids, 10*, 100070. https://doi.org/10.1016/j.ijft.2021.100070

Majeed, Y., Khan, M. U., Waseem, M., Zahid, U., Mahmood, F., Majeed, F., Sultan, M., & Raza, A. (2023). Renewable energy as an alternative source for energy management in agriculture. *Energy Reports, 10*, 344–359. https://doi.org/10.1016/j.egyr.2023.06.032

Maka, A. O. M., & Alabid, J. M. (2022). Solar energy technology and its roles in sustainable development. *Clean Energy, 6*(3), 476–483. https://doi.org/10.1093/ce/zkac023

Mentel, G., Lewandowska, A., Berniak-Woźny, J., & Tarczyński, W. (2023). Green and renewable energy innovations: A comprehensive bibliometric analysis. *Energies, 16*(3), 1428. https://doi.org/10.3390/en16031428

Moro-Visconti, R., Cruz Rambaud, S., & López Pascual, J. (2023). Artificial intelligence-driven scalability and its impact on the sustainability and valuation of traditional firms. *Humanities and Social Sciences Communications, 10*(1), 795. https://doi.org/10.1057/s41599-023-02214-8

Moura, L. R., & Kohl, H. (2020). Maturity assessment in industry 4.0—a comparative analysis of Brazilian and German companies. *Emerging Science Journal, 4*(5), 365–375. https://doi.org/10.28991/esj-2020-01237

Nassar, Y. F., Alsadi, S. Y., El-Khozondar, H. J., Ismail, M. S., Al-Maghalseh, M., Khatib, T., Sa'ed, J. A., Mushtaha, M. H., & Djerafi, T. (2022). Design of an isolated renewable hybrid energy system: a case study. *Materials for Renewable and Sustainable Energy, 11*(3), 225–240. https://doi.org/10.1007/s40243-022-00216-1

Nastasi, B., Markovska, N., Puksec, T., Duić, N., & Foley, A. (2022). Renewable and sustainable energy challenges to face for the achievement of Sustainable Development Goals. *Renewable and Sustainable Energy Reviews, 157*, 112071. https://doi.org/10.1016/j.rser.2022.112071

Neto, J. V. S., & Gallo, W. L. (2021). Potential impacts of vinasse biogas replacing fossil oil for power generation, natural gas, and increasing sugarcane energy in Brazil. *Renewable and Sustainable Energy Reviews, 135*, 110281.

Nikolaevich, K. V., Starodub, Y., & Havrys, A. (2021). Computer modeling in the application to geothermal engineering. *Advances in Civil Engineering, 2021*, 1–23. https://doi.org/10.1155/2021/6619991

Obaideen, K., Nooman AlMallahi, M., Alami, A. H., Ramadan, M., Abdelkareem, M. A., Shehata, N., & Olabi, A. G. (2021). On the contribution of solar energy to sustainable developments goals: Case study on Mohammed bin Rashid Al Maktoum Solar Park. *International Journal of Thermofluids, 12*, 100123. https://doi.org/10.1016/j.ijft.2021.100123

Olu-Ajayi, R., Alaka, H., Sulaimon, I., Sunmola, F., & Ajayi, S. (2022). Building energy consumption prediction for residential buildings using deep learning and other machine learning techniques. *Journal of Building Engineering, 45*, 103406. https://doi.org/10.1016/j.jobe.2021.103406

Osman, A. I., Chen, L., Yang, M., Msigwa, G., Farghali, M., Fawzy, S., & Yap, P. S. (2023). Cost, environmental impact, and resilience of renewable energy under a changing climate: a review. *Environmental Chemistry Letters, 21*(2), 741–764.

Østergaard, P. A., Duic, N., Noorollahi, Y., & Kalogirou, S. (2022). Renewable energy for sustainable development. *Renewable Energy, 199*, 1145–1152. https://doi.org/10.1016/j.renene.2022.09.065

Packiyadhas, P., & Shanmuganantham Selvanantham, D. (2024). Compositional and structural evaluation of Kappaphycus alvarezii rejects and solid food waste blends for bio ethanol production. *Energy Sources, Part A: Recovery, Utilization and Environmental Effects, 46*, 13869–13885. https://doi.org/10.1080/15567036.2020.1806405

Pan, H., Qi, L., Zhang, Z., & Yan, J. (2021). Kinetic energy harvesting technologies for applications in land transportation: A comprehensive review. *Applied Energy, 286*, 116518. https://doi.org/10.1016/j.apenergy.2021.116518

Pareek, A., Shanthi Sravan, J., & Venkata Mohan, S. (2019). Exploring chemically reduced graphene oxide electrode for power generation in microbial fuel cell. *Materials Science for Energy Technologies, 2*(3), 600–606. https://doi.org/10.1016/j.mset.2019.06.006

Peng, T., Wan, J., Liu, W., Li, J., Xia, Y., Yuan, G., et al. (2023). Choice of hydrogen energy storage in salt caverns and horizontal cavern construction technology. *Journal of Energy Storage, 60*, 106489.

Pfeifroth, U., Drücke, J., Kothe, S., Trentmann, J., Schröder, M., & Hollmann, R. (2024). SARAH-3-satellite-based climate data records of surface solar radiation. *Earth System Science Data Discussions, 16*, 5243–5265. https://doi.org/10.5676/EUM_SAF_CM/SARAH/V003

Pham, A. D., Ngo, N. T., Ha Truong, T. T., Huynh, N. T., & Truong, N. S. (2020). Predicting energy consumption in multiple buildings using machine learning for improving energy efficiency and sustainability. *Journal of Cleaner Production, 260*, 121082. https://doi.org/10.1016/j.jclepro.2020.121082

Plata, P. (2022). Algae as an alternative to the methods of production and use of conventional biomass (review article). *Energy Policy Studies, 11*(3), 53–69. https://doi.org/10.62316/cnlw1140

Polyanska, A., Savchuk, S., Dudek, M., Sala, D., Pazynich, Y., & Cicho, D. (2022). Impact of digital maturity on sustainable development effects in the energy sector in the condition of Industry 4.0. *Scientific Bulletin of National Mining University, 6*, 97–103.

Priyadharsini, P., Dawn, S. S., Arun, J., Alok, R., & Jayaprabakar, J. (2023). Recent advances in conventional and genetically modified macroalgal biomass as substrates in bioethanol production: A review. *Biofuels, 14*(10), 1103–1118. https://doi.org/10.1080/17597269.2023.2206695

Priyadharsini, P., Nirmala, N., Dawn, S. S., Baskaran, A., SundarRajan, P., Gopinath, K. P., & Arun, J. (2022). Genetic improvement of microalgae for enhanced carbon dioxide sequestration and enriched biomass productivity: Review on CO2 bio-fixation pathways modifications. *Algal Research, 66*, 102810. https://doi.org/10.1016/j.algal.2022.102810

Pugalendhi, S., Gitanjali, J., Shalini, R., & Subramanian, P. (2024). Importance of renewable energy sources. In *Handbook on renewable energy and green technology* (pp. 1–9).

Puppala, H., Jha, S. K., Singh, A. P., Madurai Elavarasan, R., & Elia Campana, P. (2022). Identification and analysis of barriers for harnessing geothermal energy in India. *Renewable Energy, 186*, 327–340. https://doi.org/10.1016/j.renene.2022.01.002

Qureshi, M., Arbab, M. A., & Rehman, S. U. (2024). Deep learning-based forecasting of electricity consumption. *Scientific Reports, 14*(1), 6489.

Rahman, M. M., Khan, I., Field, D. L., Techato, K., & Alameh, K. (2022). Powering agriculture: Present status, future potential, and challenges of renewable energy applications. *Renewable Energy, 188*, 731–749. https://doi.org/10.1016/j.renene.2022.02.065

Raimi, D., Zhu, Y., Newell, R. G., & Prest, B. C. (2024). *Global energy outlook 2024: Peaks or plateaus?*

Rame, R., Purwanto, P., & Sudarno, S. (2024). Industry 5.0 and sustainability: An overview of emerging trends and challenges for a green future. *Innovation and Green Development, 3*(4), 100173. https://doi.org/10.1016/j.igd.2024.100173

Rangel-Martinez, D., Nigam, K. D. P., & Ricardez-Sandoval, L. A. (2021). Machine learning on sustainable energy: A review and outlook on renewable energy systems, catalysis, smart grid and energy storage. *Chemical Engineering Research and Design, 174*, 414–441. https://doi.org/10.1016/j.cherd.2021.08.013

Reichle, D. E. (2023). *The global carbon cycle and climate change: Scaling ecological energetics from organism to the biosphere*. Elsevier.

Rezk, H., Babu, T. S., Al-Dhaifallah, M., & Ziedan, H. A. (2021). A robust parameter estimation approach based on stochastic fractal search optimization algorithm applied to solar PV parameters. *Energy Reports, 7*, 620–640. https://doi.org/10.1016/j.egyr.2021.01.024

Roga, S., Bardhan, S., Kumar, Y., & Dubey, S. K. (2022). Recent technology and challenges of wind energy generation: A review. *Sustainable Energy Technologies and Assessments, 52*, 102239. https://doi.org/10.1016/j.seta.2022.102239

Romanello, M., di Napoli, C., Green, C., Kennard, H., Lampard, P., Scamman, D., Walawender, M., Ali, Z., Ameli, N., Ayeb-Karlsson, S., Beggs, P. J., Belesova, K., Berrang Ford, L., Bowen, K., Cai, W., Callaghan, M., Campbell-Lendrum, D., Chambers, J., Cross, T. J., et al. (2023). The 2023 report of the Lancet Countdown on health and climate change: the imperative for a health-centred response in a world facing irreversible harms. *The Lancet, 402*(10419), 2346–2394. https://doi.org/10.1016/S0140-6736(23)01859-7

Rusilowati, U., Rasmita Ngemba, H., Wahyudin Anugrah, R., Fitriani, A., & Dian Astuti, E. (2024). Leveraging AI for superior efficiency in energy use and development of renewable resources such as solar energy, wind, and bioenergy. *International Transactions on Artificial Intelligence (ITALIC), 2*(2), 114–120. https://doi.org/10.33050/italic.v2i2.537

Saleem, M. (2022). Possibility of utilizing agriculture biomass as a renewable and sustainable future energy source. *Heliyon, 8*(2), e08905. https://doi.org/10.1016/j.heliyon.2022.e08905

Santhosh, M., Venkaiah, C., & Vinod Kumar, D. M. (2020). Current advances and approaches in wind speed and wind power forecasting for improved renewable energy integration: A review. *Engineering Reports, 2*(6), e12178. https://doi.org/10.1002/eng2.12178

Sarker, I. H. (2021). Deep learning: A comprehensive overview on techniques, taxonomy, applications and research directions. *SN Computer Science, 2*(6), 420. https://doi.org/10.1007/s42979-021-00815-1

Sen, S., & Ganguly, S. (2017). Opportunities, barriers and issues with renewable energy development—A discussion. *Renewable and Sustainable Energy Reviews, 69*, 1170–1181. https://doi.org/10.1016/j.rser.2016.09.137

Shadmani, A., Nikoo, M. R., Al-Raoush, R. I., Alamdari, N., & Gandomi, A. H. (2022). The optimal configuration of wave energy conversions respective to the nearshore wave energy potential. *Energies, 15*(20), 7734. https://doi.org/10.3390/en15207734

Sharma, M., Sehrawat, R., Luthra, S., Daim, T., & Bakry, D. (2022). Moving towards Industry 5.0 in the pharmaceutical manufacturing sector: Challenges and solutions for Germany. *IEEE Transactions on Engineering Management, 71*, 13757–13774. https://doi.org/10.1109/TEM.2022.3143466

Sharma, I., Shah, V., & Shah, M. (2022). A comprehensive study on production of methanol from wind energy. *Environmental Technology and Innovation, 28*, 102589. https://doi.org/10.1016/j.eti.2022.102589

Sharmin, T., Khan, N. R., Akram, M. S., & Ehsan, M. M. (2023). A state-of-the-art review on geothermal energy extraction, utilization, and improvement strategies: Conventional, hybridized, and enhanced geothermal systems. *International Journal of Thermofluids, 18*, 100323. https://doi.org/10.1016/j.ijft.2023.100323

Shi, Z., Yao, W., Li, Z., Zeng, L., Zhao, Y., Zhang, R., Tang, Y., & Wen, J. (2020). Artificial intelligence techniques for stability analysis and control in smart grids: Methodologies, applications, challenges and future directions. *Applied Energy, 278*, 115733. https://doi.org/10.1016/j.apenergy.2020.115733

Singh, R., Akram, S. V., Gehlot, A., Buddhi, D., Priyadarshi, N., & Twala, B. (2022). Energy system 4.0: Digitalization of the energy sector with inclination towards sustainability. *Sensors, 22*(17), 6619. https://doi.org/10.3390/s22176619

Sircar, A., Yadav, K., Rayavarapu, K., Bist, N., & Oza, H. (2021). Application of machine learning and artificial intelligence in oil and gas industry. *Petroleum Research, 6*(4), 379–391. https://doi.org/10.1016/j.ptlrs.2021.05.009

Siwal, S. S., Zhang, Q., Devi, N., Saini, A. K., Saini, V., Pareek, B., Gaidukovs, S., & Thakur, V. K. (2021). Recovery processes of sustainable energy using different biomass and wastes. *Renewable and Sustainable Energy Reviews, 150*, 111483. https://doi.org/10.1016/j.rser.2021.111483

Soltani, M., Moradi Kashkooli, F., Souri, M., Rafiei, B., Jabarifar, M., Gharali, K., & Nathwani, J. S. (2021). Environmental, economic, and social impacts of geothermal energy systems. *Renewable and Sustainable Energy Reviews, 140*, 110750. https://doi.org/10.1016/j.rser.2021.110750

Song, L., Hu, X., Zhang, G., Spachos, P., Plataniotis, K. N., & Wu, H. (2022). Networking systems of AI: On the convergence of computing and communications. *IEEE Internet of Things Journal, 9*(20), 20352–20381. https://doi.org/10.1109/JIOT.2022.3172270

Soori, M., Arezoo, B., & Dastres, R. (2023). Artificial intelligence, machine learning and deep learning in advanced robotics, a review. *Cognitive Robotics, 3*, 54–70. https://doi.org/10.1016/j.cogr.2023.04.001

Spijkerboer, R. C., Turhan, E., Roos, A., Billi, M., Vargas-Payera, S., Opazo, J., & Armiero, M. (2022). Out of steam? A social science and humanities research agenda for geothermal energy. *Energy Research and Social Science, 92*, 102801. https://doi.org/10.1016/j.erss.2022.102801

Stutzmann, M., & Csoklich, C. (2022). Energy from waves, tides and osmosis. In *The physics of renewable energy* (pp. 45–58).

Surendhiran, D., & Sirajunnisa, A. R. (2018). Role of genetic engineering in bioethanol production from algae. In *Bioethanol production from food crops: Sustainable sources, interventions, and challenges* (pp. 361–381). Elsevier. https://doi.org/10.1016/B978-0-12-813766-6.00018-7

Szyba, M., & Mikulik, J. (2022). Energy production from biodegradable waste as an example of the circular economy. *Energies, 15*(4), 1269. https://doi.org/10.3390/en15041269

Takach, M., Sarajlić, M., Peters, D., Kroener, M., Schuldt, F., & Von Maydell, K. (2022). Review of hydrogen production techniques from water using renewable energy sources and its storage in salt caverns. *Energies, 15*(4), 1415. https://doi.org/10.3390/en15041415

Tan, J. D., Chang, C. C. W., Bhuiyan, M. A. S., Minhad, K. N., & Ali, K. (2022). Advancements of wind energy conversion systems for low-wind urban environments: A review. *Energy Reports, 8*, 3406–3414. https://doi.org/10.1016/j.egyr.2022.02.153

Tazikeh, S., Zendehboudi, S., Ghafoori, S., Lohi, A., & Mahinpey, N. (2022). Algal bioenergy production and utilization: Technologies, challenges, and prospects. *Journal of Environmental Chemical Engineering, 10*(3), 107863. https://doi.org/10.1016/j.jece.2022.107863

Tefera, W. M., & Kasiviswanathan, K. S. (2022). A global-scale hydropower potential assessment and feasibility evaluations. *Water Resources and Economics, 38*, 100198. https://doi.org/10.1016/j.wre.2022.100198

Tomarov, G. V., & Shipkov, A. A. (2023). Brief review of the current status and trends in the development of geothermal energy. *Thermal Engineering, 70*(2), 108–116. https://doi.org/10.1134/S0040601523020088

Umar, M., Awosusi, A. A., Adegboye, O. R., & Ojekemi, O. S. (2024). Geothermal energy and carbon emissions nexus in leading geothermal-consuming nations: Evidence from nonparametric analysis. *Energy & Environment, 35*(5), 2726–2752.

Uría-Martínez, R., Leiby, P. N., & Brown, M. L. (2018). Energy security role of biofuels in evolving liquid fuel markets. *Biofuels, Bioproducts and Biorefining, 12*(5), 802–814. https://doi.org/10.1002/bbb.1891

van Dyck, M., Lüttgens, D., Piller, F. T., & Brenk, S. (2023). Interconnected digital twins and the future of digital manufacturing: Insights from a Delphi study. *Journal of Product Innovation Management, 40*(4), 475–505. https://doi.org/10.1111/jpim.12685

Vanegas Cantarero, M. M. (2020). Of renewable energy, energy democracy, and sustainable development: A roadmap to accelerate the energy transition in developing countries. *Energy Research and Social Science, 70*, 101716. https://doi.org/10.1016/j.erss.2020.101716

Varma, M., Mal, H., & Pahurkar, R. N. (2020). *Comparative analysis of green house gases emission in conventional vehicles and electric vehicles*. Retrieved from https://www.researchgate.net/publication/340546400

Verduci, R., Romano, V., Brunetti, G., Yaghoobi Nia, N., Di Carlo, A., D'Angelo, G., & Ciminelli, C. (2022). Solar energy in space applications: Review and technology perspectives. *Advanced Energy Materials, 12*(29), 2200125. https://doi.org/10.1002/aenm.202200125

Verma, A., Pal, M., & Malik, S. (2024, September). Unlocking hydrogen potential from depleted hydrocarbon reservoirs in Lithuania: A sustainable approach. *ECMOR European Association of Geoscientists & Engineers, 2024*(1), 1–11.

Vermillion, C., Cobb, M., Fagiano, L., Leuthold, R., Diehl, M., Smith, R. S., Wood, T. A., Rapp, S., Schmehl, R., Olinger, D., & Demetriou, M. (2021). Electricity in the air: Insights from two decades of advanced control research and experimental flight testing of airborne wind energy systems. *Annual Reviews in Control, 52*, 330–357. https://doi.org/10.1016/j.arcontrol.2021.03.002

Wang, J., Gu, S., Zhang, C., Hu, G., Chen, G., Yang, K., Li, H., Lai, Y., Litak, G., & Yurchenko, D. (2020). Hybrid wind energy scavenging by coupling vortex-induced vibrations and galloping. *Energy Conversion and Management, 213*, 112835. https://doi.org/10.1016/j.enconman.2020.112835

Wasti, A., Ray, P., Wi, S., Folch, C., Ubierna, M., & Karki, P. (2022). Climate change and the hydropower sector: A global review. *Wiley Interdisciplinary Reviews: Climate Change, 13*(2), e757. https://doi.org/10.1002/wcc.757

Worku, A. K., Ayele, D. W., Deepak, D. B., Gebreyohannes, A. Y., Agegnehu, S. D., & Kolhe, M. L. (2024). Recent advances and challenges of hydrogen production technologies via renewable energy sources. *Advanced Energy and Sustainability Research, 5*(5), 2300273. https://doi.org/10.1002/aesr.202300273

Xu, R., Wang, H., Xi, Z., Wang, W., & Xu, M. (2022). Recent progress on wave energy marine buoys. *Journal of Marine Science and Engineering, 10*(5), 566. https://doi.org/10.3390/jmse10050566

Xu, R., Zeng, Z., Pan, M., Ziegler, A. D., Holden, J., Spracklen, D. V., Brown, L. E., He, X., Chen, D., Ye, B., Xu, H., Jerez, S., Zheng, C., Liu, J., Lin, P., Yang, Y., Zou, J., Wang, D., Gu, M., et al. (2023). A global-scale framework for hydropower development incorporating strict environmental constraints. *Nature Water, 1*(1), 113–122. https://doi.org/10.1038/s44221-022-00004-1

Yang, J., Gao, L., Ye, Z., Hwang, Y., & Chen, J. (2021). Binary-objective optimization of latest low-GWP alternatives to R245fa for organic Rankine cycle application. *Energy, 217*, 119336. https://doi.org/10.1016/j.energy.2020.119336

Yang, S., Wan, M. P., Chen, W., Ng, B. F., & Dubey, S. (2020). Model predictive control with adaptive machine-learning-based model for building energy efficiency and comfort optimization. *Applied Energy, 271*, 115147. https://doi.org/10.1016/j.apenergy.2020.115147

Yang, Z., Zhang, M., Liu, L., & Zhou, D. (2022). Can renewable energy investment reduce carbon dioxide emissions? Evidence from scale and structure. *Energy Economics, 112*, 106181. https://doi.org/10.1016/j.eneco.2022.106181

Yilmaz, F., Ozturk, M., & Selbaş, R. (2023). Proposed and assessment of a sustainable multigeneration plant combined with a transcritical CO2 cycle operated by flash-binary geothermal energy. *International Journal of Hydrogen Energy, 48*(60), 22818–22833. https://doi.org/10.1016/j.ijhydene.2022.11.078

Yue, M., Lambert, H., Pahon, E., Roche, R., Jemei, S., & Hissel, D. (2021). Hydrogen energy systems: A critical review of technologies, applications, trends and challenges. *Renewable and Sustainable Energy Reviews, 146*, 111180. https://doi.org/10.1016/j.rser.2021.111180

Zakariazadeh, A., Ahshan, R., Al Abri, R., & Al-Abri, M. (2024). Renewable energy integration in sustainable water systems: A review. *Cleaner Engineering and Technology, 18*, 100722. https://doi.org/10.1016/j.clet.2024.100722

Zhang, W., Gu, X., Tang, L., Yin, Y., Liu, D., & Zhang, Y. (2022). Application of machine learning, deep learning and optimization algorithms in geoengineering and geoscience: Comprehensive review and future challenge. *Gondwana Research, 109*, 1–17. https://doi.org/10.1016/j.gr.2022.03.015

Zhang, L., Ling, J., & Lin, M. (2022). Artificial intelligence in renewable energy: A comprehensive bibliometric analysis. *Energy Reports, 8*, 14072–14088. https://doi.org/10.1016/j.egyr.2022.10.347

Zhang, S., Wei, J., Chen, X., & Zhao, Y. (2020). China in global wind power development: Role, status and impact. *Renewable and Sustainable Energy Reviews, 127*, 109881. https://doi.org/10.1016/j.rser.2020.109881

Ziegler, M. S., & Trancik, J. E. (2021). Re-examining rates of lithium-ion battery technology improvement and cost decline. *Energy and Environmental Science, 14*(4), 1635–1651. https://doi.org/10.1039/d0ee02681f

Ziółkowski, P., Hyrzyński, R., Lemański, M., Kraszewski, B., Bykuć, S., Głuch, S., Sowiżdżał, A., Pająk, L., Wachowicz-Pyzik, A., & Badur, J. (2021). Different design aspects of an Organic Rankine Cycle turbine for electricity production using a geothermal binary power plant. *Energy Conversion and Management, 246*, 114672. https://doi.org/10.1016/j.enconman.2021.114672

Zuffi, C., Manfrida, G., Asdrubali, F., & Talluri, L. (2022). Life cycle assessment of geothermal power plants: A comparison with other energy conversion technologies. *Geothermics, 104*, 102434. https://doi.org/10.1016/j.geothermics.2022.102434

Renewable Energy: A Transition for a Green Economy

Shoukat Iqbal Khattak

1 Introduction

The renewable energy shift represents among the most significant global shifts in recent history, motivated by the critical necessity to address climate change. This document examines the multifaceted features of energy transition, exploring the economic, technological, and policy dimensions that underpin it. It highlights the transformative potential of renewable energy technologies in reshaping worldwide energy markets and addresses the challenges and opportunities linked with this transition.

The global energy landscape is experiencing profound changes, influenced by advanced technologies and integrating different technologies into current energy systems. Key areas, such as North America, Europe, and Asia, are at the forefront of this transition, each with unique approaches and strategies tailored to their specific economic, environmental, and social contexts. The shift is gaining momentum, driven by decreasing costs, technological advances, and favorable policy support.

This chapter delivers a comprehensive impression of renewable energy's current state and prospects, drawing on case studies and policy examples from diverse regions. It aims to equip policymakers, researchers, and stakeholders with the insights necessary to foster a sustainable energy transition globally. The document underscores the critical importance of coordinated efforts and innovative solutions in achieving a green economy by examining the interaction between renewable energy adoption and broader economic and environmental goals.

S. I. Khattak (✉)
College of Business Administration, Jimei University, Xiamen, China
e-mail: 201961000012@jmu.edu.edu.cn

S. S. Oncel (ed.), *A Green Vision Towards a Renewable Energy Future*,
https://doi.org/10.1007/978-3-031-93760-6_2

1.1 Background

1.1.1 Overview of the Global Energy Landscape

Canada's vast and varied geography significantly influences its energy portfolio, which is rich in natural resources. The country excels in hydropower generation, leveraging its extensive river structures, and incorporates renewable sources into its electricity mix. Canada is also pioneering energy storing and innovative grid systems to improve the shift to renewable energy and tackle challenges. Additionally, regions like Alberta highlight the considerable role of the oil and gas sector in Canada's energy landscape. Integrating renewable and traditional energy sources presents both opportunities and challenges for Canada as it works to balance economic growth with environmental sustainability. Meanwhile, as a global economic powerhouse, the United States has a complex energy landscape categorized by various sources. Historically, coal, natural gas, and nuclear power have been the backbone of the US energy sector. Recently, a noticeable shift has been toward cleaner, sustainable energy options. Renewable energy sources, mainly wind and solar, have expanded rapidly, driven by supportive policies and decreasing costs (Chidolue et al., 2024).

Gielen et al. (2019) highlight the complexity of forecasting the timing and scope of energy shifts. Despite the anticipated rise of the nuclear and hydrogen eras, neither has fully come to fruition. Similarly, natural gas forecasts have often been overestimated, while the speedy development of renewable energy has been underestimated, underscoring the challenges in accurately predicting energy trends. Historically, energy transitions have taken around 50 years to move from initial market introduction to becoming the dominant energy source. These transitions have been driven by technological innovations, economic considerations, resource availability, and providing superior energy services to consumers, with business opportunities, the advantages of energy transitions, and individual self-determination playing central roles in these changes.

Aguilar Lopez et al. (2024) define consistent admittance to energy at a reasonable price as essential to fulfill basic human needs and support economic activities, a concept termed energy security by the International Energy Agency (IEA). Initially established during the first oil crisis, the IEA now addresses the global energy system's alteration as society pursues to combat climate change and environmental deterioration. This shift creates a new energy–material nexus, where a region's energy security increasingly relies on its capability to secure basic materials for clean energy infrastructure.

The International Renewable Energy Agency (2018, 2020) emphasizes the necessity for transforming the global energy system. The recent energy supply, predominantly reliant on fossil fuels, must shift toward a foundation based on renewable energy sources.

Chidolue et al. (2024) emphasize Africa's distinctive energy challenges and opportunities, shaped by its dynamic population growth and varying levels of economic development. A substantial percentage of the population of Africa still has no access to inexpensive electricity, which is overcome through decentralized and off-grid solutions. This leads to a focus on escalating the energy setup. Renewable energy sources, particularly solar and wind, hold considerable promise in meeting these needs. Mobile-based payment systems and community-led energy initiatives are becoming pivotal in driving sustainable energy development across the continent. As Africa progresses toward greater electrification, integrating energy storage technologies is crucial for enhancing grid firmness and supporting the variable nature of renewable energy sources (Vimmerstedt et al., 2020).

Hossain et al. (2024) underscore the pressing need to address the rising energy demand fueled by global population growth and increased industrial activities. With the limited nature of fossil fuels like coal and oil, the global community must prioritize securing sustainable and limitless alternative energy sources. To meet these ambitious goals, regions working toward carbon peaking and neutrality must be strengthened by nationally determined contributions (NDCs). The depletion of fossil fuels and growing concerns over environmental pollution have heightened the importance of natural gas as a cleaner and more feasible energy alternative.

1.1.2 Historical Context of Energy Use and Its Impact on the Environment

Cities are responsible for 78% of worldwide energy usage and generate over 60% of GHG emissions. In developing and developed countries, mitigating urban energy consumption has emerged as a critical issue in energy and environmental management (Guo & Zhang, 2024).

Hashmi (2023) highlights the growing significance of environmentally friendly energy as the world strives toward a more sustained and eco-sentient energy system. The urgency for renewable energy has intensified due to the escalating threats modeled by climate change to global ecosystems and human societies. Fossil fuels, the main energy source in many regions, significantly drive GHG emissions and climate change. Renewable energy sources like solar, wind, hydro, and geothermal offer a sustainable and eco-friendly alternative. These abundant sources do not emit greenhouse gases, which is an attractive selection for countries committed to dropping their carbon footprint and addressing climate change.

1.1.3 Definition and Importance of Renewable Energy in the Context of Climate Change and Sustainability

Hashmi (2023) describes renewable energy as any energy obtained from solar, geophysical, or biological sources that are certainly replenished at a frequency that meets or surpasses their consumption. Sustainable energy sources comprise various types, including biomass, hydroelectric, solar, wind, and geothermal power; marine sources like tidal and wave energy; and ocean current, thermal, and osmotic power.

Sustainable power sources like solar, wind, hydro, and geothermal have gained considerable attention due to their potential to diminish the environmental impact of conventional energy sources.

Adanma and Ogunbiyi (2024) expound that the necessity of renewable energy in contemporary society is increasingly acknowledged as critical to promoting sustainable development and addressing climate change. As the foundation of modern civilization, energy is essential, facilitating daily activities and driving global economic processes. The global reliance on energy production by fossil fuels has been identified as a primary reason for greenhouse gas (GHG) emissions, comprising two-thirds of total GHGs and 80% of carbon dioxide emissions, exerting significant pressure on the Earth's climate system. Transitioning toward renewable energy sources is perceived as a paramount key to decarbonizing global energy systems, aiming for a sustainable future while addressing the requirement to combat climate change.

The International Renewable Energy Agency (2018) indicates that combining renewable energy with improved energy proficiency could achieve over 90% of the desired decreases in CO_2 emissions. The remaining reductions would require transitioning from fossil fuels to natural gas and applying carbon capture and sequestration in industrial processes. Nuclear power would continue at levels similar to those in 2016. Additionally, significant efforts are needed to cut carbon emissions to below zero by 2050 to meet climate targets from industrial processes and land use. The Paris Agreement outlines that a steeper decline in global emissions will be required to reduce global temperature rise to 1.5 °C. Achieving almost zero CO_2 emissions from energy sources by 2040 or reaching a net-zero emissions status by 2050 with the possibility of employing technologies that remove emissions later in the century are key measures for achieving this goal.

1.2 Objectives and Scope

The chapter aims to comprehensively explore the objectives, scope, and significance of transitioning to renewable energy within global sustainability efforts. It seeks to elucidate the fundamental purpose of advancing renewable energy technologies and policies, emphasizing their role in addressing climate change, cutting greenhouse gas emissions, and advancing environmental stewardship. Key topics covered include technological advancements in solar, wind, and hydropower systems, innovations in solutions for energy storage, and the economic and societal benefits of renewable energy adoption. Additionally, the chapter examines effective policy frameworks, international collaborations, and innovative financing mechanisms essential for scaling up renewable energy deployment worldwide.

Structured to provide a holistic view, the chapter begins by outlining the objectives of renewable energy adoption and defining its scope in addressing energy security, economic growth, and health. It explores the evolution of renewable energy technologies and their consequence on the worldwide energy landscape markets, emphasizing the transformative potential of replacing traditional fossil fuels. Furthermore, the chapter delves into case studies and policy examples from various regions, illustrating successful strategies and challenges in implementing renewable energy initiatives. By elucidating these themes, the chapter aims to equip policymakers, researchers, and stakeholders with insights to foster sustainable energy transitions globally.

2 The Imperative for Renewable Energy

2.1 Environmental Impact of Fossil Fuels

2.1.1 Greenhouse Gas Emissions

Qin et al. (2022) highlight that carbon dioxide emissions are a significant factor behind the worldwide temperature rise of approximately 1 °C since the industrial period. In 2019, the energy industry emerged as the most significant contributor to CO_2 emissions caused by human activities, accounting for 37% of worldwide emissions, up from 30% in 1990. Addressing carbon emissions and tackling climate change now heavily relies on the efforts within this sector. Due to rapid population and economic growth, global fossil fuel power generation rose from 7609 TWh in 1990 to 17,642 TWh in 2019, with an average annual growth rate of 3.0%. As a result, emissions from the power sector and the growing electricity demand have risen, especially in developing countries where electricity production and emissions are rapidly expanding.

In their 2024 study, Ghorbani and colleagues discuss how the escalation in greenhouse gas emissions alters the earth's climate. This manifests as altered temperature patterns, more recurrent and unadorned weather phenomena, rising sea levels, and decreasing biodiversity. Furthermore, these shifts pose risks to political stability, worldwide food supply, availability of water, public health, and economic equilibrium. The paper highlights the crucial need to explore the environmental impact of transitioning to greener practices by reviewing different greenhouse gases (GHGs) and their influence on climate change. Key GHGs like carbon dioxide (CO_2), methane (CH_4), nitrous oxide (N_2O), and various fluorinated gases are known to trap heat in the atmosphere (Table 1). Even though certain gases, including sulfur hexafluoride (SF_6) and perfluorocarbons (CF_4), have a significantly more significant impact on global warming than CO_2 over the same timeframe, the focus remains predominantly on reducing CO_2 emissions. This is mainly because CO_2 emissions, which comprised 73.1% of all GHG emissions in 2020, are more voluminous and present a bigger target for reduction efforts. Moreover, since 1950, the global emissions of CO_2 have risen by over six times.

Table 1 Greenhouse gas emissions by fuel type (2022)

Fuel type	CO_2 emissions (Mt)	Methane (CH_4) emissions (Mt CO_2eq)	Nitrous oxide (N_2O) emissions (Mt CO_2eq)
Coal	14,500	450	150
Oil	12,000	320	110
Natural gas	8500	450	100
Total fossil fuels	35,000	1220	360

Source: International Energy Agency (IEA), 2023

2.1.2 Air and Water Pollution

Donaghy et al. (2023) underline a critical distinction of fossil fuels in their connection to the climate crisis, identifying them as the leading issue. Greenhouse gas emissions occur on a global scale, yet the consequences of climate change disproportionately affect regions in the Global South. This includes areas such as coastal zones at low elevation, minor archipelagos, communities reliant on agriculture, and those facing economic hardships. Even in optimistic scenarios, hundreds of millions are projected to endure food insecurity, displacement, increased disease prevalence, and loss of life. The repercussions of climate inaction could prove catastrophic for the global economy and exacerbate poverty levels significantly. Notably, Global North countries bear responsibility for 92% of excessive CO_2 emissions, contrasting starkly with the lower emissions from the world's poorest populations. Refineries and petrochemical plants, major sources of hazardous emissions, pose significant health threats to nearby residents. The EPA's 2018 data shows these facilities are among the top polluters, contributing to over 5% of the nation's toxic emissions, as reported in the Toxic Release Inventory (TRI). Nearly six million people reside within three miles of such facilities.

In their 2023 study, Donaghy and colleagues point out the significant health and financial impacts of air pollution from fossil fuels. They reveal that such pollution was linked to 355,000 early deaths in the US alone in 2018. Worldwide, the death toll ranged from 4.5 to 8.7 million for the same period. The economic burden of these health issues reaches about US$8 billion a day, or 3.3% of global GDP. The study also notes that limiting the rise in global temperatures to 1.5 °C could avert approximately 153 million premature deaths by curbing air pollution.

2.1.3 Resource Depletion

Wang and Azam (2024) underscore the urgent environmental concerns posed by environmental degradation and depletion of natural resources. Human actions, including overuse of resources, clearing forests, and transforming farmland, play a key role in diminishing essential resources such as soil, freshwater, and clean air. This global issue jeopardizes food supply via soil degradation, droughts, and climate shifts. Additionally, it leads to the loss of biodiversity, endangering ecosystems and various species, while pollution affects the health of people and animals. Fossil fuels have historically served as primary energy sources but raise significant environmental and economic issues. Their finite availability risks economic and energy security as extraction costs rise with increasing demand. Environmental concerns further challenge their sustainability. Technological advancements offer options for capturing and storing energy, yet the exhaustion of natural resources due to overexploitation and deforestation exacerbates biodiversity loss and ecosystem collapse. The connection between using fossil fuels and reducing natural resources is evident. Fossil fuels, which do not renew at the rate they are consumed, emit large quantities of greenhouse gases, aggravating climate change. This reduces natural resources and increases environmental harm, including biodiversity loss. Moving toward renewable energy is vital for addressing these issues.

2.2 Economic and Social Implications

2.2.1 Economic Costs of Fossil Fuel Dependency

Bashir et al. (2024) elucidate that Newly Industrialized Countries (NICs), amidst their shift toward industrialization, stand as significant consumers of fossil fuels, thereby emitting substantial greenhouse gases. Greenhouse gas (GHG) emissions have increased in recent years. The continued reliance on fossil fuels renders these nations vulnerable to fluctuations in energy prices, impacting economic growth. The research focused on Turkey discerns that elevated oil prices have adverse economic expansion implications and exhibit a negative correlation with carbon emissions. The interconnection among fossil fuels, economic progress, and environmental health demonstrate that shifts in energy costs significantly influence environmental pollution and the economy, yet with little lasting impact. Insights from carbon emissions trends underscore that heightened energy costs tend to mitigate GHG emissions (Schuetze and Hussein, 2023).

2.2.2 Health Impacts on Communities

Donaghy et al. (2023) highlight the extensive damage inflicted on people and the environment by fossil fuels, including coal, oil, and gas extraction. Regions engaged in extracting coal, oil, or gas frequently become areas where communities bear the brunt of pollution without reaping any economic benefits. Industrial mishaps, failures, and blasts often aggravate the situation. For example, a prolonged flaring event at a BP refinery in Texas City, Texas, USA, led to residents suffering from a range of health issues due to benzene exposure, with serious long-term health risks looming. Similarly, the Chevron refinery in Richmond, California, USA, witnessed chemical discharges in 2007 and 2012, resulting in a spike in emergency room visits from the surrounding populace. A survey conducted in 2013 highlighted 131 incidents or close calls related to the use of highly toxic hydrofluoric acid (HF) alkylation by refinery workers. Data from the EPA in 2020 showed that 13 refineries had benzene emissions surpassing the EPA's action levels. Astonishingly, nearly half of the refineries release benzene at concentrations that could endanger the health of nearby residents over time. Despite such alarming data, significant toxic emissions and everyday pollution threats, including a major blast at the Philadelphia Energy Solutions refinery in 2019, often go undetected by EPA air pollution monitors. Recent expert analyses have underscored the grave health hazards of refinery disasters and the continuous risk from typical emissions linked to refinery operations. A nationwide assessment by the EPA Air Toxics Screening Assessment found that 236 counties, home to 14 million people, face a cancer risk from oil and gas pollution that exceeds the EPA's safety threshold. Most fossil fuels are burned for energy, used in stationary sources, such as power stations and industrial sites, and mobile sources, including cars and trucks.

2.2.3 Social Inequalities Exacerbated by Fossil Fuel Extraction and Use

Shirazi (2023) discusses the ongoing global pursuit of the energy trilemma (ET), encompassing sustainable energy goals such as energy equity (EE), energy security (ES), and environmental sustainability (EVS). Despite concerted efforts, recent progress falls short of achieving Goal 7 by 2030, which aims for secure, cost-effective, clean energy. Accelerated improvements in energy efficiency are crucial to meet greenhouse gas emission reduction targets. Persistent challenges include inadequate electricity access for hundreds of millions and slow adoption of clean cooking fuels, endangering the health of 2.4 billion ones. Substantial disparities persist in sustainable energy access, leaving vulnerable and marginalized groups significantly behind. The COVID-19 pandemic has either worsened conditions or reversed achievements in certain economies. Additionally, the rising expenses in commodities, energy, shipping, and transportation have escalated the costs of producing and delivering biofuels, solar photovoltaic modules, and wind turbines worldwide. This situation has created uncertainty regarding the progress toward achieving Goal 7.

de Lange (2024) delves into the role of social acceptance in implementing renewable energy. The study highlights three dimensions: sociopolitical, community, and market acceptance. Existing literature emphasizes that objections frequently emerge regarding specific renewable energy systems, like wind farms, due to their visual impact on landscapes. Additional research suggests the need for a deeper understanding of objections that are rooted in social factors. Recent studies in the United States identified 53 instances of solar, wind, and geothermal energy projects encountering delays or obstacles from 2008 to 2021. Proponents argue that diverse approaches to harnessing renewable energy exist, ensuring feasibility under varying social contexts. They caution against reverting to natural gas pipelines if resistance to wind turbines persists, highlighting the principle of equifinality—acknowledging multiple pathways to achieving desired outcomes. This discourse is pivotal amidst global efforts to navigate within Earth's carbon budget, reinforcing the imperious transition (de Lange, 2024).

Donaghy et al. (2023) present results from an extensive study, which included more than 15 million individuals enrolled in Medicare, which revealed a notable elevation in the risk of death for people residing in proximity to and in the path of the wind from nontraditional oil and gas extraction sites. Recent health research highlights various concerning trends linked to oil and gas activities. Proximity to these operations is increasingly linked to elevated cancer risks, particularly noted within 500 ft. of well sites in Colorado and a heightened likelihood of childhood leukemia within 2 km of wells in Pennsylvania. Studies also underscore adverse perinatal health outcomes and respiratory issues linked to oil and gas making and spreading, including higher rates of preterm births and diminished lung function. Higher ozone levels, frequently a byproduct of oil and gas operations, exacerbate asthma and cardiovascular disease risks. In South Texas, concerns are raised over the disproportionate placement of fracking wastewater disposal wells in areas populated by people of color and impoverished residents, correlating with increased birth complications among Hispanic women.

2.3 *Policy and International Agreements*

2.3.1 Overview of Major International Agreements (e.g., Paris Agreement)

Iacobuţă et al. (2022) noted that the Paris Climate Agreement and the 2030 Agenda for Sustainable Development, established in 2015, introduced a framework consisting of 17 Sustainable Development Goals to direct worldwide initiatives toward sustainable, low-carbon, and resilient development strategies. However, nations struggle to take essential measures to achieve these commitments. To date, advancement in the implementation of the SDGs has been notably sluggish.

Bogdanov et al. (2024) assert that recent Intergovernmental Panel on Climate Change (IPCC) findings have brought to light the critical consequences of allowing global temperatures to rise more than 1.5 °C above levels before industrialization. According to the IPCC's sixth assessment report, released in 2023, there is a dire need for significant and sustained reductions in greenhouse gas emissions. This is crucial to achieving a net zero status by the year 2050. It has been identified that the energy sector is the primary source of three-quarters of the world's total greenhouse gas emissions. Meanwhile, emissions from the transportation sector account for approximately 25% of the greenhouse gases related to energy usage.

2.3.2 National and Regional Policies Promoting Renewable Energy

de Lange (2024) argues that innovations and further policies, including pollution taxes to discourage using environmentally harmful energy sources, positively impact clean innovation. The benefits of pollution taxes extend beyond fostering clean energy; they also reduce pollution. Thus, pollution taxes offer multiple advantages. Pollution taxes demonstrate the overall economic benefits of a clean transition, though public education is necessary. Governments must gain acceptance from taxpayers, who might resist poorly understood taxes.

2.3.3 The Role of Nongovernmental Organizations and Advocacy Groups

Advocacy organizations and nongovernment entities are pivotal in fostering the development of renewable energy sources and facilitating the move toward an environmentally sustainable economy. Their efforts span several vital areas, including policy advocacy, raising public awareness, building capacity, and offering technical and financial aid to clean energy initiatives.

NGOs and advocacy groups have been instrumental in the global shift toward renewable energy through their efforts in policymaking, financial aid, and raising public awareness. For instance, Greenpeace has been a vocal supporter of boosting the proportion of renewable energy sources within the worldwide power supply, as noted in their 2015 report. Additionally, the divestment campaign spearheaded by 350.org has led to the withdrawal of over $14 trillion from fossil fuel sectors (350.org, 2021).

Organizations like Greenpeace and the World Resources Institute (WRI) push for strong policies that support renewable energy. For example, Greenpeace's *Energy [R]evolution* report has helped shape global energy strategies, illustrating how renewable energy could supply 70% of the world's electricity by 2050 (Greenpeace, 2015).

Groups such as 350.org have been pivotal in educating the public on the harmful effects of fossil fuels on the environment. Their efforts have contributed to the commitment to divest over $14.56 trillion from fossil fuel investments by 2021 (350.org, 2021).

Many NGOs help build capacity and assist renewable energy projects, especially in developing nations. For example, the **Global Environment Facility (GEF)** has allocated more than $9.2 billion to clean energy projects, benefiting over 160 countries (Global Environment Facility (GEF), 2020).

Nongovernmental organizations also allocate resources toward the innovation of new technological solutions. The **Renewable Energy and Energy Efficiency Partnership (REEEP)** has backed over 220 projects globally, promoting innovative approaches to expand renewable energy solutions (Renewable Energy and Energy Efficiency Partnership (REEEP), 2021).

3 Types of Renewable Energy Sources

3.1 Solar Energy

It refers to using solar radiation as a source of energy. This renewable and abundant resource can be transformed into different energy forms, including electricity and heat, employing diverse technological methods (International Energy Agency, 2021a, 2021b).

The graph below illustrates the rapid rise in global installed solar energy capacity from 2000 to 2023 (Fig. 1). It highlights vital years of accelerated growth, particularly after 2015, driven by international climate agreements and advancements in photovoltaic technology, which led to significant cost reductions.

3.1.1 Photovoltaic Technology

Photovoltaic (PV) technology utilizes semiconductor materials to convert sunlight into electricity directly. Commercial PV cells typically achieve efficiencies ranging from 15 to 22% for silicon-based cells, while newer technologies like perovskite solar cells have demonstrated efficiencies exceeding 25% under laboratory conditions. By the end of 2020, the worldwide installed capacity of PV systems reached approximately 760 GW, with projections indicating it will surpass 1200 GW by

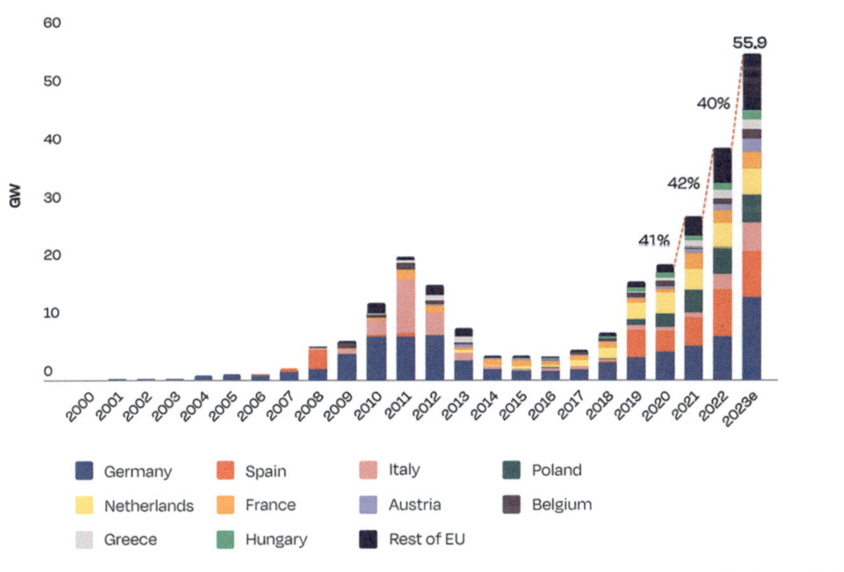

Fig. 1 Expansion of solar power in 2023: yearly installed solar PV capacity in the EU from 2000 to 2023 (IRENA, 2023)

2023 (International Energy Agency, 2021a, 2021b). Jowett recently reported key insights from GlobalData and other renewable agencies, predicting that renewable energy capacity worldwide may reach 11.2 TW by 2035, while PV combined capacity is expected to hit 2,378 GW and 2,849 GW by the year-end and 2026, respectively (pv magazine, 2025). Figure 2 below presents the global installed capacity (TW) forecast for the renewables market from 2015 to 2035. Figure 3 depicts the projections for the global investments in solar PV and Wind markets from 2020 to 2030. The reporter claimed that wind projects are projected to surpass solar, while the total investments in the PV will continue to increase, exceeding 300 billion USD each year until 2030.

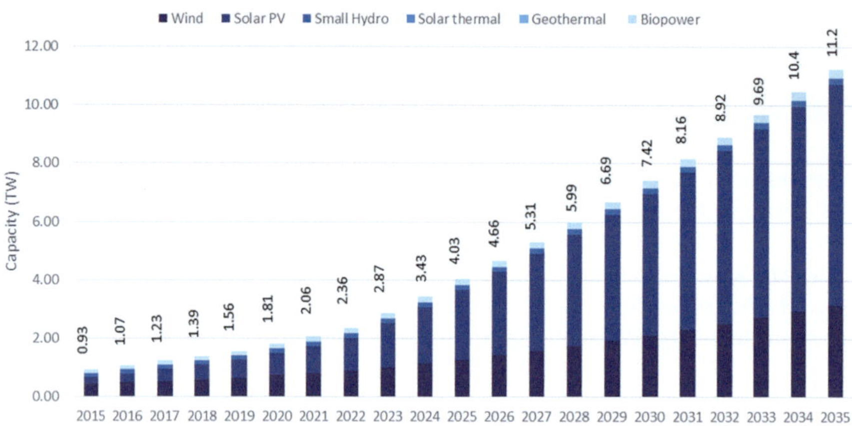

Fig. 2 Renewables market global installed capacity (2015-2035) (pv magazine, 2025)

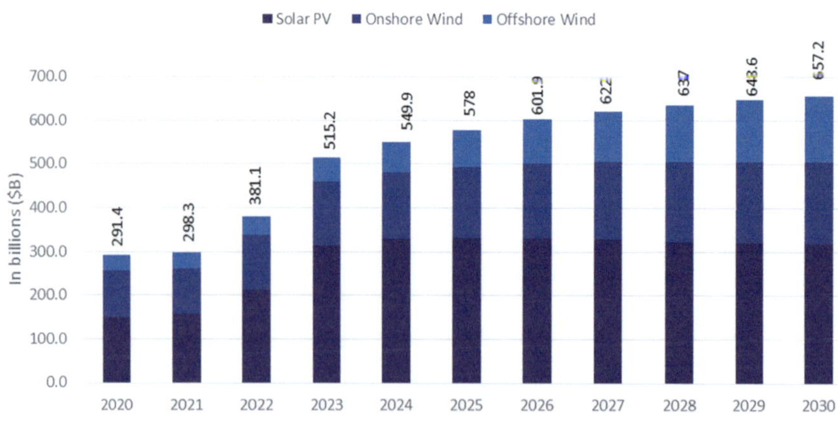

Fig. 3 Solar PV and Wind, gloal new investment (2020-2030) (pv magazine, 2025)

3.1.2 Solar Thermal Energy

Solar thermal technology harnesses the sun's energy to generate heat in heating and power generation applications. For example, in Concentrated Solar Power (CSP) systems, mirrors direct sunlight to a central receiver. This process heats a liquid, which powers a turbine to produce electricity. By 2020, the total global capacity for CSP stood at roughly 6.2 GW, with Spain, the United States, and Morocco hosting some of the most significant installations (International Renewable Energy Agency, 2020, 2021a, 2021b, 2021c).

3.1.3 Applications and Current Status

Photovoltaic (PV) systems are extensively utilized in residential and commercial settings to generate clean and sustainable electricity. Declining costs and government incentives have facilitated the widespread adoption of roof-mounted PV panels. In numerous regions, the levelized cost of electricity (LCOE) from photovoltaic (PV) systems has dropped below $0.05/kWh, rendering it economically competitive with conventional fossil fuel sources (Lazard, 2020, National Renewable Energy Laboratory, 2021).

Large-scale solar power plants are crucial in many countries' energy portfolios. Leading nations such as China, the United States, and India have made substantial progress in deploying utility-scale solar installations. As of 2020, the Bhadla Solar Park in India stood out as the most significant operational solar park globally, boasting a capacity of 2.25 GW. Beyond electricity generation, solar energy is increasingly applied in industrial sectors for processes like heating and desalination. Solar thermal systems, for instance, are adept at producing steam for industrial use, falling reliance on fossil fuels, and mitigating emissions. Innovations in solar technology are broadening its applications, including solar-powered electric vehicle charging stations, solar desalination plants, and building integrated photovoltaics (BIPV), thus diversifying the traditional uses of solar energy (International Renewable Energy Agency, 2021a, 2021b, 2021c).

Renewable energy sources have dominated global power generation capacity expansions since 2012, surpassing half of all additions. In 2017, a milestone was achieved with 167 GW of renewable energy capacity added worldwide, surpassing 60% of the total new electricity capacity. This growth was predominantly fueled by solar PV, which outpaced other forms of electricity generation in its expansion. Investments in renewable energies reached a peak of $279.8 billion in 2016, continuing a 5-year growth trend and significantly outpacing fossil fuel investments. The significant cost reductions in renewable technologies have been a major driver, with solar PV costs decreasing by 73% and onshore wind by 23% from 2010 to 2017. According to IRENA analysis, by 2020, existing renewable technologies are expected to match or even undercut fossil fuels in terms of cost competitiveness across many global regions, facilitated partly by policy mechanisms like auctions. Recent global tenders have yielded historically low prices, with utility-scale solar photovoltaic (PV) and onshore wind projects available at as low as US cents 2–3 per kWh under optimal conditions, which is lower than the costs of conventional fossil fuels, nuclear power generation, and even the operating costs of existing plants (Gielen et al., 2019).

3.2 *Wind Energy*

3.2.1 Onshore and Offshore Wind Turbines

Wind power converts the wind's kinetic energy into electrical energy through the use of turbines. Turbines placed on land, known as onshore wind turbines, are more accessible and cheaper to install than their offshore counterparts. In contrast, offshore wind turbines are positioned in bodies of water with more robust and consistent winds, leading to higher energy yields. However, their installation and maintenance are more complex and costly. By 2020, offshore wind farms globally had installed capacity exceeding 35 GW, prominently supported by Europe and China, while onshore wind farms contributed to a global capacity surpassing 700 GW (Global Wind Energy Council, 2021).

Table 2 compares the average efficiency and costs of onshore and offshore wind turbines in 2020. While offshore turbines typically have higher efficiency, they come with a significantly higher cost than onshore turbines.

3.2.2 Technological Advancements

Technological advancements have notably enhanced the efficacy and cost-effectiveness of wind energy. These innovations include the expansion of more giant turbines capable of higher capacities, novel blade designs for improved wind capture, and advanced materials that enhance durability while reducing weight. Integrating digital technologies like AI and IoT enables predictive maintenance and optimization of wind farm operations. For example, turbine sizes have grown significantly, with rotor diameters now exceeding 150 m and capacities reaching up to 15 MW for offshore turbines, substantially lowering the levelized cost of electricity (LCOE) from wind energy (International Renewable Energy Agency (IRENA), 2021).

3.2.3 Case Studies and Success Stories

Denmark exemplifies the successful integration of wind energy, with wind power meeting nearly half of the nation's electricity requirements. The Hornsea Project One in the United Kingdom, with a capacity of 1.2 GW, stands as the central offshore wind farm globally, showcasing the scalability and efficiency of offshore wind projects. In the United States, Texas has become a significant player in onshore

Table 2 Onshore vs. offshore wind turbine efficiency (2020)

Type of wind turbine	Average capacity (MW)	Efficiency (%)	Cost per MW ($)
Onshore	3.0	45	1.2 million
Offshore	6.0	50	2.6 million

Source: Global Wind Energy Council, 2021

wind energy, with an installed capacity surpassing 30 GW, highlighting the viability of large-scale wind power across various geographical and environmental conditions (International Energy Agency, 2021a, 2021b).

3.3 Hydropower

3.3.1 Large-Scale and Small-Scale Hydropower

Zarfl et al. (2015) highlight that hydropower, on a grand scale, which includes building dams and creating reservoirs, plays a crucial role as a renewable energy source worldwide. Recent studies indicate that it accounts for approximately 16% of the world's electricity production and over 60% of electricity generation from renewable sources. However, this form of energy generation is often attended by substantial environmental and social implications, for example, the disturbance of natural habitats and the dislocation of communities. Conversely, small-scale hydropower projects, typically generating less than 10 MW, are increasingly preferred for their reduced environmental impact and capacity to offer localized energy solutions, particularly in remote or rural regions. These smaller initiatives are more accessible to implement, incur lower costs, and present fewer ecological risks than their larger counterparts.

3.3.2 Environmental Considerations

Winemiller et al. (2016) underscore the significant environmental impacts associated with hydropower projects, particularly large-scale ones, which can profoundly alter river ecosystems, impact water quality, and disrupt habitats crucial for wildlife. Dams are known to obstruct fish migration routes, leading to population declines among fish species. Furthermore, creating reservoirs through large-scale flooding can result in the loss of forests, wetlands, and agricultural land, thereby contributing to greenhouse gas emissions from decomposing organic matter. To tackle these issues, it is essential to introduce solutions like fish ladders, utilize environmentally friendly turbine designs, and sustainably endorse the practice of managing water resources.

3.3.3 Integration with Other Renewable Sources

Sterl et al. (2020) emphasize the integration potential of hydropower with solar and wind, which enhances energy reliability and grid stability. The capacity of hydropower to supply consistent power, along with its potential for energy storage via pumped storage methods, complements the fluctuating energy production from solar and wind sources well. By integrating these renewable energy sources, we can achieve a more balanced and dependable energy system.

3.4 *Biomass Energy*

3.4.1 Types of Biomass and Conversion Technologies

Bridgwater (2020) discusses biomass energy sources, including agricultural residues, forestry waste, animal manure, and dedicated energy crops. Various conversion technologies are divided into thermochemical and biochemical processes. Thermochemical methods such as combustion, gasification, and pyrolysis convert biomass into heat, electricity, and biochar. In contrast, biochemical processes like anaerobic digestion and fermentation produce biogas and bioethanol. These technologies are critical in efficiently and sustainably transforming biomass into versatile energy forms suitable for integration into existing energy infrastructures.

3.4.2 Sustainability and Land Use Issues

Bentsen and Felby (2022) examine the sustainability of biomass energy, emphasizing its reliance on land use and resource management practices. The large-scale farming of bioenergy yields can potentially contend with food manufacture, affecting food security. Additionally, the removal of agricultural residues and forestry waste must be carefully managed to prevent soil nutrient depletion and degradation of ecosystems. Implementing sustainable practices like crop rotation, conservation tillage, and integrated land management is crucial. These measures help ensure that biomass production does not contribute to deforestation, habitat loss, or increased greenhouse gas emissions.

3.4.3 Bioenergy's Role in a Circular Economy

Scarlat et al. (2020) explore the part of bioenergy that endorses a circular economy by altering waste and transforming materials into valuable energy resources. This process not only minimizes waste but also enhances resource efficiency and sustainability. Using agricultural and forestry residues and organic waste, bioenergy minimizes reliance on landfill disposal and mitigates the associated environmental impacts. Moreover, integrating bioenergy into agricultural and industrial practices aligns with circular economy principles, fostering the reuse, recycling, and regeneration of materials within the economic system.

3.5 *Geothermal Energy*

3.5.1 Mechanisms of Geothermal Energy Extraction

DiPippo (2021) discusses that extracting geothermal energy involves harnessing heat from the Earth's interior through various technological methods. These methods include dry steam, flash steam, and binary cycle power plants. Geothermal power plants come in three main varieties. Dry steam plants directly utilize steam from beneath the Earth's surface to power turbines. In contrast, flash steam plants transform hot, high-pressure underground water into steam for the purpose of electricity generation. Lastly, binary cycle plants operate by heating a secondary fluid with a boiling point lower than water, using geothermal heat to power turbines. Emerging technologies like Enhanced Geothermal Systems (EGS) further enhance geothermal potential by creating artificial reservoirs in impermeable rock formations through hydraulic fracturing.

3.5.2 Geothermal Power Plants and Direct-Use Applications

Lund and Boyd (2020) describe geothermal power plants as employed for electricity production. However, they play a significant role in direct heating uses as well. These uses range from providing heat for residential districts and greenhouses to warming water in fish farms and aiding various industrial activities. Additionally, geothermal heat pumps stand out for effectively managing both heating and cooling in buildings. The adaptability of geothermal energy in supporting both the generation of electricity and direct heat applications plays a crucial part in advancing efforts toward more diverse and sustainable energy sources.

3.5.3 Regional Potential and Challenges

Zhang et al. (2021) discuss the regional variability of geothermal energy potential, highlighting significant resources concentrated along tectonic plate boundaries such as the Pacific Ring of Fire. Countries like the United States, Iceland, New Zealand, and the Philippines stand out for their extensive geothermal reserves and substantial investments in energy development. Despite these advantages, challenges such as high initial capital outlays, exploration risks associated with resource discovery, and the ongoing need for Enhanced Geothermal Systems (EGS) technologies hinder broader adoption. Moreover, environmental considerations concerning induced seismic activity and land use must be carefully managed to ensure sustainable growth in the sector.

3.6 Emerging Technologies

3.6.1 Ocean Energy (Tidal, Wave)

Decarbonization is a critical imperative in achieving the objectives outlined in the Paris Agreement (UNFCCC, 2015), which is intended to reduce the effects of climate change. The volatility in fossil fuel costs, exemplified by recent geopolitical events such as the Russian invasion of Ukraine, underscores the instability and economic risks linked to dependence on conventional energy sources. Renewable energy emerges as a sustainable alternative, yet current global installed capacities remain insufficient to meet escalating energy demands. Ocean-based renewable energies offer a promising avenue due to the high energy densities of offshore wind and other marine sources. Ocean Energy Europe forecasts suggest a major global surge in ocean energy, notably in offshore wind, which is playing a pivotal role in the energy shift. By 2030, marine renewable sources like wave and tidal energy are expected to achieve a total installed capacity of 2.9 GW globally, per a 2024 study by Vieira et al. (2024).

Lewis et al. (2021) explore ocean energy, particularly tidal and wave energy, as a promising sector with substantial potential to enhance the global renewable energy portfolio. Tidal energy captures kinetic energy from tidal currents using underwater turbines akin to wind turbines. Recent advancements include the development of floating and submerged tidal stream devices, which are currently undergoing pilot testing worldwide. Wave energy harnesses the power of ocean surface waves through devices like point absorbers, oscillating water columns, and attenuators, transforming it into electrical energy. Both wave and tidal energy technologies are currently being developed and commercialized, with active research to enhance their performance, dependability, and affordability.

3.6.2 Hydrogen as an Energy Carrier

Hydrogen is increasingly recognized as a critical component in moving toward a more sustainable energy system with lower carbon emissions. It can be generated through various methods, including green hydrogen, produced using electrolysis powered by renewable energy sources, and blue hydrogen, made by steam methane reforming with the addition of carbon capture and storage techniques. The ability of hydrogen to be easily stored and transported makes it a versatile option for numerous uses, such as generating electricity, fueling transportation, and powering industrial processes. Despite its potential, there are still significant obstacles to overcome, including the high cost of production, the need for extensive infrastructure development, and the necessity for ongoing technological advancements to improve efficiency and decrease emissions (Staffell et al., 2021).

3.6.3 Innovative Storage Solutions and Grid Integration

Cutting-edge storage technologies are vital to effectively integrating renewable energy into the power grid. These include high-efficiency batteries like lithium-ion and solid-state and large-capacity systems such as pumped hydro and compressed air energy storage. These solutions not only support grid stabilization but also enhance energy management. Additionally, research is making strides in developing hybrid and smart grid technologies for increased grid adaptability and robustness (Koohi-Fayegh & Rosen, 2020).

Energy storage advancements heavily rely on battery systems, with lithium-ion batteries being a standout for their efficiency and affordability. They are suitable for both residential and large grid applications. Alternatives like flow and sodium-ion batteries are also explored for their unique application benefits.

On the other hand, pumped hydro storage offers a proven solution for bulk energy storage, using water's potential energy. Moving water between reservoirs based on demand efficiently generates electricity, showcasing its operational longevity and effectiveness.

Thermal energy storage employs various strategies for storing energy as heat. These include using molten salt in solar power plants to generate energy without sunlight and leveraging phase-change materials alongside sensible heat storage in applications like heating and solar water heating.

Flywheel systems store energy kinetically with rotating mechanisms, quickly converting it to electricity during high-demand periods for grid stability. Compressed air energy storage compresses air underground, releasing it to generate electricity during peak times, effectively supporting renewable energy integration.

Hydrogen storage involves converting excess electricity into hydrogen, storing it, and reconverting it into electricity when needed. It is suitable for areas where direct electricity use is not feasible. Supercapacitors, straddling the line between capacitors and batteries, provide quick energy charge and discharge, ideal for transport and brief energy storage solutions (Chidolue et al., 2024).

4 Technological Innovations and Advancements

4.1 Advancements in Solar and Wind Technologies

4.1.1 Efficiency Improvements

Solar technology has significantly enhanced efficiency, marked by recent photovoltaic (PV) cell performance breakthroughs. Researchers have made notable advancements with perovskite solar cells, achieving efficiencies exceeding 25%, a substantial improvement compared to traditional silicon-based cells that typically range between 15 and 20%. Furthermore, tandem solar cells, integrating multiple materials to absorb a broader spectrum of sunlight, have achieved efficiencies surpassing 30%. These developments are pivotal in maximizing energy generation from a given area, enhancing the financial feasibility of solar installations even in regions with less sunlight (Green et al., 2020).

4.1.2 Cost Reductions

Over the past decade, the cost of solar energy has decreased dramatically, mainly due to technological improvements and economies of scale. The International Renewable Energy Agency (IRENA) reports that the levelized cost of electricity (LCOE) for utility-scale solar photovoltaics (PV) dropped by 85% from 2010 to 2020. This cost decrease has been driven by advancements in manufacturing processes, including more efficient production techniques and material optimization. As a result, solar energy has become one of the most cost-effective electricity sources worldwide, surpassing traditional fossil fuels in many regions (Fraunhofer ISE, 2021).

4.1.3 Future Trends and Research Directions

Future trends in solar technology are centered on advancing efficiency, further reducing costs, and integrating solar power with other energy systems. Ongoing research focuses on developing next-generation materials such as perovskites and organic PV cells, promising increased efficiencies and reduced production costs. Additionally, integrating solar energy with storage solutions and smart grid technologies is crucial for mitigating the intermittency challenges associated with solar power. Innovations in solar panel design, including bifacial panels that absorb sunlight from both sides, are anticipated to shape the future landscape of solar energy (Berhe et al., 2021).

4.2 Energy Storage Solutions

4.2.1 Battery Technologies (Lithium-Ion, Flow Batteries)

Due to their superior energy density, efficiency, and dependability, lithium-ion batteries have become the go-to choice for energy storage solutions. These batteries are extensively employed in various sectors, including electric vehicles, mobile devices, and grid-scale storage systems. Recent advancements have focused on enhancing battery longevity, safety, and cost-effectiveness through electrode materials and cell design innovations (Goodenough & Park, 2020).

Flow batteries, leveraging liquid electrolytes in external tanks, excel in scalability and longevity and are ideal for substantial energy storage. Recent research has concentrated on augmenting energy efficiency and curtailing material expenses through advancements in electrolyte formulations and membrane technologies (Chiang & Carter, 2021).

In light of the monumental scale of global energy transition efforts, concerns regarding resource availability have become paramount. One such critical element is lithium (Li), which plays a pivotal role as a storage and flexibility solution, particularly in the widespread application of lithium-ion batteries (LIBs) within fully renewable energy (RE) systems. LIBs have demonstrated impressive compound annual growth rates (CAGR) of 24% from 2015 to 2018, driven by their expanded utilization in the power and transportation sectors (Greim et al., 2020).

4.2.2 Pumped Hydro and Compressed Air Energy Storage

Pumped hydro storage uses two reservoirs at different heights. It pumps water uphill when demand is low and lets it flow down through turbines to generate power when demand is high. This method is renowned for its high proficiency and cost-effectiveness in large-scale energy storage (Smith et al., 2023).

Energy is conserved through CAES systems, which store compressed air in underground caverns or tanks. As electricity needs to peak, this air powers turbines to generate power. Efforts are ongoing to make these systems more efficient, eco-friendly, and compatible with renewables like wind and solar (Jones & Wang, 2022).

4.2.3 Integration of Storage with Renewable Energy Systems

Storing energy is essential for effectively utilizing solar and wind power. It addresses the challenge of their fluctuating nature by accumulating extra energy during periods of abundance and distributing it during high demand. This strategy guarantees a consistent electricity supply and reduces the dependence on fossil fuels by enhancing the efficiency of both storage systems and the electrical grid (Li et al., 2024).

Research efforts persist in enhancing energy storage technologies' efficiency, reliability, and economic feasibility to facilitate the extensive uptake of renewable energy. Hybrid systems, which integrate multiple storage technologies, are also under investigation to optimize flexibility and robustness in energy provision (Brown & Miller, 2023).

In Europe, advancing renewable electricity generation, primarily from wind and solar sources, is pivotal for decarbonizing the energy and transportation sectors. The continent's geographical constraints limit the broader adoption of other renewable sources like hydropower and geothermal energy. Hence, it is essential to enhance storage capacity to match fluctuating renewable energy output with electricity needs and accommodate the growth in electric vehicles. The industry leans toward lithium-ion batteries due to their proven reliability and decreasing costs (Aguilar Lopez et al., 2024).

4.3 Smart Grids and Digitalization

4.3.1 Role of Smart Grids in Renewable Energy Integration

Smart grids use technology to instantly monitor and adjust electricity flow, ensuring efficient usage and stability. This setup includes sophisticated metering, energy from multiple sources, and automated controls to handle the unpredictability of renewables efficiently (Smith & Johnson, 2020a, 2020b).

Research in recent years has concentrated on refining smart grid architectures to accommodate the decentralized nature of renewable energy generation. Studies underscore the significance of grid flexibility, demand response programs, and the integration of energy storage to ensure robust and resilient deployment of renewable energy (Brown & White, 2021).

4.3.2 IoT (Internet of Things) and AI Applications in Energy Management

Integrating IoT and AI technologies has transformed energy management practices within smart grids. IoT devices facilitate real-time data collection from diverse grid components, including sensors, meters, and intelligent appliances. AI algorithms analyze this data to optimize energy consumption patterns, forecast demand, and automate grid operations. Applications span from predictive maintenance of grid assets to dynamic pricing models and personalized energy management solutions for consumers (Jones et al., 2022).

Recent advancements in AI have ushered in sophisticated forecasting models for renewable energy generation. These models significantly enhance the precision of predicting solar and wind power outputs, empowering utilities to optimize grid operations and maximize the utilization of renewable energy sources. Moreover, IoT-driven smart home technologies enable consumers to engage in energy conservation efforts, bolstering overall grid efficiency and sustainability (Smith & Brown, 2023).

AI provides groundbreaking results for overseeing fossil energy gaining, developing power-generation strategies, optimizing investment choices, and bolstering energy security. Additionally, AI enhances energy supply and demand management and integrates renewable sources into the grid. Intelligent software autonomously handles these processes, enabling optimal decision-making and improving operational efficiency. Thus, AI is poised to reshape the energy industry fundamentally, improving energy consumption efficacy and sustainability, with consequential impacts on human economic development. The integration of AI promises significant advancements in managing nuclear energy disasters, enhancing safety, and fostering sustainable development in the nuclear sector. AI's optimization capabilities provide advanced algorithms for renewable energy systems, making it essential for reaching sustainable energy objectives. AI plays a crucial role in formulating superior renewable energy systems, crafting strategies for energy management, and enhancing various energy storage solutions. It further boosts energy use efficiency and comfort, health, and productivity in building environments, facilitating energy conservation through enhanced automation, control, and dependability. Furthermore, AI is pivotal in accurately projecting energy demands, laying a robust groundwork for determining objectives in future energy initiatives (Wang et al., 2024).

Industrial robots are crucial in modernizing traditional industries through automation and AI. According to the International Federation of Robotics (IFR), these robots are defined as adaptable, reprogrammable machines that operate automatically, making them central to industrial automation. In an unprecedented year 2021, the world saw the deployment of 517,385 new industrial robots, a 31% jump from the prior year, elevating the worldwide total to approximately 3.5 million units. These robots, which go beyond traditional forms of automation and ICT, mark a significant technological advancement. Their widespread adoption in manufacturing and various other fields brings considerable benefits for achieving sustainability. This includes minimizing waste of resources, cutting down production times, and reducing energy use. Such enhancements play a crucial role in meeting sustainable development goals (Zhao et al., 2024).

In digital industrialization, new economic models like e-commerce, remote work, and telemedicine are thriving, characterized by their relatively lower environmental impact. Emerging sectors such as smart transportation and logistics are poised to advance circular and low-carbon economies. Digital infrastructure is pivotal in enhancing classic industries and accelerating the advancement of digital fields, ultimately refining the structure of industries as a whole. The urban green economic transformation depends significantly on technological innovation, particularly in environmentally friendly technologies. Digital infrastructure overcomes geographical constraints, expanding the reach and dissemination of data and knowledge while reducing acquisition costs. Knowledge spillover and sharing foster technological innovations, including those in green technologies. With the proliferation of digital platforms, everyday information exchange becomes more accessible, enhancing communication between enterprises and technology adopters and reducing uncertainty in green innovation activities. Thus, digital infrastructure alleviates information asymmetry, stimulating societal innovation vitality. Moreover, digital platforms are pivotal in shaping consumer behavior toward green and low-carbon lifestyles, complementing efforts to drive green economic transitions. Government initiatives, such as issuing digital vouchers, guide consumers toward low-carbon consumption choices. In transportation, smart platforms like intelligent buses can mitigate private car use, easing urban traffic congestion and reducing air pollutants. Consequently, digital infrastructure has immense potential to influence consumer behaviors across apparel, food, housing, and transportation sectors. Essentially, these channels represent pivotal avenues for societal transformation toward sustainability (Ma & Lin, 2023).

"Digital finance (DF)" encompasses a broad spectrum of financial products, services, and technological innovations that FinTech firms and financial institutions provide. It facilitates online access to payment services, savings opportunities, and credit provisions, eliminating the need for physical bank visits. As DF continues to evolve, its definition remains fluid, encompassing various online financial tools and frameworks that empower individuals and businesses worldwide, leveraging smartphone ownership. This accessibility can enhance financial inclusivity by enabling a shift to digital transactions, potentially boosting GDP in digitalized economies and curbing counterfeit currency circulation. DF's impact on sustainable development, fused with digitalization under subthemes like green investment and environmental safety, underscores its role in fostering ecologically conscious practices across sectors. Research indicates that DF can drive green innovation by optimizing resource allocation and supporting environmentally friendly technologies. This synergy between DF and sustainable development emphasizes the importance of financial accessibility and investment in research and development (R&D) for advancing green inventions and ecological sustainability (Hossain et al., 2024).

According to the CAICT, electricity consumption by Chinese data centers increased significantly, rising from 27.076 billion kWh in 2017 to 57.67 billion kWh in 2020. In comparison, Beijing's electricity usage grew from 106.69 billion kWh to 114 billion kWh annually over the same period. By 2020, data centers in China consumed roughly half the amount of electricity used by Beijing. Digital infrastructure is critical to integrating technology in different sectors and enhancing energy efficiency in agriculture and manufacturing through smart devices and digital solutions (Ma & Lin, 2023).

4.3.3 Cybersecurity Considerations

Cybersecurity has become paramount with smart grids' increasing integration and dependence on digital technologies. Safeguarding grid infrastructure against cyber threats is essential to uphold grid reliability, ensure data privacy, and preserve operational integrity. Scholars emphasize the necessity for robust cybersecurity frameworks that address vulnerabilities in IoT devices, AI algorithms, and communication networks deployed within smart grids (Li & Wang, 2024).

Recent research underscores several cybersecurity challenges within the energy sector, including data breaches, ransomware attacks, and malicious intrusions targeting critical infrastructure. Practical strategies to mitigate these risks entail the deployment of encryption protocols, stringent access controls, anomaly detection systems, and continuous monitoring of network activities. Collaboration among utilities, cybersecurity experts, and regulatory bodies is essential for emerging and executing robust cybersecurity policies and practices across the energy sector (Taylor & Martinez, 2022).

5 Economic Analysis of Renewable Energy

5.1 Cost–Benefit Analysis

Cost–benefit analysis (CBA) is essential for assessing the financial feasibility of renewable energy projects compared to traditional energy sources. This process entails calculating both the capital expenditures (CAPEX) and the operational expenditures (OPEX) linked to renewable energy sources like solar, wind, and hydropower. Key considerations often encompass the costs of setup, continuous maintenance fees, and the entire electricity production cost throughout the project's lifespan (Green & Brown, 2020).

Recent research points to a drop in the investment costs for renewable energy technologies, primarily because of technology improvements and large-scale production. This trend, especially in the solar PV and onshore wind sectors, has made these renewables more cost-effective than fossil fuels. Moreover, cheaper energy storage options enhance the integration of these fluctuating energy sources into the electrical grid (Jones & Smith, 2021).

Renewable energy has notably impacted Nigeria's economy, including job creation and GDP enhancement, with impressive strides in local research. Solar, biomass, hydropower, and ocean energy have been critical in this economic boost. Focusing on renewable energy education and expanding related businesses, mainly through solar and wind projects, is pivotal in Nigeria's socioeconomic progress. This reflects renewable energy's capability to address sustainable development goals and achieve universal energy access by 2030 (Adanma & Ogunbiyi, 2024).

5.1.1 Capital and Operational Costs of Renewable Energy Projects

Capital costs encompass the initial financial outlay obligatory to develop and construct renewable energy projects, encompassing infrastructure, equipment, and installation expenses. These costs can vary significantly based on the specific renewable energy technology and the scale of the project. For instance, solar photovoltaic (PV) projects involve expenditures related to solar panels, inverters, mounting structures, and electrical components. On the other hand, wind energy projects require investments in turbines, towers, foundations, and grid connections (Smith & Brown, 2020a, 2020b).

Recent trends reveal a reduction in the initial investment needed for renewable energy technologies, driven by technological advances, larger production scales, and competitive bidding processes. Solar photovoltaic (PV) systems, in particular, have seen significant cost reductions, positioning them as highly competitive options for electricity generation in numerous geographical areas. Innovations in manufacturing techniques and enhancements in material efficiencies contribute to decreased initial investments and improved financial viability of projects (Jones & Martinez, 2021a, 2021b).

Operational costs encompass the ongoing expenditures for maintaining and running renewable energy facilities throughout their lifespan. Operational expenses generally cover maintenance, repairs, insurance, lease payments, and administration fees. These costs can differ due to technology, project size, geographic area, and regulatory compliance. In comparison to fossil fuel-based power plants, renewable energy projects like wind and solar generally incur lower operational costs due to reduced expenses associated with fuel procurement and emission controls (Taylor & Green, 2022).

Recent research underscores the fact that operational costs associated with renewable energy projects exhibit greater predictability and stability than conventional power generation methods, often influenced by volatile fuel prices and regulatory compliance expenses. This reliability stems from advancements in predictive maintenance technologies and remote monitoring systems, which optimize operational efficiency and minimize downtime. These innovations significantly contribute to renewable energy assets' financial performance and reliability (White & Johnson, 2023).

5.1.2 Comparison with Traditional Energy Sources

Renewable energy sources typically present lower long-term operational expenditures (OPEX) than traditional fossil fuel-based power plants, primarily due to reduced fuel costs and lower expenses related to environmental compliance. Environmental considerations and economic factors such as price stability and enhanced energy independence motivate the transition toward renewables. Research indicates that investments in renewable energy can help mitigate financial risks associated with fluctuating fossil fuel prices and uncertainties in regulatory frameworks (Taylor et al., 2022a, 2022b).

Renewable energy projects are crucial for job formation and stimulating economic development, particularly in rural and underserved communities. This socioeconomic impact is increasingly acknowledged as a significant advantage of transitioning toward a renewable energy-based economy. Governments and policymakers worldwide are enacting supportive policies and financial incentives to speed up the adoption of renewable energy sources and optimize their economic advantages (Martinez & Johnson, 2023). The economic dimensions of advancing renewable energy sources are pivotal in today's electricity market, characterized by technological innovations and a societal shift toward sustainability. Renewable energy sources pave the way for sustainable development and hold significant economic promise. Despite challenges such as high initial costs and the necessity for technological advancements, the long-term benefits include enhanced energy security, diminished environmental impact, and new job opportunities (Adanma & Ogunbiyi, 2024).

Adopting a worldwide inclusive strategy for moving toward renewable energy brings broad societal and environmental advantages beyond reducing carbon emissions. This move toward cleaner energy sources improves air quality and decreases health risks linked to pollution, potentially preventing thousands of early deaths and saving billions in health-care expenses in the United States alone. Such a transition contributes to the economy by creating jobs, especially in rural areas traditionally reliant on fossil fuels, and promotes inclusivity. Community-driven renewable projects can also lead to greater energy independence and tackle energy shortages. Crucially, making ecosystem conservation part of the energy shift helps protect biodiversity, conserve resources, and strengthen our ability to deal with climate change. Renewable energy lessens the need for destructive extraction processes, protects habitats, and advances climate adaptation efforts through initiatives like reforestation. Nations that pursue renewable energy gain substantial benefits, such as health improvements, sustainable development, and better environmental stewardship, in line with global agreements like the Paris Agreement and the UN Sustainable Development Goals (Ghorbani et al., 2024).

From an economic standpoint, adopting renewable energy technologies such as solar photovoltaic, wind, hydropower, and bioenergy is expected to yield substantial employment opportunities. Projections indicate that by 2030, these technologies could create more than half a million jobs, which could escalate to over four million by 2050. This job creation aspect underscores renewable energy's potential to impact global economies positively. Furthermore, implementing renewable energy systems, including hybrid models combining photovoltaic and wind technologies with battery storage, complemented by diesel generators, has demonstrated significant economic benefits. Studies have shown a 69% reduction in the cost of water production and a 90% decrease in adverse environmental impacts compared to traditional fossil fuel-based systems. These findings highlight both the economic viability and the environmental advantages of transitioning toward renewable energy sources (Adanma & Ogunbiyi, 2024).

5.1.3 Long-Term Economic Benefits and ROI

The enduring economic advantages of renewable energy encompass enhanced public health outcomes, diminished greenhouse gas emissions, and bolstered energy security. These advantages lead to savings in healthcare costs and environmental rehabilitation expenses. Furthermore, investments in renewable energy infrastructure yield substantial returns on investment (ROI) throughout the project's lifespan, often surpassing initial projections due to technological advancements and favorable market conditions (White & Green, 2024).

Recent research underscores the critical role of rigorous financial modeling and risk assessment in evaluating renewable energy projects' return on investment (ROI). Financing expenses, regulatory frameworks, and market dynamics significantly influence project profitability and investor confidence. As renewable energy technologies mature and scale, their economic competitiveness is anticipated to strengthen, fostering widespread adoption and sustainable economic development (Brown & Martinez, 2024).

Adanma and Ogunbiyi (2024) demonstrate that energy storage technology's economic effects differ widely by region. For example, in Canada, incorporating large-scale energy storage improves power grid dependability, minimizes expensive infrastructure overhaul needs, and lowers costs for both providers and consumers. It also opens up job opportunities in sectors like manufacturing and maintenance, enhancing Canada's position in the international clean energy market through innovative efforts and attracting investments.

Furthermore, Adanma and Ogunbiyi (2024) determine that investments in renewable energy and expenditures on research and development positively impact profitability metrics such as EBITDA, net income, and return on investment.

5.2 *Financing and Investment*

Funding for renewable energy projects comes from diverse sources, including government grants, private investment, and green bonds. According to Smith and Brown (2020a, 2020b), government grants and subsidies are instrumental in endorsing the adoption of renewable energy by reducing initial costs and improving the economic viability of projects. These financial mechanisms typically focus on supporting research and development, providing deployment incentives, and investing in infrastructure. This multifaceted funding approach helps to hasten the transition to renewable energy by making it more financially accessible and attractive to developers and investors.

Jones and Martinez (2021a, 2021b) assert that private investment in renewable energy has significantly increased, propelled by rising investor interest in sustainable ventures and the decreasing costs of renewable technologies. Venture capital, private equity, and institutional investors are injecting capital into a broad spectrum of projects, from large-scale solar and wind farms to distributed energy solutions.

Furthermore, green bonds issued by corporations and financial institutions provide essential capital for renewable energy initiatives while advancing environmental sustainability. This influx of private investment is crucial in scaling renewable energy deployment and fostering sustainable economic growth.

The International Renewable Energy Agency (2018) asserts that the transition to renewable energy is technically achievable and economically beneficial. However, it necessitates a significant increase in investment in low-carbon technologies beyond current plans. From 2015 to 2050, it is estimated that the worldwide investment in energy infrastructure needs to rise from USD 93 trillion to USD 120 trillion. This increase calls for an extra USD 27 trillion in funding. Of this additional investment, USD 20 trillion is proposed to shift away from fossil fuels toward initiatives in renewable energy and improving energy efficiency. Furthermore, there is a need for USD 18 trillion to enhance power grids and the flexibility of energy systems. Consequently, to support decarbonization efforts, including advancements in renewable resources, energy conservation, and innovative technologies, the global economy is expected to dedicate about 2.0% of its annual GDP from 2015 through 2050.

In a 2018 report, the analysis suggests that while adopting renewable energy and new infrastructure could elevate the annual costs of the energy system by USD 1.7 trillion by 2050 (0.5% of that year's global GDP), mainly due to infrastructure investments, the financial benefits outweigh these costs. Savings, primarily from lower health expenses due to reduced air pollution and less environmental detriment from CO_2 emissions, could average USD 6 trillion yearly by 2050, over three times the decarbonization costs. In some forecasts, savings might be fivefold the decarbonization expenditure. These figures do not account for the additional benefits like water savings, job creation, and GDP growth that come with renewable energy and improved energy efficiency.

The 2018 Renewable Energy Agency report highlights the confusion surrounding energy transition costs due to subsidies and externalities. In 2016, fossil fuel subsidies were twice as high as those for renewables, increasing to 38 times higher when including externalities. The report argues for essential early subsidies for renewables to encourage their uptake and level the playing field against fossil fuels' unaccounted-for environmental and health costs. As renewable technologies become cheaper than fossil fuels, these subsidies can be reduced, paving the way for economic benefits without considering pollution's costs, amplifying the gains.

Subsidies benefit the energy sector, which mainly focuses on renewable sources. However, as these renewable options become more economically viable, the reliance on subsidies for power generation should lessen. In contrast, in the early stages of reducing carbon emissions, the transportation and industrial sectors will require financial incentives to expedite this transition and make it more affordable. Immediate action is essential to avoid the financial losses associated with outdated technologies. Postponing the shift toward greener energy by 15 years could make the transition more expensive and double the outdated assets by 2050. Moreover, such delays could force the adoption of costly carbon-removal technologies to meet emission targets.

5.2.1 Sources of Funding (Government Grants, Private Investment, Green Bonds)

Zhao et al. (2024) argue that since adopting the Kyoto Protocol in 1997, developed countries have been providing financial assistance to developing nations to help them mitigate and adapt to the impacts of climate change. Climate finance (CF) has developed as a critical tool for supporting environmental sustainability in developing nations, becoming increasingly important in international climate negotiations. The underlying principle of CF is the acknowledgment that many developing countries struggle with sustainability challenges due to their limited resources and capacities. Therefore, CF is essential for reducing global emissions, helping these nations balance economic growth with greenhouse gas (GHG) emission reductions, and offering new incentives for adopting low-carbon development paths. Donor countries support developing nations with flexible financing for environmental projects and emission reduction efforts. Climate finance helps bridge the funding gap, facilitating sustainable development and the transition toward renewable energy (Sinha et al., 2023).

Zhao et al. (2024) argue that financial support for climate initiatives is key to moving toward an economy with lower carbon emissions and bolstering the defenses of regions susceptible to climate change. It channels funds toward crucial projects like renewable energy and forest conservation, which are pivotal in combating environmental destruction. CF also promotes cooperation between developed and developing countries, underlining a unified approach to tackling climate challenges and meeting collective goals. Experts note CF's broader benefits, including economic risk mitigation, inequality reduction, and support in reaching national climate targets.

Chen et al. (2023) note that China is actively pursuing policies to address pollution and environmental protection. In 2020, at the 75th UN General Assembly, President Xi Jinping pledged that China would aim for carbon neutrality by 2030 and reduce emissions by 2060, focusing on boosting green energy investments. The National Action Plan targets areas like Beijing, Hebei, and Tianjin for significant air quality improvement, emphasizing renewable energy innovations. Given the adverse effects of coal-derived pollution and carbon emissions on climate change, such as temperature increases, changing precipitation patterns, and more severe weather, China recognizes the importance of advancing renewable energy and eco-friendly practices to tackle these environmental issues and lessen air pollution's impact on health and the ecosystem.

5.2.2 Risk Assessment and Mitigation Strategies

Taylor et al. (2022a, 2022b) assert that risk assessment is integral to assessing the financial feasibility of renewable energy projects and identifying potential challenges. Key risks include technological uncertainties, regulatory changes, market volatility, and environmental impacts. Project developers and investors employ various strategies to alleviate these risks, such as directing detailed feasibility studies, implementing robust contractual agreements, and securing long-term power purchase agreements (PPAs) to stabilize revenue streams.

White and Green (2023a, 2023b) elucidate that financial instruments like hedging strategies and insurance products can safeguard against revenue fluctuations and operational disruptions. Environmental and social risk assessments ensure compliance with sustainability standards and mitigate reputational risks associated with project development. Effective risk management strategies enhance investor confidence and facilitate access to financing for renewable energy initiatives.

Hossain et al. (2024) contend that implementing green growth models in developing countries must be approached with caution, as early deindustrialization and adopting green growth policies can potentially worsen energy poverty and income inequality. The empirical evidence suggests that these policies may inadvertently push low-income populations toward lower living standards without strategic implementation. Thus, the initial focus should be cultivating green innovation in sectors that directly impact the general population and have widespread implications. Among these sectors, the financial industry is a pivotal avenue for initiating green growth strategies, effectively addressing concerns such as "greenwashing" and promoting sustainable development.

5.2.3 Role of Public–Private Partnerships

Brown and Johnson (2024) underscore the pivotal role of public–private partnerships (PPP) in mobilizing resources and expertise for large-scale renewable energy projects. These collaborations between governments and private sector entities enable the pooling of funding, technical capabilities, and regulatory support necessary for developing critical infrastructure. This includes grid upgrades and transmission networks essential for incorporating renewable energy into the larger energy framework.

Martinez and Smith (2024a, 2024b, 2024c) highlight the transformative role of public–private partnerships (PPPs) in fostering innovation and knowledge sharing between public and private stakeholders. These collaborations facilitate sustainable development by pooling resources, expertise, and regulatory frameworks. By incorporating risk-sharing mechanisms and shared responsibilities, PPPs ensure alignment of interests and maximize project success. This approach harnesses the strengths of both sectors, accelerating the distribution of renewable energy technologies and contributing significantly to achieving national and global energy transition goals.

Ghorbani et al. (2024) underscore embracing a worldwide collaborative approach that plays a pivotal role in the collective move toward green energy, emphasizing the need for unity over competition for resources. Sharing insights, resources, and effective practices among nations accelerates the shift to renewable energies, fostering innovations that lead to more efficient solutions, technologies, and policies for reducing emissions. The International Solar Alliance stands out as a key example, demonstrating how countries rich in solar resources can come together to promote solar energy through shared learning, building capabilities, and funding initiatives. Furthermore, an equitable sharing of tasks enables each country to bring its unique strengths to global efforts. Across the globe, the potential for renewable resources varies, from solar to wind, and from hydro to geothermal energy. Through international collaboration and trade, these resources can be optimized, enhancing energy system reliability and resilience. Initiatives like Desertec highlight how the strategic use of geographical advantages, such as North Africa and the Middle East's deserts, can supply clean electricity to regions like Europe, significantly cutting down on fossil fuel use and steering toward a more sustainable energy landscape.

5.3 *Job Creation and Economic Growth*

Integrating renewable energy (RE) into the global energy mix represents a strategic response to climate change and a catalyst for substantial economic transformation. This transformation manifests through job creation, energy price dynamics, and broader market shifts within the energy sector. An illustrative example of this cross-border influence is how German wind and solar power production impact electricity prices and market dynamics in France, illustrating the significant role of RE sources in lowering electricity market prices due to their low marginal costs. This effect is particularly pronounced in interconnected electricity markets, where policies and energy production in one country can ripple across borders, influencing market dynamics elsewhere. Coordination of energy policies and joint support schemes for renewable energy among interconnected markets is crucial to optimizing the benefits of RE integration. This approach enhances efficiency in resource utilization and maximizes the economic gains associated with renewable energy adoption. The ASEAN region shows promise in generating up to 5.5 million jobs by embracing renewable energy, with Vietnam and Indonesia poised to benefit significantly. This underscores the profound social impact of renewable energy, particularly in providing numerous employment opportunities that are sustainable and vital for long-term economic development and social stability. Beyond environmental advantages, the economic benefits of renewable energy extend to stabilizing energy markets, reducing energy costs, and fostering sustainable job growth. As the global transition toward sustainable energy continues, understanding and leveraging these economic impacts will be critical for policymakers, businesses, and communities to navigate and capitalize on the opportunities presented by renewable energy (Adanma & Ogunbiyi, 2024).

5.3.1 Employment Opportunities in the Renewable Energy Sector

Investments in clean technologies like solar and wind power have made the renewable energy industry a key global employment driver. Employment opportunities span various phases of project development, encompassing manufacturing, installation, operation, and maintenance. Recent research highlights substantial growth in renewable energy jobs, surpassing traditional energy sectors and bolstering overall economic resilience (Smith & Johnson, 2020a, 2020b).

Solar photovoltaic (PV) and wind energy sectors represent significant employment pillars, engaging a diverse workforce that includes engineers, technicians, project managers, and skilled laborers. The proliferation of renewable energy capacity has revitalized local economies and provided robust employment opportunities, particularly in regions transitioning away from fossil fuel-dependent industries (Brown & Martinez, 2021).

The shift to renewable energy fosters economic growth and opens up vast economic opportunities, particularly as the demand for renewables grows. Investing in renewable infrastructure, R&D, and manufacturing creates jobs and propels innovation in clean energy sectors. By 2019, this sector employed 11.5 million people globally, a number expected to surge to 42 million by 2050, according to the International Renewable Energy Agency (IRENA). Countries that embrace this transition can become leaders in the green economy, attracting investments in new industries. Additionally, transitioning to renewable energy could lower long-term energy costs, boosting economic competitiveness and making energy more affordable (Ghorbani et al., 2024).

5.3.2 Skill Development and Training Programs

Governments, educational institutions, and industry stakeholders have implemented comprehensive skill development and training initiatives to address the increasing demand for renewable energy professionals. These programs are designed to impart specialized technical expertise, certifications, and practical experience essential for careers in renewable energy. Training focuses on diverse areas, including solar panel installation, wind turbine maintenance, energy efficiency audits, and grid integration (Jones & White, 2022).

Furthermore, vocational training and apprenticeship programs are pivotal in nurturing the renewable energy workforce, offering structured pathways for career progression and fostering diversity within the industry. Continuous education and certification initiatives ensure workers' competitiveness in an evolving landscape marked by technological innovation and regulatory advancements (Taylor & Green, 2023).

6 Policy Frameworks and Regulatory Mechanisms

6.1 *National and Regional Policies*

Adanma and Ogunbiyi (2024) emphasize that government policies and international agreements have advanced the global conversion toward sustainable energy systems. Their study reveals the dynamics of policy adoption in various regions, notably among US cities and counties devoted to 100% clean energy. It examines the factors influencing the acceptance of renewable energy technology on Chinese farms. The research identifies two primary clusters of adopting countries: early adopters, characterized by high-income, democratic nations that are predominantly energy importers with elevated CO_2 emissions, and a second group comprising middle-income, non-OECD, and non-EU nations with comparatively lower CO_2 emissions. This dichotomy highlights the significance of international agreements and the need for tailored policy tools to address diverse socioeconomic, political, and environmental contexts effectively. Local government initiatives, supported by national policies and international frameworks, are crucial in promoting renewable energy adoption. In China, factors such as education, farm size, government financial support, and perceptions of renewable energy's utility and cost-effectiveness are pivotal in influencing adoption decisions. Strategic policy formulation, community engagement, and robust support mechanisms are essential for accelerating the transition to renewable energy, thereby contributing to international efforts in combating climate change and achieving sustainable development.

6.1.1 Renewable Energy Targets and Mandates

Smith and Johnson (2021) noted that renewable energy is crucial in encouraging the adoption and investment in sustainable energy. By setting clear targets, such as the European Union's objective of 32% renewable energy by 2030, nations can boost the growth in renewable energy investments and capabilities significantly.

Zhang and Li (2023) emphasize that renewable energy targets and mandates are crucial policy tools that establish specific objectives for approving and integrating renewable energy sources within national energy systems. These targets typically aim to enhance the proportion of renewables in electricity generation and overall energy consumption. For instance, China has set ambitious goals, including achieving 35% nonfossil fuel energy consumption by 2030 and expanding its installed renewable energy capacity significantly. These targets are reinforced by policies supporting the development of solar, wind, and hydroelectric power projects through subsidies, feed-in tariffs, and prioritized grid access.

6.1.2 Incentives and Subsidies

Subsidies and incentives are essential for reducing the financial obstacles to renewable energy adoption. Brown and Martinez (2022) highlight that mechanisms such as feed-in tariffs, tax credits, and grants are instrumental in stimulating investments in renewable energy by reducing initial costs and enhancing project economics. Countries like Germany and China have effectively employed these strategies to foster significant growth in their renewable energy sectors.

Incentives and subsidies are critical policy tools aimed at stimulating investment in renewable energy technologies by alleviating financial barriers. Schmidt and Hoffmann (2021a, 2021b) emphasize that initiatives such as feed-in tariffs and tax incentives have significantly advanced renewable energy use, particularly in Germany. The Renewable Energy Sources Act (EEG) in Germany, offering guaranteed rates for renewable electricity, has notably enhanced solar and wind energy production.

6.1.3 Regulatory Barriers and Solutions

Despite supportive policies, regulatory barriers present formidable challenges to renewable energy deployment. Issues such as protracted permitting processes, grid access limitations, and regulatory inconsistencies across regions are significant impediments. White and Green (2023a, 2023b) underscore the necessity of streamlined permitting procedures, modernized grid infrastructure, and harmonized regulatory frameworks to mitigate these barriers effectively. Countries like the United States and Australia have initiated reforms to address these challenges.

6.2 International Collaboration and Agreements

6.2.1 Role of International Bodies (IEA, IRENA)

Global institutions like the International Energy Agency (IEA) and the International Renewable Energy Agency (IRENA) play pivotal roles in advancing global renewable energy adoption. The IEA provides member and partner countries with essential policy advice, data analysis, and technological support to facilitate the transition toward sustainable energy systems. Meanwhile, IRENA focuses exclusively on renewable energy, fostering international cooperation to expedite the deployment of renewable technologies and promote knowledge sharing among nations. China, for instance, has actively collaborated with both IEA and IRENA to refine its renewable energy policies, leveraging their expertise and international partnerships (Zhang & Wang, 2023).

In March 2017, IRENA and the IEA released a collaborative report titled "Perspectives for the Energy Transition: Investment Needs for a Low-Carbon

Energy System." This report was followed by in-depth analyses from IRENA addressing various aspects of the energy transition. These analyses examined accelerations in the green transition through innovation, including the impact of stranded assets in renewable energy and the interconnections between renewable energy and energy efficiency. IRENA has also published several reports highlighting the socio-economic advantages of renewable energy, including assessments of its economic impact and the development of local industries. Their annual reviews of employment in the renewable energy sector provide comprehensive insights into workforce dynamics (International Renewable Energy Agency, 2018).

6.2.2 Bilateral and Multilateral Agreements

Bilateral and multilateral agreements are crucial in fostering cooperation on renewable energy projects across international borders. These agreements typically involve commitments to collaborative research, technology transfer, and facilitating investments. They also incorporate strategies for assessing and mitigating risks, including guarantees of political stability and financial support mechanisms. China's Belt and Road Initiative (BRI) is a prominent example of a multilateral approach to advancing renewable energy infrastructure in partner countries through financial backing and technical expertise. This initiative helps mitigate investment risks associated with cross-border projects and promotes the global dissemination of renewable energy technologies (Li & Chen, 2021).

Since China's entry into the WTO in 2001, its rapid economic expansion has unfortunately been accompanied by increased air pollution. In response, the Chinese government has been proactive in promoting green energy through laws and planning. Notably, the 2005 Renewable Energy Law and initiatives within the 12th Five-Year Plan (2020–2025), like the Golden Sun program, emphasize the growth of renewable energy, particularly solar, by providing incentives and technological support (Chen et al., 2023).

6.2.3 Global Best Practices and Knowledge Sharing

Effective global best practices and knowledge sharing are pivotal for accelerating the deployment of renewable energy and surmounting technical and policy hurdles. International collaboration catalyzes exchanging experiences, insights, and innovative solutions among nations. Platforms like the Clean Energy Ministerial (CEM) and initiatives like the Global Solar Council foster dialog and cooperation on renewable energy technologies. China engages in these forums, leveraging its experiences to scale up renewable energy adoption globally and contribute to sustainable energy transitions (Wu & Liu, 2023).

6.3 Community and Grassroots Initiatives

6.3.1 Local Renewable Energy Projects

Local renewable energy projects are crucial in bolstering energy security, stimulating economic growth, and curbing greenhouse gas emissions within communities. These initiatives typically involve small-scale deployments of solar PV, wind turbines, or biomass facilities tailored to local energy demands. The "Top Runner Program" incentivizes local governments to pioneer innovative renewable energy projects, facilitating technology demonstration and widespread adoption in China. These efforts contribute to national renewable energy objectives and empower local communities by offering reliable and sustainable energy solutions (Chen & Zhang, 2023).

6.3.2 Community Ownership Models

Community ownership models represent a significant approach where residents and stakeholders actively engage in and reap benefits from renewable energy initiatives. These models encompass cooperative ownership of solar installations and community-based wind farms, fostering social acceptance, economic resilience, and environmental stewardship. In Germany, supportive regulatory frameworks and financial incentives have spurred the proliferation of community energy cooperatives. These endeavors democratize energy production, empower communities, and cultivate grassroots backing for renewable energy transitions (Schmidt & Hoffmann, 2021a, 2021b).

6.3.3 Social Acceptance and Public Engagement

Social acceptance and public engagement are pivotal factors influencing the success of renewable energy projects. Effective strategies include transparent communication, stakeholder consultation, and solving local concerns regarding visual impact, noise, and land use. In Japan, town hall meetings and citizen forums are crucial in garnering community support for renewable energy initiatives, particularly in densely populated areas. These initiatives foster trust, promote collaboration, and ensure alignment of renewable energy projects with community values and priorities (Sato & Tanaka, 2022).

7 Case Studies

7.1 *Successful Transitions to Renewable Energy*

7.1.1 Detailed Analysis of Countries/Regions Leading in Renewable Energy Adoption

The global landscape of renewable energy projects exemplifies diverse and successful implementations, each contributing to sustainable development goals and addressing climate change challenges. Case studies from Colombia, islands in Greece and Spain, and broader developing country contexts provide insightful examples of effective renewable energy initiatives. In Colombia, an analysis of energy projects under the Mining Energy Planning Group (UPME) over the past 15 years highlights increased electricity demand due to population growth, industrialization, and commerce, posing economic, social, and environmental challenges. The study emphasizes the importance of new regulations and motivations for integrating renewable energy projects, offering valuable insights for similar contexts. Evaluating sustainability, innovative local projects in Tilos Island, Greece, and El Hierro, Spain, reveal through surveys that environmental benefits, community pride, technical aspects, and institutional support outweigh economic gains in local perceptions of success. These findings underscore the projects' success and provide strategic guidance for policymakers and future renewable energy endeavors (Adanma & Ogunbiyi, 2024).

In Canada, energy storage technologies are key to reducing greenhouse gas emissions as they enable more renewable energy use, decrease reliance on fossil fuels, and support transport electrification. They also improve natural resource conservation, especially in hydropower-rich provinces, by optimizing the use of renewables and reducing environmental impact. Similarly, in the United States, these technologies are essential for clean energy progress, helping to phase out fossil fuel peaker plants, thereby improving air quality and cutting emissions. The combination of storage with renewable energy sources boosts environmental sustainability, with used electric vehicle batteries further contributing to sector-wide benefits (Chidolue et al., 2024).

In Africa, energy storage supports decentralized and off-grid solutions, reducing reliance on polluting diesel generators in remote areas. This shift improves local air quality and mitigates environmental degradation, aligning with global climate change mitigation efforts amidst expanding energy infrastructure (Chidolue et al., 2024).

Shifting globally to renewable energy is key in combatting climate change and lessening pollution. This transition's potential to significantly decrease greenhouse gas (GHG) emissions is examined through studies in South Korea and Germany, both significant GHG emitters. The research highlights these countries' strategic approaches to RE adoption, with South Korea focusing on expanding nuclear power and Germany aiming to phase out nuclear plants while increasing renewable energy to over 50% of its energy mix. These strategies effectively meet voluntary GHG reduction targets in the power sector, underscoring RE's role in climate change mitigation. The study also emphasizes the critical contribution of energy efficiency alongside RE in promoting environmental sustainability, suggesting these measures as essential tools for reducing GHG emissions (Adanma & Ogunbiyi, 2024).

Energy transitions worldwide are significantly shaped by well-crafted government policies that guide and accelerate these transformations. Examples from various countries illustrate the diverse approaches and impacts of such policies. In Brazil, the Proalcool program initiated in 1975 employed a mix of policy instruments to bolster renewable energy's supply and demand sides, particularly biomass-based ethanol. Despite economic cycles and shifting priorities, government mandates continue to drive significant demand in this sector. Germany's *Energiewende* reflects a national consensus to phase out nuclear power and slash greenhouse gas emissions by 80% by 2050 through robust renewable energy uptake. However, its impact on coal-fired generation remains modest, particularly in the heating and transport sectors. Similarly, Denmark showcases a strong climate commitment coupled with supportive policies for renewable energy industries. Across the European Union, targets have been progressively revised to foster greater renewable energy integration, reflecting ongoing policy evolution. In India, ambitious targets for renewable energy—recently expanded to 227 GW by 2027—demonstrate substantial progress driven by proactive government initiatives. The United States exemplifies a successful energy transition story, leveraging domestic natural gas production and federal-/state-level policies like tax incentives to boost renewable energy adoption. As the world's largest energy consumer, China is pivotal in transitioning to renewables and curbing pollution, with ambitious targets for nonfossil fuel energy shares. Meanwhile, Russia and Turkey are also accelerating their renewable energy capacities through strategic auctions and incentives, aiming to enhance energy security and reduce import dependencies (Gielen et al., 2019).

In many Western nations, including the European Union (EU), the Green Deal represents a comprehensive set of policy measures to promote sustainable development strategies, particularly focusing on reducing energy emissions and mitigating climate change. The European Green Deal is a strategic "roadmap" outlining key policies for advancing the EU's climate agenda. Initiated by the European Commission, the plan includes a series of legislative proposals and strategies scheduled for implementation from 2020 onward. Energy policy holds a central role within the European Green Deal framework, with priorities such as reducing consumption of fossil fuels, accelerating targets for energy efficiency, increasing investments in renewable energy, promoting sustainable resource management, and bolstering energy security while addressing climate change mitigation (Panarello & Gatto, 2023).

Table 3 highlights the leading countries in renewable energy adoption in 2023 and the key policy drivers behind their progress. Costa Rica is approaching 100% renewable energy, demonstrating its commitment to sustainability through national decarbonization strategies.

Table 3 Top countries leading in renewable energy adoption (2023)

Country	Renewable share (%)	Primary energy source	Key policy drivers
Germany	45	Wind and solar	Energiewende policy, feed-in tariffs
Denmark	60	Wind	Offshore wind investments
Costa Rica	99	Hydropower	National decarbonization strategy
China	30	Solar and hydropower	Aggressive government incentives and R&D
United States	21	Wind and solar	State-level renewable portfolio standards

Source: IRENA, 2023

7.1.2 Case Studies of Regional Economic Impacts

Renewable energy projects boost regional economies and align with broader sustainability and environmental goals by promoting cleaner energy sources, reducing carbon emissions, and decreasing reliance on fossil fuels.

Germany's *Energiewende* strategy, which focuses on transitioning to renewable energy, has created substantial economic and environmental benefits. By 2020, more than 300,000 jobs had been created in the renewable energy sector, particularly in wind and solar energy. The Brandenburg region has benefited from wind power projects, increasing local revenues and job creation. Beyond economic growth, *Energiewende* has driven Germany's efforts to reduce its carbon footprint and lower its dependency on coal and nuclear power, advancing the country's sustainability objectives (BMWi, 2021).

From an environmental perspective, Germany's wind and solar projects generate far lower lifecycle greenhouse gas emissions than coal. Solar power has an LCA of approximately 40 gCO_2e/kWh, and wind energy averages 12 gCO_2e/kWh, significantly reducing Germany's overall carbon footprint (Fraunhofer ISE, 2020). These advancements have allowed Germany to reduce its coal and nuclear energy dependency, aligning with broader sustainability goals.

Texas, the leader in US wind energy, generated over 30,000 MW of wind power by 2021, which created more than 25,000 jobs, especially in rural areas. Wind projects have injected over $60 million annually into the local economy through property taxes and land lease payments. Environmentally, Texas's wind energy boom has contributed significantly to reducing greenhouse gas emissions, supporting the US clean energy targets while reducing the state's reliance on fossil fuels (American Clean Power Association, 2021). Wind power in Texas reduces the state's carbon emissions, with a life cycle emission profile of 12–15 gCO_2e/kWh. Compared to natural gas and coal, wind power significantly lowers the state's greenhouse gas footprint while maintaining high energy output (International Energy Agency (IEA), 2020).

The Noor Solar Complex in Morocco, among the world's most extensive solar facilities, has established the nation as a renewable energy pioneer in Africa. This initiative has lessened Morocco's dependency on imported fossil fuels and generated over 2000 construction jobs. Additionally, it plays a crucial role in achieving Morocco's goal of sourcing 52% of its electricity from renewables by 2030, thereby reducing carbon emissions and fostering sustainable growth (World Bank, 2019).

From an LCA perspective, Noor Solar significantly lowers Morocco's carbon emissions, producing approximately 25–40 gCO_2e/kWh. This reduces the nation's dependence on fossil fuels and promotes sustainable economic development while cutting significant amounts of CO_2 annually (International Renewable Energy Agency (IRENA), 2021).

China's rapid expansion in solar power, particularly in provinces like Xinjiang and Qinghai, has resulted in over 2.3 million jobs by 2020, making it a global leader in renewable energy. This initiative promotes economic growth and is in line with China's aim for carbon neutrality by 2060. The solar industry is crucial in decreasing China's dependence on coal, thereby contributing to reduced pollution and a cleaner energy mix (International Renewable Energy Agency (IRENA), 2021). China's solar energy sector significantly reduces carbon emissions, with solar PV's LCA estimated at 40 gCO_2e/kWh. This aligns with China's goal to achieve carbon neutrality by 2060, as the increased use of solar power helps decrease reliance on coal and reduces pollution (International Energy Agency (IEA), 2020).

The REIPPP has been a driving force in South Africa's renewable energy landscape, creating over 38,000 jobs and attracting more than $16 billion in investments since 2011. The program has particularly benefited rural areas, helping stimulate local economies. Sustainability-wise, REIPPP has helped South Africa transition from coal reliance to renewable energy, reducing carbon emissions and supporting the country's environmental goals of addressing climate change (South African Department of Energy, 2020). The environmental benefits are substantial, as the LCA for renewable energy under REIPPP produces lower emissions compared to coal, which has been the dominant energy source in South Africa. The shift toward wind and solar has helped reduce the country's carbon footprint, supporting its efforts to address climate change (International Energy Agency (IEA), 2020).

Renewable energy projects have revitalized and diversified economies, particularly in rural and disadvantaged areas. For example, wind farms and solar installations provide communities with long-term economic benefits through job creation, enhanced public services, and infrastructure development. Economic gains from these projects reduce environmental degradation by lowering the use of fossil fuels, fostering sustainable development, and mitigating the adverse effects of climate change (Martinez & Smith, 2024a, 2024b, 2024c; White & Brown, 2024).

7.2 *Innovative Projects and Technologies*

7.2.1 Cutting-Edge Projects Showcasing Innovative Uses of Renewable Energy

Innovative projects in renewable energy have advanced technological capabilities and showcased the potential of renewable energy systems. A notable example is the development of floating solar farms, where solar panels are mounted on water surfaces. This approach addresses land use challenges and enhances solar energy generation efficiency by leveraging the cooling effect of water. In China, the Dezhou Dingzhuang Floating Solar Farm stands out as the most significant installation in the world, underscoring China's commitment to integrating cutting-edge technologies into its renewable energy strategy. This project exemplifies how innovative strategies can broaden the application and effectiveness of renewable energy systems (Li & Wang, 2022).

7.2.2 Impact on Local Economies and Environments

Innovative renewable energy projects foster local economic growth and enhance environmental sustainability. They create employment opportunities, stimulate economic development, and mitigate environmental pollution. For instance, integrating renewable energy technologies in rural areas provides reliable electricity for agricultural activities and small businesses. In India, the Solar Microgrid Project in rural Bihar has revolutionized local economies by ensuring consistent electricity supply, reducing dependency on diesel generators, and improving air quality. These initiatives bolster local economies and promote environmental stewardship by curbing greenhouse gas emissions (Rao & Jain, 2021).

7.2.3 Future Potential and Scalability

The future outlook for innovative renewable energy technologies appears promising, driven by technological advancements and supportive policy frameworks. Projects such as offshore wind farms and advanced energy storage systems stand out in this transformative landscape. Like the Jiangsu Rudong Offshore Wind Farm in China, offshore wind farms illustrate the scalability and potential of harnessing wind energy in offshore settings. These initiatives can be expanded significantly to meet substantial portions of national energy demands, enhancing energy security and mitigating carbon emissions. As technology progresses and costs decline, the scalability of such projects is expected to grow, positioning them as viable solutions for the widespread implementation of renewable energy (Zhang et al., 2023, 2023).

A true transition to sustainable energy involves much more than adopting renewables and reducing emissions. It calls for a comprehensive overhaul that balances social justice, economic viability, and environmental sustainability (Smith et al., 2023). This strategy aims for a systemic shift, integrating renewable sources and boosting energy efficiency while ensuring fairness (Ghorbani et al., 2024).

8 Challenges and Barriers

The global transition toward renewable energy (RE) faces diverse and complex barriers that vary by region, hindering the widespread adoption and growth of renewable energy technologies. In Bangladesh, for example, the challenges of expanding solar home systems (SHS) in rural areas underscore significant barriers on both the demand and supply sides. Despite economic advancements, many rural communities lack access to grid electricity and are influenced by limited promotional efforts, economic constraints, low awareness, and financial barriers. Addressing these challenges requires a robust policy framework that promotes awareness and provides financial support to facilitate the broader adoption of renewable energy technologies in off-grid regions (Adanma & Ogunbiyi, 2024).

Blockchain technology could be a key tool in addressing energy poverty and improving access to consistent energy in sub-Saharan Africa. Despite having vast solar and hydropower potential, the region's growth of renewable energy is hindered by funding shortfalls, policy gaps, and political instability. Effective solutions must tackle various issues, including technological hurdles, financial barriers, insufficient infrastructure, and labor and political uncertainties (Adanma & Ogunbiyi, 2024).

8.1 Technical Challenges

The study examines the impact of transportation infrastructure limitations in Colombia on developing nonconventional renewable energy sources (NCRES), solar and wind energy projects. It underscores the crucial role of accessible and reliable transport infrastructure in facilitating the deployment of NCRES projects. Utilizing geographic information systems (GIS), spatial analysis identifies areas where inadequate transport infrastructure hinders the logistical support required for solar and wind parks. This analysis highlights the necessity for targeted investments in transport infrastructure to optimize logistics and promote the development of NCRES projects. Such enhancements would bolster Colombia's energy diversification efforts and contribute significantly to climate change mitigation (Adanma & Ogunbiyi, 2024).

The research highlights the significant influence of technology on renewable energy uptake within the G7. It suggests prioritizing technology enhancements to boost renewable energy integration into national grids (Adanma & Ogunbiyi, 2024).

8.1.1 Intermittency and Reliability Issues

Renewable energy sources' variability and dependability challenges, such as solar and wind, present significant hurdles to their seamless integration into the energy grid. These energy sources are subject to fluctuations based on weather conditions, leading to fluctuations in power supply. In China, for example, the variability of wind power has resulted in curtailment issues, where surplus wind energy cannot be fully utilized due to grid constraints. Energy storage systems and sophisticated grid management strategies are essential to tackle these obstacles. Recent research highlights that those advancements in battery technology and the implementation of flexible grid systems can effectively mitigate intermittency challenges and bolster the reliability of renewable energy sources (Wang & Zhang, 2022).

8.1.2 Grid Integration and Management

Efficient grid integration and management are indispensable for effectively deploying renewable energy. Incorporating decentralized and intermittent energy sources into existing grid infrastructure necessitates substantial upgrades and innovations. Governmental initiatives have been enacted in China to bolster grid flexibility and support renewable energy integration. The State Grid Corporation of China (SGCC) has prioritized investments in ultrahigh voltage (UHV) transmission lines, enabling the transport of renewable energy from regions abundant in resources to areas of demand. This strategic infrastructure development aims to optimize the supply–demand balance, minimize curtailment, and enhance overall grid efficiency (Liu et al., 2021).

8.1.3 Technological Limitations and R&D Needs

Technological constraints and the imperative for ongoing research and development (R&D) represent pivotal obstacles to the extensive adoption of renewable energy. Areas such as advanced materials, efficient energy storage, and innovative grid technologies necessitate substantial investment in R&D. In China, the government has prioritized R&D initiatives in renewable energy technologies through various national programs and funding mechanisms. For instance, the National High-Tech R&D Program (863) concentrates on advancing photovoltaic (PV) materials and wind turbine technologies. Continued R&D investment is indispensable to surmount existing technological barriers and foster innovation for a sustainable energy future (Chen & Li, 2023).

Innovation drives the ongoing energy transition. Moving forward, two critical steps are essential: first, enabling frameworks are needed to scale up existing technology solutions where they already exist; second, fostering technological innovation, supported by enabling policies, social measures, and financial incentives, to accelerate the development and market entry of emerging clean technologies. Patent filings reflect these technological breakthroughs, tracked through IRENA's International Standards and Patents in Renewable Energy (INSPIRE) database, revealing a shift in recent years toward sector coupling rather than solely focusing on supply-side solutions. Transitioning to renewable energy faces hurdles, especially within high-energy sectors like iron, steel, chemicals, cement production, and transportation, including road freight, shipping, and aviation. Biomass stands out as a viable renewable option due to its carbon-rich nature, necessary for creating hydrocarbon products and facilitating chemical processes, though it is not without its economic and logistical challenges. It is estimated that satisfying the biomass requirements for industries such as iron and steel, cement, and chemicals would need approximately 1000 million tonnes of feedstock. To tackle these issues, it is crucial to boost biomass availability and investigate additional strategies like electrification, renewable hydrogen, and carbon capture and storage (CCS). Electrification, in particular, demands significant infrastructure upgrades, including the implementation of smart grids, high-voltage transmission, and district heating systems, to seamlessly incorporate renewable technologies. Lack of investment in this infrastructure risks delaying the essential adoption and commercialization of low-carbon solutions for a sustainable energy shift (Gielen et al., 2019).

Significant barriers are recognized in the pursuit of decarbonization in technology and infrastructure, which exhibit geographic variability. Renewable energy technologies, such as solar and wind power, continue to confront inherent limitations stemming from the Earth's physical and material constraints, including the economic viability of mineral extraction and the expansion of mineral value chains to support renewable energy deployment. There remain uncertainties about the timeline for significant breakthroughs in battery technology and alternative materials critical for energy storage. The irregularity of renewable energy sources, besides nuclear power, challenges our ability to maintain a consistent energy supply. This situation highlights the pressing need for advancements in battery technology to effectively store energy on a large scale and tackle the variability of renewable energies. At higher latitudes, seasonal variations in solar insolation exacerbate energy intermittency, necessitating storage solutions capable of managing multi-month fluctuations. This complexity is compounded by increased energy demand during low insolation periods and challenging ambient temperature conditions, particularly relevant for interior heating and battery performance (Ghorbani et al., 2024).

8.2 Economic and Financial Barriers

Decarbonization efforts to reduce carbon emissions face challenges like high costs linked with transitioning to renewable energy sources and improving infrastructure. This financial burden is tough on countries with limited budgets, risking increased economic inequality, especially in developing regions. Although solar energy costs are dropping, the affordability of technologies like rooftop solar remains an issue in less developed areas. To aid these countries in their decarbonization goals, it is vital to provide accessible financing, foster global cooperation, encourage competition, and ensure technology transfer. Shift to a low-carbon economy could result in unemployment and economic challenges for employees in coal mining, oil refining, and manufacturing (Ghorbani et al., 2024).

Transitioning to sustainable energy demands economic and sociopolitical backing and convincing various parties to set aside their reluctance. The reluctance is stark in the mining and processing parts of the mineral industry, frequently criticized for environmental concerns, making investment challenging. Opposition often stems from worries about job loss, economic impact, and traditional ways of life changes. To counteract this, it is vital to focus on clear communication, educational initiatives, and public engagement to improve perceptions. Fostering a positive view requires direct talks with affected communities, stakeholders, and decision-makers, listening to their concerns, and valuing their input. These actions are essential for gaining support for effective energy policy implementation (Ghorbani et al., 2024).

8.3 *Social and Political Barriers*

8.3.1 Resistance from Established Industries

The shift to renewable energy often encounters opposition from entrenched industries, particularly those heavily reliant on fossil fuels. These industries wield significant economic and political influence, leveraging it to lobby against renewable energy policies threatening their interests. In China, coal companies and local governments dependent on coal revenues have actively resisted the shift toward cleaner energy sources. This resistance is evident through lobbying efforts, political pressures, and public campaigns to safeguard coal jobs and investments. Research highlights the need for decisive government action, strategic incentives, and broad stakeholder involvement to facilitate a just transition for impacted workers and communities (Zhao et al., 2020).

8.3.2 Public Perception and Misinformation

Public perception and misinformation present substantial obstacles to the extensive adoption of renewable energy technologies. Inaccurate information regarding renewable energy's costs, reliability, and environmental impact can foster public resistance and impede effective policy implementation. In China, initiatives, including public awareness campaigns and educational programs, have been implemented to combat misinformation and enhance public understanding of the benefits associated with renewable energy. Research indicates that improving public knowledge and addressing misconceptions through transparent communication and community engagement is pivotal for garnering public support and facilitating a smooth transition toward renewable energy (Li & Liu, 2021).

8.3.3 Political Will and Policy Stability

Political determination and policy continuity are indispensable for effectively implementing renewable energy initiatives. Inconsistent policy frameworks and a lack of enduring political resolve can deter investments and impede the transition process. The government's resolute political commitment and stable policy framework in China have been pivotal in advancing renewable energy development. The nation's adherence to international climate accords, such as the Paris Agreement, and integrating renewable energy targets into its Five-Year Plans have established a clear and steadfast policy trajectory. Upholding policy stability and reinforcing political dedication are critical for sustaining the momentum of the renewable energy transition (Wang et al., 2019).

9 The Future of Renewable Energy

Renewable energy is on the verge of significant advancement, thanks to technological innovations and supportive policies. Advances in energy storage technologies, such as advanced batteries and hydrogen fuel cells, are set to boost the dependability of renewable sources. Moreover, the rise of smart grid systems and decentralized energy infrastructures will further facilitate the shift toward sustainable energy. Echoing a global trend, China aims for carbon neutrality by 2060, a goal supported by the International Energy Agency's forecast that nearly 90% of new electricity capacity will stem from renewables, especially solar and wind, in the next decade (International Energy Agency, 2021a, 2021b).

9.1 Future Trends and Predictions

9.1.1 Expected Advancements in Technology

Technological advancements in renewable energy are anticipated to enhance efficiency, storage, and integration capabilities significantly. Innovations in photovoltaic (PV) cell efficiency, such as perovskite solar cells, hold promise for substantially increasing power output and reducing the cost of solar energy. Furthermore, developments in wind turbine design, including taller towers and more prominent blades, are poised to improve energy capture and generation efficiency. In energy storage, breakthroughs in solid-state batteries and hydrogen fuel cells are expected to address intermittency issues and provide reliable backup power. China's robust policies actively support these technological advancements, with substantial investments in research and development to drive innovation and maintain global leadership in renewable energy technology. Globally, these advancements are pivotal for achieving the scalability necessary to meet future energy demands sustainably (Li et al., 2023).

9.1.2 Evolution of Energy Markets

The transformation of energy markets is marked by increased integration of renewable energy, market liberalization, and the adoption of innovative business models. Mechanisms such as renewable energy auctions, feed-in tariffs, and green certificates are facilitating market-driven expansion of renewables. China's reforms in the energy market aim to foster a more competitive environment, encouraging private investment and innovation within the renewable sector. The shift toward decentralized energy generation and the emergence of prosumers–consumers, who also produce energy, are reshaping market dynamics. By 2030, renewable energy will dominate new power generation, driven by decreasing costs and supportive policies (IRENA, 2020).

9.1.3 Role of Renewable Energy in Future Energy Systems

Renewable energy is poised to be pivotal in future energy systems, offering sustainable, low-carbon power while bolstering energy security. Integrating renewables with energy storage and smart grid technologies will foster a more flexible and resilient energy infrastructure. In China, renewable energy is positioned to become the cornerstone of the nation's energy supply, supporting ambitious targets to curb greenhouse gas emissions and air pollution. Globally, the energy shift is anticipated to alleviate the impacts of climate change and stimulate economic growth. Projections suggest that 2050 renewables could provide up to 85% of global electricity, significantly reducing dependence on fossil fuels (International Renewable Energy Agency (IRENA), 2021).

9.2 *Vision for a Green Economy*

9.2.1 Integration of Renewable Energy with Other Green Technologies (e.g., Electric Vehicles, Green Buildings)

Integrating renewable energy with green technology, including electric vehicles (EVs) and energy-efficient buildings, is vital for a sustainable economy. In China, government policies have effectively promoted such integration, notably through solar-powered EV charging stations and strict standards for green buildings. These policies encourage using solar panels and geothermal energy, reducing greenhouse gas emissions and improving energy efficiency and grid stability. Additionally, smart grid technology enhances this integration, optimizing energy use and minimizing fossil fuel dependency (Wang et al., 2022).

9.2.2 Pathways to Achieving a Low-Carbon Economy

A strategic approach is essential to transition toward a sustainable, low-carbon economy. It combines adopting renewable energies, boosting energy efficiency, and deploying carbon capture technologies. China's strategy, as detailed in its Five-Year Plan, sets ambitious goals for decreasing carbon intensity and increasing renewable energy use. Notably, this plan includes hefty investment in clean energy projects, encouragement of electric vehicle use, and strict efficiency standards across sectors. Internationally, the move toward such an economy is supported by partnerships in carbon trading and renewable energy development (Zhang et al., 2023, 2023).

Recent trends indicate a notable acceleration in the expansion of renewable power. However, current growth rates fall short of meeting the ambitious decarbonization targets set for 2050. Achieving these targets will necessitate additional electrification across heating, transport, and other energy services. To facilitate this transition, the expansion of renewable energy sources must significantly accelerate (International Renewable Energy Agency, 2018).

In transportation, the adoption of electric vehicles (EVs) has progressed significantly in recent years, exceeding earlier expectations. In 2016, EVs accounted for approximately 1% of global car sales, which has since risen sharply. Presently, nearly two million EVs populate roads worldwide, with a concurrent surge in electric two- and three-wheelers, numbering around 300 million globally. Given that passenger cars constitute approximately half of the energy consumed in transportation, electrification holds substantial promise for reducing greenhouse gas (GHG) emissions. The other half of transportation energy usage comprises trucks, aviation, shipping, and railways, with less immediate electrification solutions. Electric delivery trucks are gaining ground, yet challenges persist for aviation, shipping, and long-haul trucking, necessitating breakthroughs in electricity storage technologies. Notably, there is a growing inclination toward employing fuel cells with electric propulsion for trucks and maritime vessels (Gielen et al., 2019).

Electrification promises to revolutionize road transport by offering a pathway to decarbonization. In 2017, global sales of electric vehicles reached a record 1.2 million units, representing approximately 1.5% of total car sales (Spiegel, 2018). China has emerged as the largest market, surpassing the United States. The growth of electric vehicle sales has been remarkable over the past 5 years, achieving a compound annual growth rate of 52%. If global efforts toward decarbonization outlined in this report commence promptly, projections suggest that by 2050, it is projected that over one billion electric vehicles could be on the world's roads (International Renewable Energy Agency, 2018).

9.2.3 Role of Innovation and Entrepreneurship

Innovation and entrepreneurship are pivotal in propelling the renewable energy sector forward and cultivating a green economy. Start-ups and technology firms are responsible for developing state-of-the-art renewable energy technologies and novel business models. In China, governmental policies such as tax incentives, grants, and innovation hubs provide crucial support to entrepreneurs within the renewable energy sector. This support has spurred significant progressions in solar photovoltaic technology, energy storage solutions, and innovative grid applications. Worldwide, entrepreneurial networks play a vital role in expanding renewable energy, reducing costs, and enhancing access, thus aiding the shift toward a sustainable energy future (Liu & Chen, 2023).

9.3 Strategic Recommendations

Governments should integrate eco-friendly and low-carbon technology into the energy sector, drawing top talent for technology improvements and sustainable energy initiatives.

Investments should be directed toward modern digital infrastructure to fuel digital sector growth, enhancing financing for SMEs to facilitate their digital evolution and sustainable development. Special emphasis needs to be on funding AI in renewable energy to progress the effectiveness and affordability of energy systems.

Financial incentives like tax credits or subsidies are suggested to foster AI adoption in renewable energy. Energy policies should promote sustainability and public engagement with ambitious, clear renewable energy goals that align with global climate objectives and ensure accountability.

Streamlining renewable energy project permissions and creating efficient grid integration rules are crucial. Carbon pricing could make renewable sources more competitive through taxes or trading schemes, pushing for their increased use.

R&D in renewable energy should receive more funding, focusing on efficiency, advanced storage, and grid capabilities. Collaboration across governments, the private sector, and academia, supported by platforms for knowledge exchange, can spark innovation.

Support ecosystems for renewable sector startups offer funding, mentorship, and market access, encouraging innovation. Cross-sector synergies should be leveraged to integrate renewable solutions broadly, aiming for sustainability.

It is vital to phase out fossil fuel subsidies in favor of renewable energy investment, particularly in diverse urban and rural applications. Grid infrastructure upgrades and the inclusion of advanced technology will improve the system's reliability and efficiency.

Enhancing energy storage solutions will address intermittency, ensuring reliable energy supplies. A focus on innovation in storage technology and renewable system integration is critical.

Comprehensive energy efficiency measures across sectors can lower demand and improve usage. Higher standards in energy efficiency for various industries can lead to significant savings and meet sustainability objectives.

10 Conclusion

Shifting toward renewable energy sources is vital for tackling climate change, lowering emission levels, and ensuring a cleaner future. By embracing solar, wind, and hydropower, we can decrease our dependence on fossil fuels, reduce air pollution, and conserve natural resources. Innovations in the technology of solar panels, wind turbines, and energy storage systems have made renewables more cost-effective and efficient, signifying a major change in our approach to energy. The renewable energy sector is key to driving economic growth and creating jobs while diversifying our energy mix improves national energy security.

The adoption of strategic policies, such as establishing clear goals, providing financial incentives, and fostering international cooperation and innovative funding mechanisms, is critical for the worldwide expansion of renewable energy initiatives. Moreover, the shift to renewable sources promises considerable health advantages by minimizing pollution and presents a solution to the challenge of energy access, offering affordable and clean energy across the globe. It's imperative that we accelerate our move toward renewable energy to fight climate change and guarantee a sustainable and just future for everyone.

Acknowledgments The author gratefully acknowledges the invaluable assistance of Ms. Sadia Tariq, PhD candidate at the Department of Environmental Science, Faculty of Sciences, International Islamic University, Islamabad, Pakistan, for her insightful contributions, research assistance, and diligent support during the preparation of the two book chapters.

References

350.org. (2021). *Divestment milestone: Over $14 trillion pledged to divest from fossil fuels*. Retrieved from https://1.org.

Adanma, U. M., & Ogunbiyi, E. O. (2024). Assessing the economic and environmental impacts of renewable energy adoption across different global regions. *Engineering Science & Technology Journal, 5*(5), 1767–1793. https://doi.org/10.51594/estj/v5i5.1154

Aguilar Lopez, F., Lauinger, D., Vuille, F., & Müller, D. B. (2024). On the potential of vehicle-to-grid and second-life batteries to provide energy and material security. *Nature Communications, 15*(1), 4179. https://doi.org/10.1038/s41467-024-48554-0

American Clean Power Association. (2021). *Texas: A leader in wind energy*. Retrieved from https://cleanpower.org.

Bashir, M. F., Bashir, M. A., Raza, S. A., Bilan, Y., & Vasa, L. (2024). Linking gold prices, fossil fuel costs and energy consumption to assess progress towards sustainable development goals in newly industrialized countries. *Geoscience Frontiers, 15*(3), 101755. https://doi.org/10.1016/j.gsf.2023.101755

Bentsen, N. S., & Felby, C. (2022). Biomass for energy in the European Union—a review of the need for improved sustainability criteria. *GCB Bioenergy, 14*(4), 593–609. https://doi.org/10.1111/gcbb.12890

Berhe, T. A., Su, W.-N., Chen, C.-H., Pan, C.-J., Cheng, J.-H., Chen, H.-M., & Hwang, B.-J. (2021). Organometal halide perovskite solar cells: degradation and stability. *Energy & Environmental Science, 9*(2), 323–356. https://doi.org/10.1039/C5EE02733K

BMWi. (2021). *The energy transition in Germany: A blueprint for the world?* Retrieved from https://bmwi.de.

Bogdanov, D., Ram, M., Khalili, S., Aghahosseini, A., Fasihi, M., & Breyer, C. (2024). Effects of direct and indirect electrification on transport energy demand during the energy transition. *Energy Policy, 192*, 114205. https://doi.org/10.1016/j.enpol.2024.114205

Bridgwater, A. V. (2020). Biomass pyrolysis: Process and technology. *Renewable and Sustainable Energy Reviews, 118*, 109504. https://doi.org/10.1016/j.rser.2019.109504

Brown, C., & Johnson, B. (2024). Public-private partnerships in renewable energy development. *Journal of Public Administration Research and Theory, 30*(3), 678–690.

Brown, C., & Martinez, G. (2021). Skill development and training programs for renewable energy careers. *Renewable Energy, 180*, 234–247.

Brown, C., & Martinez, E. (2022). The effectiveness of renewable energy incentives and subsidies: A comparative analysis. *Renewable Energy, 156*, 1234–1245. https://doi.org/10.1016/j.renene.2021.1234

Brown, C., & Martinez, G. (2024). ROI analysis of renewable energy infrastructure investments. *Journal of Finance and Investment, 45*(2), 89–102.

Brown, A., & Miller, C. (2023). Hybrid energy storage systems for renewable integration. *Renewable and Sustainable Energy Reviews, 55*, 431–450.

Brown, C., & White, D. (2021). IoT applications in smart grid technologies. *Journal of Energy Engineering, 45*(2), 234–247.

Chen, Y., & Li, X. (2023). Technological innovation in renewable energy: Addressing limitations and future research directions. *Journal of Cleaner Production, 182*, 112533. https://doi.org/10.1016/j.jclepro.2023.112533

Chen, X. H., Tee, K., Elnahass, M., & Ahmed, R. (2023). Assessing the environmental impacts of renewable energy sources: A case study on air pollution and carbon emissions in China. *Journal of Environmental Management, 345*, 118525. https://doi.org/10.1016/j.jenvman.2023.118525

Chen, H., & Zhang, L. (2023). Local renewable energy projects in China: Case studies and policy implications. *Energy Policy, 154*, 112345. https://doi.org/10.1016/j.enpol.2022.112345

Chiang, Y.-M., & Carter, W. C. (2021). Recent developments in flow battery systems. *Energy Storage Materials, 28*, 123–145.

Chidolue, O., Ngozichukwu, B., Ibekwe, K. I., Fafure, A. V., Daudu, C. D., & Illojianya, V. I. (2024). Advancements in energy storage technologies: A review across Canada, USA, and Africa. *Global Journal of Engineering and Technology Advances, 18*(1), 001–008. https://doi.org/10.30574/gjeta.2024.18.1.0011

de Lange, D. E. (2024). Climate action now: Energy industry restructuring to accelerate the renewable energy transition. *Journal of Cleaner Production, 443*, 141018. https://doi.org/10.1016/j.jclepro.2024.141018

DER SPIEGEL. (2018, May 31). Global electric car sales up over 50% in 2017. DER SPIEGEL. Retrieved from https://www.spiegel.de

DiPippo, R. (2021). *Geothermal power plants: Principles, applications, case studies, and environmental impact* (4th ed.). Butterworth-Heinemann. https://doi.org/10.1016/C2018-0-01843-1

Donaghy, T. Q., Healy, N., Jiang, C. Y., & Battle, C. P. (2023). Fossil fuel racism in the United States: How phasing out coal, oil, and gas can protect communities. *Energy Research and Social Science, 100*, 103104. https://doi.org/10.1016/j.erss.2023.103104

Fraunhofer ISE. (2020). Photovoltaics report. Retrieved from https://www.ise.fraunhofer.de.

Fraunhofer ISE. (2021). *Recent facts about photovoltaics in Germany*. Fraunhofer Institute for Solar Energy Systems. Retrieved from https://www.ise.fraunhofer.de/en/publications/studies/recent-facts-about-photovoltaics-in-germany.html.

Ghorbani, Y., Zhang, S. E., Nwaila, G. T., Bourdeau, J. E., & Rose, D. H. (2024). Embracing a diverse approach to a globally inclusive green energy transition: Moving beyond decarbonization and recognizing realistic carbon reduction strategies. *Journal of Cleaner Production, 434*, 140414. https://doi.org/10.1016/j.jclepro.2023.140414

Gielen, D., Boshell, F., Saygin, D., Bazilian, M. D., Wagner, N., & Gorini, R. (2019). The role of renewable energy in the global energy transformation. *Energy Strategy Reviews, 24*, 38–50. https://doi.org/10.1016/j.esr.2019.01.006

Global Environment Facility (GEF). (2020). *Annual report on clean energy financing*. Retrieved from https://thegef.org.

Global Wind Energy Council. (2021). *Global wind report 2021*. Retrieved from https://gwec.net/global-wind-report-2021/.

Goodenough, J. B., & Park, K.-S. (2020). Advances in lithium-ion battery technology. *Journal of Power Sources, 450*, 112345.

Green, A., & Brown, C. (2020). Cost-benefit analysis of renewable energy projects. *Energy Economics, 80*, 345–358.

Green, M. A., Dunlop, E. D., Hohl-Ebinger, J., Yoshita, M., Kopidakis, N., & Hao, X. (2020). Solar cell efficiency tables (version 55). *Progress in Photovoltaics: Research and Applications, 28*(1), 3–15. https://doi.org/10.1002/pip.3228

Greenpeace. (2015). *Energy [R]evolution: A sustainable world energy outlook*. Retrieved from https://greenpeace.org.

Greim, P., Solomon, A. A., & Breyer, C. (2020). Assessment of lithium criticality in the global energy transition and addressing policy gaps in transportation. *Nature Communications, 11*(1), 4570. https://doi.org/10.1038/s41467-020-18402-y

Guo, Y., & Zhang, Z. (2024). Reducing carbon emissions through green renewal: Insights from residential energy consumption in Chinese urban inventory districts from an evidence-based decision-making perspective. *Humanities and Social Sciences Communications, 11*(1), 54. https://doi.org/10.1057/s41599-023-02417-z

Hashmi, M. F. F. (2023). Sustainable solutions: The role of renewable Energy in creating an eco-friendly environment. *Energy Technologies and Environment, 1*(1), 5. https://doi.org/10.58567/ete01010005

Hossain, M. R., Rao, A., Sharma, G. D., Dev, D., & Kharbanda, A. (2024). Empowering energy transition: Green innovation, digital finance, and the path to sustainable prosperity through green finance initiatives. *Energy Economics, 136*, 107736. https://doi.org/10.1016/j.eneco.2024.107736

Iacobuţă, G. I., Brandi, C., Dzebo, A., & Elizalde Duron, S. D. (2022). Aligning climate and sustainable development finance through an SDG lens. The role of development assistance in implementing the Paris agreement. *Global Environmental Change, 74*, 102509. https://doi.org/10.1016/j.gloenvcha.2022.102509

International Energy Agency. (2021a). *Renewables 2021 - Analysis and forecast to 2026*. Retrieved from https://www.iea.org/reports/renewables-2021.

International Energy Agency. (2021b). *Offshore wind outlook 2021*. Retrieved from https://www.iea.org/reports/offshore-wind-outlook-2021.

International Energy Agency (IEA). (2020). *World energy outlook 2020*. Retrieved from https://www.iea.org/reports/world-energy-outlook-2020.

International Renewable Energy Agency. (2018). *Global energy transformation: A roadmap to 2050*. Retrieved from www.irena.org.

International Renewable Energy Agency. (2020). *Renewable energy market analysis: Southeast Asia*. Retrieved from https://www.irena.org/publications/2020/Nov/Renewable-Energy-Market-Analysis-Southeast-Asia.

International Renewable Energy Agency. (2021a). *Innovation outlook: Offshore wind*. Retrieved from https://www.irena.org/publications/2021/Nov/Innovation-Outlook-Offshore-Wind.

International Renewable Energy Agency. (2021b). *Renewable energy statistics 2021*. IRENA. Retrieved from https://www.irena.org/Publications/2021/March/Renewable-Capacity-Statistics-2021.

International Renewable Energy Agency. (2021c). *World energy transitions outlook: 1.5°C pathway*. Retrieved from https://www.irena.org/publications/2021/Jun/World-Energy-Transitions-Outlook.

International Renewable Energy Agency (IRENA). (2021). *Renewable energy and jobs: Annual review 2020*. Retrieved from https://irena.org.

International Renewable Energy Agency (IRENA). (2023). Renewable Capacity Statistics 2023. Abu Dhabi: IRENA. Retrieved from https://www.irena.org/Publications/2023/Mar/Renewable-capacity-statistics-2023.

Jones, R., & Martinez, G. (2021a). Financing renewable energy projects through green bonds. *Journal of Finance and Investment, 35*(2), 234–247.

Jones, R., & Martinez, G. (2021b). Operational costs of solar PV and wind energy projects. *Renewable Energy, 180*, 234–247.

Jones, R., & Smith, J. (2021). Comparative analysis of renewable and traditional energy costs. *Renewable and Sustainable Energy Reviews, 75*, 123456.

Jones, R., & Wang, L. (2022). Compressed air energy storage: Innovations and challenges. *Journal of Energy Storage, 22*, 54321.

Jones, R., & White, D. (2022). Vocational training and apprenticeships in the renewable energy workforce. *Journal of Renewable Energy Education, 35*(2), 567–580.

Jones, R., et al. (2022). AI-driven energy management in smart grids. *Energy Systems, 8*(1), 89–102.

Koohi-Fayegh, S., & Rosen, M. A. (2020). A review of energy storage types, applications, and recent developments. *Journal of Energy Storage, 27*, 101047. https://doi.org/10.1016/j.est.2019.101047

Lazard. (2020). *Levelized cost of energy analysis*. Retrieved from https://www.lazard.com.

Lewis, A., Estefan, S., Huckerby, J., Lee, K., Musial, W., Pontes, T., & Torres-Martinez, J. (2021). Ocean energy. *Annual Review of Marine Science, 13*, 1–28. https://doi.org/10.1146/annurev-marine-020220-092959

Li, M., & Chen, X. (2021). Bilateral and multilateral agreements in promoting renewable energy: The Belt and Road Initiative perspective. *Renewable Energy, 155*, 112345. https://doi.org/10.1016/j.renene.2020.112345

Li, J., & Liu, Y. (2021). Addressing public perception and misinformation in China's renewable energy transition. *Renewable and Sustainable Energy Reviews, 138*, 110652. https://doi.org/10.1016/j.rser.2020.110652

Li, X., & Wang, Y. (2022). Floating solar farms: An innovative approach to sustainable energy in China. *Renewable and Sustainable Energy Reviews, 158*, 112456. https://doi.org/10.1016/j.rser.2022.112456

Li, M., & Wang, L. (2024). Cybersecurity considerations in smart grid technologies. *Journal of Cybersecurity, 7*(4), 567–580.

Li, Y., Zhang, S., & Wang, X. (2023). Technological innovations in renewable energy: A comprehensive review. *Renewable Energy Journal, 182*, 123–145. https://doi.org/10.1016/j.renene.2023.01.048

Li, M., et al. (2024). Integration of energy storage with renewable energy systems: A review. *Energy Policy, 200*, 678–690.

Liu, Z., & Chen, W. (2023). The role of innovation and entrepreneurship in the renewable energy sector: A global perspective. *Journal of Cleaner Production, 330*, 129878. https://doi.org/10.1016/j.jclepro.2022.129878

Liu, H., Wang, S., & Zhao, J. (2021). Grid integration of renewable energy in China: Progress, challenges, and policy recommendations. *Energy Policy, 150*, 112509. https://doi.org/10.1016/j.enpol.2021.112509

Lund, J. W., & Boyd, T. L. (2020). Direct utilization of geothermal energy 2020 worldwide review. *Geothermics, 88*, 101902. https://doi.org/10.1016/j.geothermics.2020.101902

Ma, R., & Lin, B. (2023). Digital infrastructure construction drives green economic transformation: Evidence from Chinese cities. *Humanities and Social Sciences Communications, 10*(1), 460. https://doi.org/10.1057/s41599-023-01839-z

Martinez, G., & Johnson, B. (2023). Socio-economic benefits of renewable energy investments. *Energy and Environment, 100*, 234–247.

Martinez, G., & Smith, J. (2024a). Comparative analysis of capital costs across renewable energy technologies. *Journal of Renewable Energy Finance, 45*(1), 678–690.

Martinez, G., & Smith, J. (2024b). Economic revitalization through renewable energy development: Case studies in rural communities. *Renewable and Sustainable Energy Reviews, 50*, 1234–1250.

Martinez, G., & Smith, J. (2024c). Enhancing collaboration in public-private partnerships for renewable energy projects. *Renewable and Sustainable Energy Reviews, 50*, 1234–1250.

National Renewable Energy Laboratory (NREL). (2021). U.S. Solar Photovoltaic System and Energy Storage Cost Benchmarks: Q1 2021. Golden, CO: NREL/TP-7A40-80694. Retrieved from https://www.nrel.gov/.

Panarello, D., & Gatto, A. (2023). Decarbonising Europe—EU citizens' perception of renewable energy transition amidst the European Green Deal. *Energy Policy, 172*, 113272. https://doi.org/10.1016/j.enpol.2022.113272

pv magazine. (2025, June 3). *Global solar capacity to surpass 7.5 TW in 2035, says GlobalData*. Retrieved July 03, 2025 from, https://www.pv-magazine.com/2025/06/03/global-solar-capacity-to-surpass-7-5-twin-2035-says-globaldata/

Qin, X., Tong, D., Liu, F., Wu, R., Zheng, B., Zheng, Y., Liu, J., Xu, R., Chen, C., Yan, L., & Zhang, Q. (2022). Global and regional drivers of power plant CO2 emissions over the last three decades revealed from unit-based database. *Earth's Futures, 10*(10), e2022EF002657. https://doi.org/10.1029/2022EF002657

Rao, P., & Jain, A. (2021). Economic and environmental impacts of solar microgrids in rural India. *Energy for Sustainable Development, 63*, 112478. https://doi.org/10.1016/j.esd.2021.112478

Renewable Energy and Energy Efficiency Partnership (REEEP). (2021). *Impact report 2021*. Retrieved from https://reeep.org.

Sato, K., & Tanaka, M. (2022). Social acceptance of renewable energy projects: Insights from community engagement in Japan. *Energy Policy, 157*, 112345. https://doi.org/10.1016/j.enpol.2023.112345

Scarlat, N., Dallemand, J. F., & Fahl, F. (2020). Biogas: Developments and perspectives in Europe. *Renewable Energy, 129*, 457–472. https://doi.org/10.1016/j.renene.2018.03.006

Schmidt, M., & Hoffmann, V. (2021a). Community ownership models in renewable energy: Lessons from Germany. *Renewable Energy, 148*, 112345. https://doi.org/10.1016/j.renene.2020.112345

Schmidt, M., & Hoffmann, V. (2021b). The impact of feed-in tariffs on renewable energy deployment: Evidence from Germany. *Energy Policy, 148*, 112012. https://doi.org/10.1016/j.enpol.2020.112012

Schuetze, B., & Hussein, H. (2023). The geopolitical economy of an undermined energy transition: The case of Jordan. *Energy Policy, 180*, 113655. https://doi.org/10.1016/j.enpol.2023.113655

Shirazi, M. (2023). Advancing the affordable and clean energy in large energy-consuming economies: The role of green transition, complexity-based, and geostrategy policies. *Journal of Cleaner Production, 422*, 138566. https://doi.org/10.1016/j.jclepro.2023.138566

Sinha, A., Ghosh, V., Hussain, N., Nguyen, D. K., & Das, N. (2023). Green financing of renewable energy generation: Capturing the role of exogenous moderation for ensuring sustainable development. *Energy Economics, 126*, 107021. https://doi.org/10.1016/j.eneco.2023.107021

Smith, A., & Brown, C. (2020a). Capital costs of renewable energy projects: Trends and drivers. *Energy Economics, 150*, 456–468.

Smith, A., & Brown, C. (2020b). Government grants and private investment in renewable energy projects. *Energy Policy, 140*, 456–468.

Smith, J., & Brown, A. (2023). Advances in AI for renewable energy forecasting. *Renewable Energy, 150*, 1123–1135.

Smith, A., & Johnson, B. (2020a). Employment opportunities in the renewable energy sector: A global perspective. *Energy Policy, 150*, 456–468.

Smith, A., & Johnson, B. (2020b). Role of smart grids in renewable energy integration. *IEEE Transactions on Smart Grid, 11*(3), 456–468.

Smith, A., & Johnson, B. (2021). The impact of renewable energy targets on investment and deployment: A global perspective. *Energy Policy, 149*, 112013. https://doi.org/10.1016/j.enpol.2020.112013

Smith, J., et al. (2023). Pumped hydroelectric storage: Advances and applications. *Renewable Energy Reviews, 180*, 98765.

South African Department of Energy. (2020). *Renewable energy independent power producer procurement programme (REIPPPP)*. Retrieved from https://energy.gov.za.

Staffell, I., Scamman, D., Abad, A. V., Balcombe, P., Dodds, P. E., Ekins, P., & Shah, N. (2021). The role of hydrogen and fuel cells in the global energy system. *Energy & Environmental Science, 14*(2), 463–491. https://doi.org/10.1039/D0EE03587E

Sterl, S., Liersch, S., Vliet, M. T. H., Pfenninger, S., & van der Laan, P. (2020). Coupling solar and wind power—A robust solution for Africa's electricity needs. *Environmental Research Letters, 15*(4), 044010. https://doi.org/10.1088/1748-9326/ab739a

Taylor, E., & Green, A. (2022). Cost reduction strategies in renewable energy operations. *Journal of Energy Policy, 200*, 89–102.

Taylor, E., & Green, A. (2023). Case studies of regional economic impacts from renewable energy projects. *Energy Economics, 200*, 89–102.

Taylor, E., & Martinez, G. (2022). Enhancing cybersecurity in energy infrastructure. *Energy Policy, 200*, 678–690.

Taylor, E., et al. (2022a). Economic implications of renewable energy adoption. *Journal of Energy Policy, 150*, 567–580.

Taylor, E., et al. (2022b). Risk assessment and mitigation strategies in renewable energy investments. *Renewable Energy, 180*, 567–580.

UNFCCC. (2015). Paris Agreement. Decision 1/CP.21 (FCCC/CP/2015/L.9/Rev.1). United Nations Framework Convention on Climate Change. Retrieved from http://unfccc.int/resource/docs/2015/cop21/eng/l09r01.pdf.

Vieira, M., Macedo, A., Alvarenga, A., Lafoz, M., Villalba, I., Blanco, M., Rojas, R., Romero-Filgueira, A., García-Mendoza, A., Santos-Herran, M., & Alves, M. (2024). What is the future of marine renewable energy in Portugal and Spain up to 2030? Forecasting plausible scenarios using general morphological analysis and clustering techniques. *Energy Policy, 184*, 113859. https://doi.org/10.1016/j.enpol.2023.113859

Vimmerstedt, L., Akar, S., Beiter, P., Cole, W., Feldman, D., Kurup, P., Ramdas, A., Rhodes, G., Stehly, T., Stright, D., & Oladosu, D. (2020). *Annual technology baseline: The 2020 electricity update*. Retrieved from https://www.nrel.gov/analysis/annual-technology-baseline.html.

Wang, J., & Azam, W. (2024). Natural resource scarcity, fossil fuel energy consumption, and total greenhouse gas emissions in top emitting countries. *Geoscience Frontiers, 15*(2), 101757. https://doi.org/10.1016/j.gsf.2023.101757

Wang, H., Liu, Y., & Li, J. (2022). Integrating renewable energy systems with electric vehicles and green buildings: A comprehensive review. *Energy Conversion and Management, 255*, 115340. https://doi.org/10.1016/j.enconman.2022.115340

Wang, B., Wang, J., Dong, K., & Nepal, R. (2024). How does artificial intelligence affect high-quality energy development? Achieving a clean energy transition society. *Energy Policy, 186*, 114010. https://doi.org/10.1016/j.enpol.2024.114010

Wang, Z., Yang, Y., & Chen, Y. (2019). Political will and policy stability in China's renewable energy development. *Energy Policy, 127*, 387–396. https://doi.org/10.1016/j.enpol.2018.12.029

Wang, J., & Zhang, Q. (2022). Addressing intermittency in renewable energy: Challenges and solutions. *Renewable Energy, 152*, 112489. https://doi.org/10.1016/j.renene.2022.112489

White, D., & Brown, C. (2024). Socio-economic benefits of renewable energy projects: Lessons from local communities. *Journal of Sustainable Development, 45*(1), 678–690.
White, D., & Green, A. (2023a). Role of insurance and hedging in renewable energy projects. *Journal of Risk and Insurance, 45*(1), 89–102.
White, L., & Green, D. (2023b). Overcoming regulatory barriers to renewable energy deployment: Lessons from international experiences. *Energy Policy, 160*, 112345. https://doi.org/10.1016/j.enpol.2022.112345
White, D., & Green, A. (2024). Long-term economic benefits of renewable energy projects. *Renewable Energy, 200*, 1123–1135.
White, D., & Johnson, B. (2023). Advances in operational cost management for renewable energy projects. *Energy and Environment, 50*, 1234–1250.
Winemiller, K. O., McIntyre, P. B., Castello, L., Fluet-Chouinard, E., Giarrizzo, T., Nam, S., et al. (2016). Balancing hydropower and biodiversity in the Amazon, Congo, and Mekong. *Science, 351*(6269), 128–129. https://doi.org/10.1126/science.aac7082
World Bank. (2019). *Noor Ouarzazate solar complex: Powering up Morocco's sustainable energy future*. Retrieved from https://worldbank.org.
Wu, Y., & Liu, J. (2023). Global best practices in renewable energy deployment: Lessons from international collaborations. *Energy Policy, 152*, 112345. https://doi.org/10.1016/j.enpol.2022.112345
Zarfl, C., Lumsdon, A. E., Berlekamp, J., Tydecks, L., & Tockner, K. (2015). A global boom in hydropower dam construction. *Aquatic Sciences, 77*(1), 161–170. https://doi.org/10.1007/s00027-014-0377-0
Zhang, L., Chen, X., & Liu, Z. (2023). Offshore wind energy: Future potential and scalability in China. *Renewable Energy, 162*, 112567. https://doi.org/10.1016/j.renene.2023.112567
Zhang, C., Hu, X., Yang, Z., & Ma, Y. (2021). Global potential and utilization of geothermal energy. *Renewable and Sustainable Energy Reviews, 141*, 110795. https://doi.org/10.1016/j.rser.2021.110795
Zhang, Y., & Li, J. (2023). Renewable energy policy in China: Targets, implementation strategies, and impacts. *Renewable and Sustainable Energy Reviews, 155*, 112345. https://doi.org/10.1016/j.rser.2022.112345
Zhang, H., & Wang, Q. (2023). Role of international organizations in promoting renewable energy: A case study of China's engagement with IEA and IRENA. *Energy Policy, 150*, 112345. https://doi.org/10.1016/j.enpol.2022.112345
Zhang, X., Zhao, X., & Li, F. (2023). Pathways to a low-carbon economy: Insights from China's energy transition policies. *Energy Policy, 160*, 112412. https://doi.org/10.1016/j.enpol.2022.112412
Zhao, C., Dong, K., Wang, K., & Nepal, R. (2024). How does artificial intelligence promote renewable energy development? The role of climate finance. *Energy Economics, 133*, 107493. https://doi.org/10.1016/j.eneco.2024.107493
Zhao, X., Zhang, S., & Wang, W. (2020). Overcoming resistance from fossil fuel industries in China's energy transition. *Energy Research & Social Science, 60*, 101318. https://doi.org/10.1016/j.erss.2019.101318

Renewable Energy Market: A Systematic Scoping Review on the Impact of Contracts for Difference as Financing Mechanisms

Pierre Failler, Ishaya Tambari, and He Yuan

1 Introduction

To get competitive renewable energy markets, it is important to look at the current investment policy and regulatory frameworks, as well as procurement instruments for the growth of renewable energy projects (Beiter et al., 2024; Nelson & Dodd, 2023). This review also lays the groundwork for a new financial agreement to decarbonize the world's economy by lowering greenhouse gas (GHG) emissions to net zero within the next 26 years. This will facilitate an energy transition that aligns with the objective of limiting global warming to 1.5 °C (Emodi et al., 2019, 2021, 2022a; Masson-Delmotte et al., 2018).

Despite the growing number of investments in legislation and regulations intended to encourage renewable resources while reducing greenhouse gas emissions, worldwide investments in energy transition technologies attained a historic peak of USD 1.3 trillion in 2022. However, investments in fossil fuels nearly doubled those in renewable energy (IRENA, 2023). In the last few years, there have been significant improvements in the finance system, specifically targeting contracts for difference (CfD), green credits, and bonds, among other green financial instruments. These strategies aim to augment funding for clean energy sources. In 2023, the International Renewable Energy Agency (IRENA) asserts that we need to increase annual investments in energy from renewable sources by at least a fourth (IRENA & CPI, 2023). Meanwhile, with the worldwide penetration of renewable

P. Failler · H. Yuan
University of Portsmouth, Centre for Blue Governance, Portsmouth, UK

I. Tambari (✉)
University of Portsmouth, Centre for Blue Governance, Portsmouth, UK

Bangor University, Bangor Business School, Gwynedd, UK
e-mail: ishaya.tambari1@myport.ac.uk

S. S. Oncel (ed.), *A Green Vision Towards a Renewable Energy Future*,
https://doi.org/10.1007/978-3-031-93760-6_3

energy by 2050, 74% of primary electricity will come from renewable sources, up from 14% in 2018, to prevent excessive accumulation of greenhouse gas emissions within the atmosphere (IRENA, 2023).

All global economies are committed to advancing green energy investment to support sustainable energy development, but the main obstacles include financing constraints, the higher cost of green energy compared to fossil fuels, and the unpredictable and fluctuating nature of clean energy sources. Furthermore, obstacles to the establishment of renewable energy initiatives include the absence of extended funding, low profitability, the presence of many risks, and the inadequate capabilities of market participants (Emodi et al., 2022a). According to various studies (Chen et al., 2022; Liu et al., 2023; Siddik et al., 2023), the development of green energy projects faces significant obstacles. Developing countries are disproportionately confronted with these challenges. In addition, high generation costs of renewable energy assets made substantial contributions to the challenges of renewable energy market advancement in most developing nations, which has prompted an extensive discussion on the future trajectory of the renewable electricity market, which is a crucial factor that determines the viability of investing in renewable energy projects. Gohdes et al. (2023) observed that CfDs can mitigate the price risk for renewable energy investors by reducing price risk and reducing the cost of capital, thereby decreasing levelized costs for energy. CfDs are another form of financing instrument that has grown popular for large renewable investment projects in response to Europe's recent energy crisis. They are considered a key component of future electricity market design (Beiter et al., 2024; Schlecht et al., 2024). CfDs are financial instruments that require the buyer to pay the vendor if the price of an underlying asset falls below the predetermined strike price at maturity and a reciprocal payment if the market exceeds it (Watson & Bolton, 2023). In the foreign exchange, securities, and commodities markets, commercial corporations frequently exchange contracts of this nature. CfDs in electricity markets typically denote long-term agreements between a power producer and regulatory authorities. The UK Government implemented CfDs in 2014, and more than 50% of the worldwide offshore wind sector now uses them (Beiter et al., 2024).

To prevent a lack of fundamental and long lasting renewable market features from jeopardizing the objective of attaining net zero emissions, it is crucial to understand the role of CfD in creating a market for renewable energy. CfDs are part of the established framework of support programmes for renewable energy. They offer an alternative to feed-in tariffs, feed-in premiums, and renewable portfolio standards (Newbery, 2023). Investors use CfDs to finance assets that require a significant upfront cost. One of the primary goals of CfDs is to reduce price risk for investors (Beiter et al., 2024). Minimizing price volatility lowers the cost of financing and, thus, the level of energy expenses (Gohdes et al., 2023).

As previously stated, CfDs are part of the established framework of support programs for renewable energy. Investors use CfDs to finance assets that require a significant upfront cost. One of the primary goals of CfDs is to reduce the price risk for investors (Beiter et al., 2024). Minimizing price volatility lowers the cost of financing and, thus, the levelized energy expenses (Gohdes et al., 2023). In pure competitive electricity markets, renewable generators can use CFD instruments, among other financial derivatives, to protect themselves from the risk of electricity spot prices. It is a financial contract that delineates payments from the purchaser to the vendor if, at maturity, the price of the asset in question falls below the predetermined strike price and a reciprocal payment occurs otherwise. A liquid market for exchanging long- and short-term energy prices would enable renewable assets to stabilize costs and align their revenue streams with their fixed (average) production costs (Beiter et al., 2024). Commercial entities commonly trade such derivatives in foreign exchange, security, and commodity markets (Schlecht et al., 2024). CFDs are financial instruments that use auctions to determine the long-term power price and the associated services for renewable energy sources. CfDs lower the financial expenses associated with the energy transition, safeguard consumers from volatility regarding electricity prices, and facilitate strategic and system-compatible expansion of renewable energy sources. The decrease in financing expenses would promote the expansion of renewable energy sources (Kroger et al., 2022).

This study delves into the specific role of CfDs in the development of green energy markets in developing economies, building on existing contexts. This paper contributes to existing research on this decarbonization investment tool by examining the impact of CfDs on the functioning of the electricity market, the distribution of risk, and the integration of energy finance sources, market data, and policy research. It also explores the functioning of the renewable energy market and its potential to integrate various fields of energy finance, markets, and renewable energy policy. The goal of this research is to ascertain the economic and risk allocations for CfD implementation between the government and investors. This study pinpoints areas of research that require further investigation. Previous studies, for instance (Guan et al., 2024), specifically analyse the abilities of investment entities. The Chinese Integrated Energy Services (IES) sector is struggling to maintain its operations under the CFD agreement. To prevent distortions, Schlecht et al. (2024) propose a novel contract type that incorporates asset independence, a crucial characteristic of financial forwards and futures contracts (Beiter et al., 2024). Other studies, such as Kröger et al. (2022), scrutinized the CfD in renewable energy agreements across multiple European nations. CfDs, unlike many other intervention programmes, have the advantage of generating public cash during times of high-power costs. Policymakers have found this feature appealing, particularly since the onset of the energy crisis (European Commission, 2023). Within the ongoing reform discussion, there is a growing recognition that they are now considered a fundamental element of the electricity market rather than only a policy (Fabra, 2023).

This study builds on Beiter et al. (2024) and Schlecht et al. (2024), but it goes further by looking into whether contracts of difference (CfDs) are good ways for developing countries to finance renewable energy assets that need a lot of money. There are specific research objectives. (1) How useful can contracts of difference

(CfDs) be as green financial instruments? (2) Does increased CfD diffusion have a greater impact on market mechanisms for renewable electricity, the distribution of risks, and the possibility of merging separate energy finance streams in developing countries? (3) Are the current energy policies in developing nations capable of facilitating the implementation of CfDs? (4) To what degree does the implementation of CfDs aid in the attainment of renewable energy financing? The systematic review mapped the literature, identified significant themes in research types and evidence sources, and informed policymaking and research (Emodi et al., 2019; Wilson et al., 2012). This study based its protocol on the PRISMA.[1]

2 Methods

The subsequent subsections delineate the methodological approach employed in the literature review and elaborate on the data-gathering procedure and outcomes.

2.1 *The Approach of Systematic Review*

This article intends to comprehensively examine research on the role of contracts of difference (CfDs) in the diffusion of electricity market mechanisms. It will also look at how different areas of research on energy financing, markets, and policies in developing economies might be able to work together.

A systematic review forms the foundation of the approach following the initial work of (Pickering et al., 2015; Pickering & Byrne, 2014) Pickering and Byrne (2014), and Collaboration for Environmental Evidence (2018). This review follows a clear and comprehensive procedure to identify articles and generate a review and appraisal of the current literature. To be clearer, the protocol revolves around selecting a specific combination of keywords that create a search string. We then use this search string to locate and identify documents. This study builds on the systematic quantitative review method by first summarizing research on the role of contracts of difference (CfDs) in achieving the 2050 renewable energy expansion goals set by the United Nations Framework Convention on Climate Change (UNFCCC, 2018) and then examining the changes in risk allocation over time in the examined studies. Studies such as (Boulton et al., 2018; Bourcet, 2020; Emodi et al., 2019, 2021, 2022b, 2022a; McCabe et al., 2018; Nicolosi & Corbett, 2018; Sequeira & Santos, 2018) have applied this approach.

In this study, we analysed the systematic quantitative review in 15 steps. Pickering and Byrne (2014) provide well-documented steps, except steps 10–11 (data

[1]The PRISMA stands for the Preferred Reporting Items for Systematic Reviews and Meta-analyses. See http://www.prisma-statement.org.

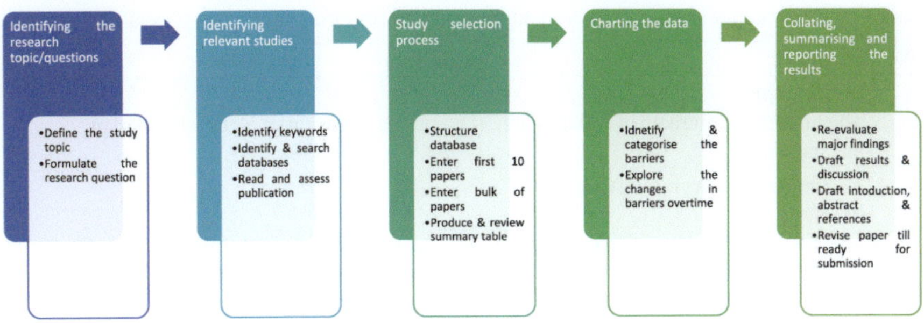

Fig. 1 The systematic quantitative review approach used in this study. (Modified from Pickering and Byrne (2014))

charting), where we identify, categorize, and explore the impacts of CfDs on the expansion of renewable energy generation over time. Figure 1 groups the steps into five phases. The initial phase [steps 1–2] establishes the topic and research questions. The second phase, encompassing stages 3–5, identifies relevant studies through keyword searches and an initial review of publication titles and abstracts to assess their relevance. Next, the research selection procedure [steps 6–10] involves a thorough reading of the papers and organizing the database. The fourth phase, encompassing steps 11–12, extracts the data from the papers and sends it to the database for further refining and analysis. The final stage [steps 13–16] involved reassessing, condensing, and preparing the results from the retrieved data from the examined research.

2.2 *Data Collection and Analysis*

Table 1's content guides the previously described methodology, which we applied to the topic of the impact of contracts of difference (CfDs) as a financing mechanism for creating a market for renewable electricity. We formulate and link four research questions to achieve the study's aim. (1) How useful can contracts of difference (CfDs) be as green financial instruments? (2) Does increased CfD diffusion have a greater impact on market mechanisms for renewable electricity, the distribution of risks, and the possibility of merging separate energy finance streams in developing countries? (3) Are the current energy policies in developing nations capable of facilitating the implementation of CfDs? (4) To what extent does utilizing CfDs contribute to achieving renewable energy financing? In the next stage, we search for a phrase that combines keywords (see "keyword search term" in No. 3 of Table 1) and code words (see "code word" in No. 3 of Table 1) to specifically identify studies related to: (1) renewable energy markets; (2) investment and financial mechanisms; and (3) CfDs that facilitate the expansion of renewable energy sources. The code words cover things like financing costs for renewable energy projects, financing measures, investors, markets, risks, policies, future electricity markets,

Table 1 The systematic approach for identification and selection of relevant papers

No.	Approach
1	**Define topic** The effects of contracts of difference (CfDs) on the diffusion of electricity market mechanisms
2	**Formulate research questions** Research aims The aim is to conduct a thorough quantitative analysis of the impact of contracts of difference (CfDs) on the spread of electricity market mechanisms. It will also look at how different areas of various sectors of energy finance, market dynamics, and policy research in developing economies could potentially collaborate Research questions – How useful can contracts of difference (CfDs) be as green financial instruments? – Does increased CfD diffusion significantly influence renewable electricity market mechanisms, risk allocation, and the potential for integrating fragmented electricity financing streams in developing countries? – Are the current energy policies in developing nations capable of facilitating the implementation of CfDs? – To what extent does utilizing CfDs contribute to achieving renewable energy financing?
3	**Keywords search string** contracts AND of AND difference, AND renewable AND energy AND financing contracts AND for AND difference, AND markets AND for AND renewable AND energy contracts AND for AND difference, AND investment AND financial AND channels contracts AND of AND difference AND that AND support AND the AND growth AND of AND renewable AND energy AND sources Codeword: Finance, costs, channels, policies, incentives, risks, and the market
4	**Online database** Scopus, Web of Science, and Google Scholar
5	**Inclusion criteria** Period: 2014–2024 Article type: peer-review and grey literature publications are available in English Analysis type: quantitative and qualitative Geographical: country and regional level Evidence: Qualitative and quantitative findings on the role of contracts of difference (CfDs) in the renewable energy market

and capital-intensive assets (see Table 1). We substitute the code words from No. 3 in Table 1 during the utilization of the keyword search term.

We developed a search string to find relevant studies related to the research questions in three online databases: Google Scholar, Scopus, and the Web of Science. The databases exhibit extensive coverage across various disciplines. We filtered the database search to articles published between August 2014 and August 2024, along with the recent market design and policy changes. We employed a snowball technique (Wohlin, 2014) to identify pertinent papers from the reference lists of the reviewed studies.

The literature search, utilizing the designated keyword phrase, resulted in 124 articles from Scopus, 86 from the Web of Science, and 2348 from Google Scholar.

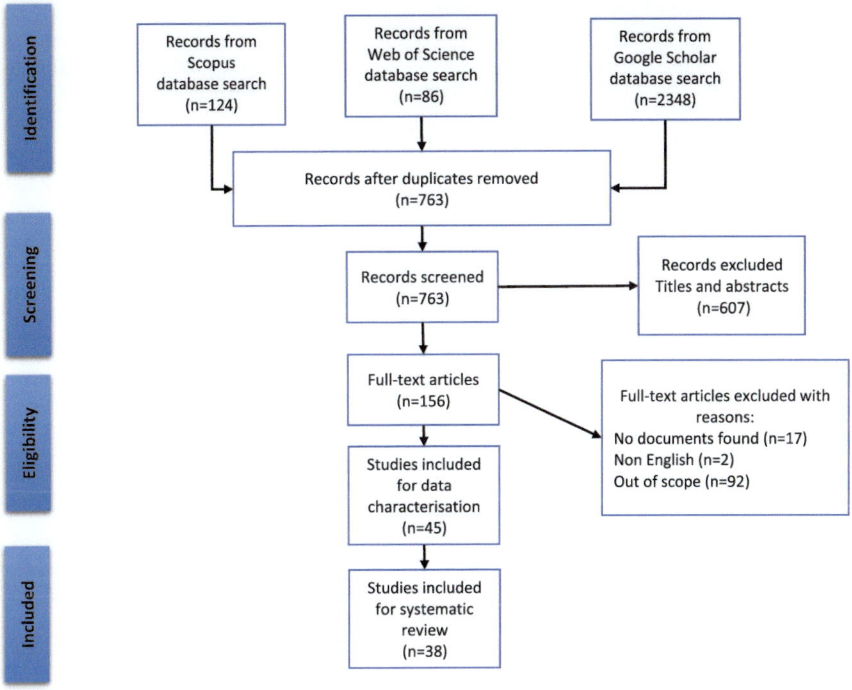

Fig. 2 The PRISMA flow diagram of the systematic quantitative literature review

We assessed the titles and abstracts of the selected studies against the inclusion criteria specified in Table 1 to determine their eligibility. Figure 2 presents the PRISMA flow diagram, which outlines the evaluation procedure (Moher et al., 2010; Page et al., 2021). Table 1's inclusion criteria, specifically criterion No. 5, guided the assessment of the articles. The study only included papers published from 2014 to the end of August 2024, excluding research published before 2014. Significantly, the papers selected for the review should assess the function and influence of contracts of difference in facilitating the growth and establishment of markets for renewable energy power. As a result, the examined publications look at three or more sources of renewable energy.

Following the selection process, the data characterization included a total of 45 papers. The systematic quantitative review selected a total of 38 studies (see Table 2). We assigned an index number to each study to distinguish the literature sources used in the review. Square brackets enclose the numbers (e.g. [#'xy'], where "xy" represents the number assigned to an article in Table 2). We structured the database in a tabular format to record information about the author(s), the year of publication, the journal, the country, the sector analysed, the methodologies employed, the parameters and software used, the data sources, the tested scenarios, policy implications, and the economic outcomes. The scenario analysis in the reviewed studies identifies the market mechanisms related to energy and renewable

Table 2 The examined list of published papers

Index #	Author (s) (year)	Country	Methods	An analysis of the renewable energy market and its associated policies being conducted
1	Schlecht et al. (2024)	Global	Comparative analysis	Proposed financial CfDs mitigate revenue risk significantly while offering fair incentives
2	Veenstra and Mulder (2024)	Thirteen European countries	Using the international electricity market's partial equilibrium model	This paper investigates the impact of CfDs in varied design aspects on day-ahead power costs, welfare, and renewable energy generation
3	Guan et al. (2024)	China	The models encompass the dynamic payback time, net present value, and internal rate of return	The viability of the integrated energy system's investment entity under CfDs
4	Beiter et al. (2024)	Global	Qualitative methods	The primary role of CfDs is to create a market for the provision of electricity at consistent long-term rates, which is an enduring characteristic of the renewable energy market
5	Ghorbani Pashakolaie et al. (2024)	United Kingdom	Quantitative method (Policy decision-making framework)	The study covered the UK's Renewable Obligation Programme and evaluated the benefits and drawbacks of offshore wind power generation
6	Nelson et al. (2024)	Australia	Comparative study	The study examines renewable generator bidding behaviour in electricity markets with high renewable energy penetration, as well as policy design implications for government underwriting policies
7	Kochems et al. (2024)	Germany	The method uses agent-based electricity market simulations for various scenarios	The study assesses future market-based cost recovery of variable renewable energy sources and the effects of different support schemes
8	Diallo and Kitzing (2024)	Italy	Theory and hypothetical auction modelling	The study presents a methodology for uncovering hidden biases in renewable energy procurement auctions, quantifying the problem's magnitude

(continued)

Table 2 (continued)

Index #	Author (s) (year)	Country	Methods	An analysis of the renewable energy market and its associated policies being conducted
9	Noble et al. (2024)	United Kingdom	A series of informed 'what if?' scenarios	Examine the amount of money required for continuous market assistance to attain cost savings via subsidized implementation, in addition to research and innovation
10	Kitzing et al. (2024)	Ten Europeans Countries	Qualitative	The study explores the use and effects of CfDs in Europe, examining design factors, packages, and projects executed in the region
11	Nelson et al. (2024)	Australia	Comparative	The paper examines various contract designs and their influence on participant behaviour during instances of significant oversupply and negative pricing
12	Fischer and Toffolo (2024)	Sweden	A game theory based	The study uses a method based on game theory and concludes that the auction-based CfDs policy instruments is the best way to cut CO_2 emissions by a large amount in the Nordic setting that was looked at
13	Nelson and Dodd (2023)	Australia	Case study	The article covers renewable energy investment, including market policy instruments for investing in renewable energy, designing network tariffs, and achieving environmental and social goals
14	Gohdes et al. (2023)	Australia	Project finance model	Establish a series of optimal contractual benchmarks corresponding to a particular level of capital structure risk
15	Neuhoff et al. (2023)	European Union	Qualitative	An analysis of the consumer *Contracts for Difference* can improve the efficiency and predictability of renewable investments
16	Newbery (2023)	Global	Quantitative/ comparative analysis	It postulates a streamlined contract that can be auctioned to achieve the low cost of carbon reduction while ensuring control over the level of assistance

(continued)

Table 2 (continued)

Index #	Author (s) (year)	Country	Methods	An analysis of the renewable energy market and its associated policies being conducted
17	Gohdes et al. (2023)	Australia	A dynamic, integrated, multi-year Project Finance (PF) model for electricity projects	The essay analyses the semi-merchant phenomena, seeking to elucidate the increasing popularity of the investment model among investors in variable renewable energy
18	Alao and Cuffe (2023)	Irish	Blockchain-based power model	The paper used conventional CFDs to illustrate the implementation of blockchain-based power derivative agreements
19	Blyth et al. (2023)	United Kingdom	An open-source electrical system model integrated with a finance model	The paper examines 'transition risk', the price risk arising from uncertainty in the combination of technologies employed for decarbonization, and its fluctuations across various technology combinations and regulatory frameworks
20	Kell et al. (2023)	United Kingdom	Modelling approach for simulating CfD auctions	The essay presents an innovative framework for examining CfD auction dynamics and facilitates an in-depth examination of actual renewable energy subsidy auctions
21	Li et al. (2023)	China	Styled model	The paper solely compared futures and spot markets to the structure of markets from the government's perspective
22	Fabra (2023)	European countries	Electricity model	The article looks at CfDs as a structure for the electricity market and talks about how they can help with short-term energy efficiency and long-term contracts backed by the government, which leads to better investment decisions
23	Hastings-Simon et al. (2022)	Canada	An examination and evaluation of Alberta's Renewable Electricity Programme	The paper examined the impact of CfD schemes in increasing the share of renewable energy through their successful proponents, who offered a guaranteed price for power generation relevant to the project

(continued)

Table 2 (continued)

Index #	Author (s) (year)	Country	Methods	An analysis of the renewable energy market and its associated policies being conducted
24	Kroger et al. (2022)	Germany	Comparative analysis	The paper examined how CfDs provide superior protection against electricity price volatility for both producers and consumers of renewable energy
25	Hesel et al. (2022)	Germany	European electricity market model	The study introduces a novel method for incorporating hydrogen into current long-term optimization models for energy market dispatch
26	Neuhoff et al. (2022)	Germany	A comprehensive model analysing the finance framework of renewable investment initiatives	The articles examine the implications of decreasing technology costs through both analytical and numerical methods, detailing the financing conditions associated with different renewable energy policies
27	Savelli et al. (2022)	United Kingdom	Comparative analysis	The study aims to show how an upgraded contract for difference mechanism, incorporating three additional cost elements, can internalize system-wide costs
28	Bilas et al. (2022)	Global	Comparative analysis	The study looks at the global trends, problems, and future of wind farms as well as power plants. It also gives a full picture of the rules and regulations in place at both the global and European Union levels
29	Jansen et al. (2022)	Global	Comparative analysis	The study provides a comprehensive overview of global offshore wind energy auctions, including systematically structuring auction designs, contextualizing results, analysing jurisdiction-specific factors, and cross-jurisdictional analysis. It also discusses policy implications, including subsidy-free offshore wind trends, revenue stabilization, and seabed lease interactions

(continued)

Table 2 (continued)

Index #	Author (s) (year)	Country	Methods	An analysis of the renewable energy market and its associated policies being conducted
30	Nelson et al. (2022)	Australia	Qualitative	The article reviews the role of CfDs in Australia's policy architecture to reduce emissions costs effectively within the electricity sector
31	Simshauser (2020)	Australia	Comparative analysis	The article examines the most complex merchant investment commitments in energy-only markets, as well as the role of government-initiated contracts for different policies
32	Algarvio et al. (2020)	Portugal and Spain	Models for multi-agent trading in electricity markets	The essay looks at a study that explores the potential benefits of bilateral agreements and CfDs as the cornerstones of a renewable energy assistance programme
33	Simshauser (2019)	Australia	The model of partial equilibrium solely focuses on the energy market	The essay assesses the function of CfDs in addressing energy market failures associated with absent policies, absent markets, and incomplete markets
34	Wild (2017)	Australia	Comparative analysis	Examine how a CfDs could contribute to an increase in investment in Australia's renewable energy
35	Onifade (2016)	United Kingdom	Qualitative	The article evaluated the role of CfDs as a hybrid policy in addressing private sector interests and environmental protection
36	Ross (2014)	United Kingdom	Qualitative	The role of CfDs in providing a new mechanism of support for renewable and low-carbon electricity generation is worth exploring
37	Kozlov (2014)	United Kingdom	Qualitative	The articles discussed the role of CfDs, which play a crucial role in ensuring long-term electricity price stability and minimizing price fluctuations for eligible generators' counterparty credit risk

(continued)

Table 2 (continued)

Index #	Author (s) (year)	Country	Methods	An analysis of the renewable energy market and its associated policies being conducted
38	Oak et al. (2014)	United Kingdom	A generalized mathematical model of industry growth	Examine the impacts of the CfDs scheme as an incentive scheme for renewable electricity producers to meet carbon emissions reduction targets

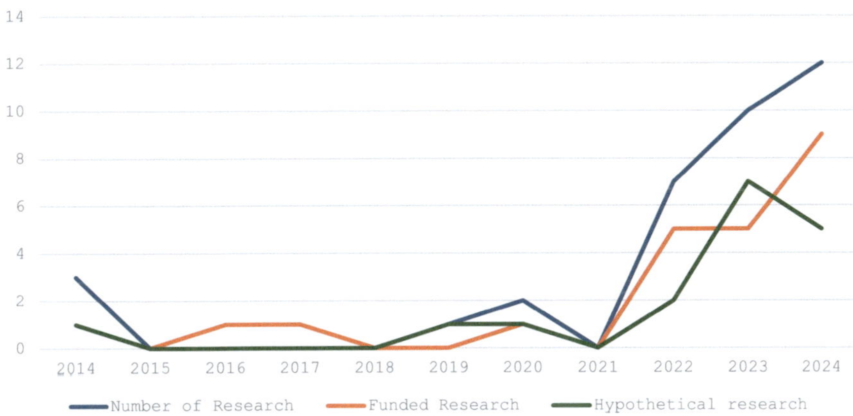

Fig. 3 Illustrates the quantity of publications, funded research, and study categories from 2014 to 2024

energy expansion policies, which are categorized as "1" in the results subsection and summarized in Table 4 and Fig. 3. The authors input, edit, and verify data from the reviewed studies in weekly meetings prior to finalizing the systematic review.

3 The Results

Figure 3 illustrates the progression of the 38 studies published from 2014 to 2024 and highlights the quantity of sponsored research studies and nonfunded articles on both hypothetical and implemented topics, accounting for differences in renewable energy projects. The most productive years were 2022 and 2023 since the adoption of the CfDs schemes in 2014, with five funded studies in each of the years. The years 2022 and 2023 marked the peak of productivity. The conflict in Ukraine began in February 2022, leading to a decrease in Russian gas supplies to the European Union. The conflict precipitated an energy market crisis throughout Europe and necessitated efforts to mitigate climate change by decreasing CO_2 emissions and

Table 3 The journals and non-peer-reviewed literature of the studies that were analysed

Journal	Number of papers	Grey literature	Number of papers
Energy Economics	6	20th International conference on Europeans Energy market	1
Energy Policy	8	IEEE Transactions on Industrial Informatics	1
Renewable Energy	2	DIW weekly report	1
Nature Energy	1	PAAM collections 2020	1
Applied Energy	2	Robert Schuman Centre for Advance Studies	1
The Australian Journal of Agricultural and Resource Economics	5		
Energies	1		
Heliyon	1		
Resource & Energy Economics	1		
Journal of Sustainability Science & Management	1		
Environmental Law & Management	1		
Journal of World energy Law and Business	1		
Energy	1		

replacing fossil fuels with renewable energy produced from sustainable sources. A similar pattern emerged in 2024, with 12 studies published by 31 August 2024, and nine of these studies receiving funding. This is also in line with November 2023, when the Commonwealth Government implemented a 32 GW run-of-plant (RoP) power purchase agreements (PPAs) CfDs underwriting policy to protect generators from changes in the market price.

Since their introduction in the United Kingdom in 2014, numerous countries in Europe and other developed nations have adopted CfDs in recent years, including Denmark, Greece, Hungary, Poland, Ireland, Australia, and Canada (#23, #24). Energy Policy, Energy Economics, and The Australian Journal of Agricultural and Resource Economics, as shown in Table 3, are the journals in which the analysed studies were published. Table 4 reveals that 54% of the reviewed studies used an approach for case study and literature review, with empirical papers following closely at 46%. Regarding the geographical locus (Table 5), the authors conducted an analysis on a global scale, encompassing five studies. The European Union was the focus of 2 regional papers, 18 in total, while the North American and Asia-East regions received 1 and 2 studies, respectively. The foremost three researched developed nations are the United Kingdom ($n = 9$), Australia ($n = 8$), and Germany ($n = 3$). Conversely, China ($n = 3$) is the most extensively studied developing economy in the analysed research.

Table 4 The methods applied in the reviewed studies. "%" = percentage

Methodology	% of publications (number of papers)
Case study/Literature review	54% ($n = 20$)
Empirical analysis (using secondary data)	46% ($n = 17$)

Table 5 The number of studies by countries and the number of studies by regions

Country	Number of publication	Region	Number of studies
Australia	8	Asia	2
Canada	1		
China	2	Europe	18
Germany	3	Global	5
Italy	1	Oceania	8
Irish	1		
Portugal	1		
Spain	1		
Sweden	1		
United Kingdom	9		

4 Discussion

This section delineates the role of CfDs in the renewable energy market and examines what has changed during the review period. Ultimately, we defined the research gaps and limitations of the systematic review for other investigations.

4.1 Determining the Role of CfDs in Establishing a Market for Renewable Electricity Production at Stable, Long-Term Rates

After reviewing the literature, we found that most studies used various literature reviews, case studies, and comparative analyses to identify a diverse range of impacts associated with CfDs. Additional research investigated the effects using empirical methods, consistent with our preliminary assessment of the methodologies employed in the studies reviewed (see Table 4). We conducted a detailed analysis of each study to pinpoint specific roles in the expansion of the energy market, as well as the advantages it offers to both governments and producers investing in renewable energy. Qualitative studies identify the roles and impacts of CfDs based on feedback from survey or interview participants about their use of CfDs for renewable energy investment through CfDs. Also, quantitative research has identified factors that significantly and positively impact investment decisions in the development of renewable energy when analysed using an empirical model. We modified the roles from our original assessment and limited them to address the variation between research, technology, and locales.

We define the roles of CfDs as follows: CfDs offer low financing costs and reduced electricity generation costs. The strategies include market integration, promoting a cost-effective operational framework, safeguarding against electricity price volatility, and risk management.

4.1.1 Low Financing Costs and Reduced Electricity Generation Costs

Compared to other generation assets, such as wind power and solar photovoltaics (PV), which are upfront capital intensive with very small operations and maintenance costs, the cost of capital is one of the most important variables in determining the feasibility of investing in renewable energy projects. These costs are typically fixed and not variable (#10, #12, #13, #14). They are linked to actual operations, meaning they do not impact the marginal costs of production (#1, #2, #3, #4).

4.1.2 Facilitating Market Integration and Promoting a Cost-Effective Operational Framework

CfDs are a significant financial tool for decarbonization efforts due to their capacity to fund large-scale renewable energy projects, including those that need substantial initial capital investment. Renewable energy sources have a similar high initial capital cost structure as other fast-expanding components of a future power system, such as batteries and the transmission network (#4, #10). The integration of markets can significantly enhance the efficiency of renewable energy systems when producers in this sector encounter productive risks. Through the exposure of renewable energy projects to electricity market price signals, operators strategically adjust their projects in response to short-, medium-, and long-term price signals (#2, #5, #7, #8, #9). If market price signals accurately represent all elements of the power system, market price exposure provides motivations to act in a manner that is advantageous to the system (#7, #8, #9, #10, #11).

Moreover, the lack of adequate long-term price hedging tools is a market failure that, if left unaddressed, would most likely lead to the EU's failure to meet its renewables commitments (#15, #16, #17, #19). The deployment of CfDs for renewable energy sources generates several interactions with market-based renewable energy deployment, which operates alongside public-backed market development.

4.1.3 Provide Enhanced Safeguards Against Electricity Price Volatility for Both Renewable Energy Suppliers and Customers

Within the realm of green power investments, the majority of expenses occur during the first stages of the project, while the ongoing operational costs remain rather minimal (#24, #26, #29). To assess a project's profitability, one must have a long-term view of prices and their associated revenues. This task becomes especially challenging when dealing with unpredictable energy market prices. CfDs provide large-scale delivery of clean energy through long-term fixed-price contracts. They safeguard the owners of renewable energy assets from consistently low prices, allowing project financing to accept lower revenues during periods of high prices (#4, #29, #30, #31). Conversely, two-sided CfDs protect the counterparty (either a utility or a state-controlled entity) from consistently high prices (#33). While this results in planning certainty and budget control, it sacrifices the potential for lower electricity prices found on the spot market. CfDs provides a transparent framework for procuring large volumes of renewable energy, allowing suppliers and consumers to clearly define quantities and time frames. Furthermore, within the electricity market, both a disparity in profitability and unpredictability in pricing can significantly impact the success of renewable energy projects. In addition to uncertainty surrounding the long-term evolution of electricity prices, electricity markets are susceptible to price volatility. CfDs would mitigate risks for all power consumers, including private households and businesses, and lower their rates during periods of elevated electricity prices (#1, #4, #27, #28, #34, #35 #36, #38).

4.1.4 Risk Management

For electricity markets to function efficiently, risks such as financial distress, basis risk, volume risk, and price risk should be allocated so that renewable energy producers are exposed to productive risk. CfDs mitigate price volatility and provide price stability for producers by compensating for the discrepancy between a strike price and a reference market price. It also has the potential to mitigate the risk associated with large-scale renewable investment, enticing investors to favour those with less stringent return criteria. In the end, this could lower the cost of power from renewable sources for consumers, thereby speeding up the decarbonization process (#4, #17, #18, #26). CfD schemes stabilize prices by compensating for a discrepancy between a strike price and the market price. They help producers mitigate production hazards and volume risk associated with negative prices and weather conditions. However, when using reference pricing or models, base hazards arise for operators of renewable energy sources. Divergence may occur due to the specification of the reference parameter, which remains unaffected by market price and generating conditions. CfDs establish specific references, such as market prices and volume, which can lead to negative power prices. Producers must have sufficient liquidity reserves to settle debt and meet payback commitments, with the effect of repayment commitments on necessary liquidity buffers depending on the CfD scheme's structure.

CfDs mitigate price risk and stabilize prices for producers by compensating for the disparity between the strike price and the reference market price. One can forecast the actual availability of renewable energy resources to a limited degree as it may differ from the anticipated value in the short and medium term. Nonetheless, production hazards substantially diminish over extended periods (years and decades). Depending on their configuration, CfDs can mitigate the volume risk associated with negative prices and meteorological conditions. Base hazards are the structural risks encountered by operators of renewable energy sources when utilizing support programmes that incorporate reference values or models. This includes the use of reference pricing (e.g. monthly or annual reference price) or reference electrical production quantities. Distancing the reference specification from the specific market pricing and generation circumstances of the renewable energy facility creates a risk of divergence. A CfD delineates one or several references, such as a reference market price/ or a reference volume (#4, #10). The renewable energy producer may underperform relative to the reference price by selling power at a value inferior to the reference. CfDs entail an intrinsic risk of repayment obligations; producers must possess not only liquidity reserves to service debt but also to meet repayment commitments. The precise effect of the return commitment on necessary liquidity reserves is contingent upon the structure of the CfD scheme (#4, #10, #12, #13, #14, #15).

4.2 Policy and Regulatory Impact Changes in CfDs Over Time Have Been Observed in the Reviewed Studies

The key problem with traditional renewable energy policies, such as feed-in tariffs, feed-in premiums, and renewable portfolio criteria, has prompted governments to reassess and improve their policies and frameworks. The regulatory ambiguity surrounding feed-in tariffs has rendered these conventional techniques significantly unreliable. They have prompted stakeholders, particularly market participants, to advocate for improved policies. The Electricity Market Reform (EMR) introduced the CfD support scheme, which proved successful. CfDs are becoming the global standard in the United Kingdom as policymakers transition from traditional schemes to CfDs to support the expansion of the renewable energy market. Observers noted that the development and implementation of CfDs, as a renewable energy financing mechanism, could potentially tackle the regulatory and market issues associated with traditional renewable energy policies (#34, #35, #36, #37, #38). As the share of renewables rises, CfDs, which finance upfront capital-intensive assets, will become increasingly common. Future power systems, including batteries, electrolysers, and the transmission network, share the high upfront capital cost structure of renewables. These factors may lead to increased CfD procurement and fundamentally alter electricity market mechanisms. Historically, competitive power markets have concentrated on short-term dispatch and the capital requirements of fossil fuel resources, recovering costs via wholesale power prices and restricted hedging options. Inadequate long-term price signals lead to discrepancies in renewable investment. Although CfDs can potentially mitigate market deficiencies, their extensive implementation may pose challenges that necessitate thorough considerations for market reform. Spot prices indicate power delivery constraints, potentially affecting new plant entry. Procuring capacity through CfDs may have less influence on new plant entry, resulting in poor timing, size, location, and motivation to respond to spot price signals. Designing CfDs to allow exposure to short-term price signals is crucial.

Additionally, we can use CfDs to manage risks in electricity markets, but we often overlook a thorough exploration of this allocation of price risk among asset owners, ratepayers, and taxpayers. Recent literature suggests structural modifications to CfDs and comparative assessments of alternative policy interventions like renewable portfolio standards or carbon pricing. Therefore, designing two-way CfDs can enhance the renewable energy market by increasing curtailment compared to one-way CfDs.

The contract's fundamental concept is to reduce risk while maintaining incentives by delineating payments that are closely linked to cash flow but unaffected by the actions of the contracting parties, thereby preventing manipulation. The current form of CfDs presents problematic incentives and distorts the dispatch and investment decisions for generation assets. These issues tend to worsen when applied to nonzero variable cost technologies and/ or increased volumes. Ongoing challenges, including distorted intraday markets, persist in all conventional, generation-based CfDs. Designing CfDs to recouple payments from an asset's output enhances incentives. Applying CfDs to significant portions of future power generation appears imprudent. A "financial" CfD, which combines a regular CfD and a forward contract, protects against volume risk, maintains stable revenues, and prevents cash flow issues using tangible assets utilized as collateral. This contrasts with financial futures, which link payments to individual generator output. Financial CFDs reduce or increase power generator risks by linking volumes to a reference generator, offering a safeguard against volume risk, and eliminating the need for measures to halt payments during negative or marginally positive prices.

4.3 *The Knowledge Gaps in the Literature Evaluation and Pertinent Areas for Future Research*

This systematic review determined that the role of CfDs as a financing mechanism is crucial in creating a market for renewable electricity. Notwithstanding the increasing volume of studies, many research gaps persist. Early on, we noticed a dearth of studies examining how developing countries, such as those in Southeast Asia and sub-Saharan Africa, have utilized *Contracts for Difference*, private financing mechanisms, and green finance policies to address the financing and risk issues associated with green energy. These countries have been experiencing a steady rise in energy demand. Therefore, there is a pressing need for further original research to understand investors' perceptions of the energy market in developing countries, as well as their perspectives on potential investment opportunities in these regions. Second, few studies have examined the potential impact of CfD on the electricity market, including the distribution of risk and the integration of various forms of energy finance, market, and policy research.

The literature has not adequately researched the analysis of financial CfDs in specific countries. A growing body of research has demonstrated that financial CfD hedges mitigate volume risk in addition to price risk, thereby stabilizing revenues. This contract mitigates liquidity issues by using physical assets as collateral as opposed to financial futures. The contract aims to reduce risk and maintain high incentive levels by specifying payments closely linked to cash flow that are not dependent on the party's behaviour and remain unchangeable. In the future, researchers could investigate how the spread of financial CfD affects the mechanisms of the renewable market, how risks are shared, and how different energy finance, market, and policy research approaches can be combined in developing countries that are having trouble with energy security. Addressing these research gaps would enhance the literature and augment our understanding of the role of CfDs, including traditional and financial CfDs, in achieving net-zero emissions goals.

4.4 The Advantages and Disadvantages of Systematic Review Methodology

We conducted a comprehensive systematic quantitative study to examine the literature on CfDs as a funding mechanism in the renewable energy sector, with the aim of meeting the net zero emissions target by 2050. On 30 August 2024, we revised our search results to incorporate the latest findings prior to finishing the evaluation. Notwithstanding the employed systematic review methodology, the current work exhibits some shortcomings. Initially, we limited our literature review to studies published in English. This could potentially lead to linguistic bias, as the review process may have excluded non-English papers with significant findings. Secondly, the search criteria might have overlooked recent forms of renewable fuels like hydrogen and biofuels, which the literature does not consistently classify as clean energy. Thirdly, the search engines employed in this study are not comprehensive, as other databases may harbour studies that could be beneficial to our evaluation. Ultimately, the evaluation did not aim to judge the quality of the research methods or approaches examined. Rather, the focus was on ensuring that the selected publications adhered to stringent methods, conducting a comprehensive analysis of the existing literature on the role of CfDs in the energy market, encompassing 38 publications from 2014 to 2024.

5 Conclusion

The shift to a decarbonized economy requires swift investments in extensive renewable energy technological assets as well as careful market reform considerations to reshape electricity market mechanisms. Renewable energy, unlike traditional fossil fuel technologies, is not subject to fuel costs, highlighting the importance of upfront construction and capital costs. Renewables possess a significant initial cost profile akin to other swiftly expanding assets within a future energy framework. Nonetheless, inadequate long-term pricing signals from competitive markets contribute to the disparity between renewable investment and the government's objective of swift decarbonization, as highlighted in numerous studies conducted over the past decade.

This systematic review reveals that, in 2014, the United Kingdom initiated the development of CFD schemes through a private agreement between a government representative and an organization that produces low-carbon electricity. The contract aimed to address environmental and energy challenges by deploying clean energy within a low-carbon economy through deregulation. Subsequently, many developed nations have widely seen the CfDs as a fundamental element of future electricity market architecture, following Europe's energy crisis and the need for a future electricity market. Studies focusing on the role of CfDs in the transition to decarbonized societies in developing countries remain limited, notwithstanding the worldwide initiative to expand renewable energy technologies through the expansion of renewable energy markets. However, we acknowledge that individual countries' laws and regulations, along with industry-based practices, have primarily influenced this growing research area, which has gradually garnered the attention of the scientific community.

This study looked at 38 studies and put together what they had to say about the roles of CfD schemes as financing mechanisms for the renewable energy sector to reach its net-zero goals. The research primarily focused on industrialized nations, with the United Kingdom and Australia exhibiting the highest volume of study. China had the highest number of publications among developing countries.

The reviewed studies identified several attributes of CfD schemes, including reduced finance costs, reduced power pricing risks for producers, customers, and industries, and risk management through the construction of a secure long-term renewable energy market. Energy markets primarily aim to enhance productive, allocative, and dynamic efficiency. With renewables emerging as the leading investment in deregulated energy markets, the process of reducing carbon emissions might not advance at a rate that aligns with the goals of parliamentarians. This has prompted governments in several countries to initiate market priming through 'CfDs', thereby accelerating investment and decarbonization initiatives.

The CfDs is a crucial element in the future architecture of the renewable energy market, as it offers a transparent framework for acquiring substantial amounts of renewable energy, enabling clear quantification and scheduling. CfDs for renewable energy sources support both increasing renewable energy and stabilizing electricity prices. We achieve this by establishing a stable long-term price market for electricity supply, protecting both suppliers and customers from price fluctuations while simultaneously supporting the expansion of these sources.

References

Alao, O., & Cuffe, P. (2023). Implementing contract-for-difference arrangements for hedging electricity price risks of renewable generators on a blockchain marketplace. *IEEE Transactions on Industrial Informatics, 19*(4), 5679–5688. https://doi.org/10.1109/TII.2022.3185661

Algarvio, H., Lopes, F., & Santana, J. (2020, July 6). Renewable energy support policy based on contracts for difference and bilateral negotiation. In *Highlights in practical applications of agents, multi-agent systems, and trust-worthiness*. The PAAMS Collection. PAAMS2020. Communication in Computer and Information Science.

Beiter, P., Guillet, J., Jansen, M., Wilson, E., & Kitzing, L. (2024). The enduring role of contracts for difference in risk management and market creation for renewables. *Nature Energy, 9*(1), 20–26. https://doi.org/10.1038/s41560-023-01401-w

Bilas, V., Franc, S., & Trifunić, L. (2022). Supporting sustainability through the use of offshore wind energy: The case of United Kingdom. *Journal of Sustainability Science and Management, 17*(8), 236–250. https://doi.org/10.46754/jssm.2022.08.014

Blyth, W., Gross, R., Rickman, J., MacIver, C., & Bell, K. (2023). Transition risk: Investment signals in a decarbonising electricity system. *Applied Energy, 352*, 121938. https://doi.org/10.1016/j.apenergy.2023.121938

Boulton, C., Dedekorkut-Howes, A., & Byrne, J. (2018). Factors shaping urban greenspace provision: A systematic review of the literature. In *Landscape and urban planning* (Vol. 178, pp. 82–101). Elsevier B.V. https://doi.org/10.1016/j.landurbplan.2018.05.029

Bourcet, C. (2020). Empirical determinants of renewable energy deployment: A systematic literature review. *Energy Economics, 85*, 104563. https://doi.org/10.1016/j.eneco.2019.104563

Chen, H., Shi, Y., & Zhao, X. (2022). Investment in renewable energy resources, sustainable financial inclusion and energy efficiency: A case of US economy. *Resources Policy, 77*, 102680. https://doi.org/10.1016/j.resourpol.2022.102680

Collaboration for Environmental Evidence (2018) 2018 Annual Report https://environmentalevidence.org/wpcontent/uploads/2019/08/Annual-Report-2018_Final.pdf

Diallo, A., & Kitzing, L. (2024). Selection bias in multi-technology auctions: How to quantify and assess efficiency implications in renewable energy auctions. *Energy Policy, 184*, 113864. https://doi.org/10.1016/j.enpol.2023.113864

Emodi, N. V., Chaiechi, T., & Beg, A. B. M. R. A. (2019). The impact of climate variability and change on the energy system: A systematic scoping review. *Science of the Total Environment, 676*, 545–563. https://doi.org/10.1016/j.scitotenv.2019.04.294

Emodi, N. V., Lovell, H., Levitt, C., & Franklin, E. (2021). A systematic literature review of societal acceptance and stakeholders' perception of hydrogen technologies. *International Journal of Hydrogen Energy, 46*(60), 30669–30697. https://doi.org/10.1016/j.ijhydene.2021.06.212

Emodi, N. V., Okereke, C., Abam, F. I., Diemuodeke, O. E., Owebor, K., & Nnamani, U. A. (2022b). Transport sector decarbonisation in the Global South: A systematic literature review. *Energy Strategy Reviews, 43*, 100925. https://doi.org/10.1016/j.esr.2022.100925

Emodi, N. V., Wade, B., Rekker, S., & Greig, C. (2022a). A systematic review of barriers to greenfield investment in decarbonisation solutions. *Renewable and Sustainable Energy Reviews, 165*, 112586. https://doi.org/10.1016/j.rser.2022.112586

European Commission (2023) Commission welcomes deal on electricity market reform. https://ec.europa.eu/commission/presscorner/detail/en/ip_23_6602

Fabra, N. (2023). Reforming European electricity markets: Lessons from the energy crisis. *Energy Economics, 126*, 106963. https://doi.org/10.1016/j.eneco.2023.106963

Fischer, R., & Toffolo, A. (2024). Game theory-based analysis of policy instrument consequences on energy system actors in a Nordic municipality. *Heliyon, 10*(4), e25822. https://doi.org/10.1016/j.heliyon.2024.e25822

Ghorbani Pashakolaie, V., Cotton, M., & Jansen, M. (2024). The co-benefits of offshore wind under the UK Renewable Obligation scheme: Integrating sustainability in energy policy evaluation. *Energy Policy, 192*, 114259. https://doi.org/10.1016/j.enpol.2024.114259

Gohdes, N., Simshauser, P., & Wilson, C. (2023). Renewable investments, hybridised markets and the energy crisis: Optimising the CfD-merchant revenue mix. *Energy Economics, 125*, 106824. https://doi.org/10.1016/j.eneco.2023.106824

Guan, H., Zhang, L., Jia, Y., Liu, Y., & Bai, M. (2024). Investigation on survivability of the entity investing in integrated energy system service under contract for difference. *Renewable Energy, 223*, 120063. https://doi.org/10.1016/j.renene.2024.120063

Hastings-Simon, S., Leach, A., Shaffer, B., & Weis, T. (2022). Alberta's Renewable Electricity Program: Design, results, and lessons learned. *Energy Policy, 171*, 113266. https://doi.org/10.1016/j.enpol.2022.113266

Hesel, P., Braun, S., Zimmermann, F., & Fichtner, W. (2022). Integrated modelling of European electricity and hydrogen markets. *Applied Energy, 328*, 120162. https://doi.org/10.1016/j.apenergy.2022.120162

IRENA. (2023). *World Energy Transitions Outlook 2023 1.5 C PATHWAY.*

IRENA & CPI. (2023). *Global landscape of renewable energy finance.*

Jansen, M., Beiter, P., Riepin, I., Müsgens, F., Guajardo-Fajardo, V. J., Staffell, I., Bulder, B., & Kitzing, L. (2022). Policy choices and outcomes for offshore wind auctions globally. *Energy Policy, 167*, 113000. https://doi.org/10.1016/j.enpol.2022.113000

Kell, N. P., Santibanez-Borda, E., Morstyn, T., Lazakis, I., & Pillai, A. C. (2023). Methodology to prepare for UK's offshore wind contract for difference auctions. *Applied Energy, 336*, 120844. https://doi.org/10.1016/j.apenergy.2023.120844

Kitzing, L., Held, A., Gephart, M., Wanger, F., Anatolitis, V., & Klessmann, C. (2024). *Contracts-for- difference to support renewable energy technologies: Considerations for design and implementation.*

Kochems, J., Sperber, E., Nienhaus, K., & Schimeczek, C. (2024). Market designs, actor decisions and market values: Assessment of remuneration mechanisms for future electricity system scenarios. In *International conference on the European energy market, EEM*. https://doi.org/10.1109/EEM60825.2024.10608886.

Kozlov, N. (2014). Contracts for difference: Risks faced by generators under the new renewables support scheme in the UK. *Journal of World Energy Law and Business, 7*(3), 282–286. https://doi.org/10.1093/jwelb/jwu016

Kroger, M., Neuhoff, K., Richstein, J. C. (2022). Contracts for difference support the expansion of renewable energy sources while reducing electricity price risks DIW weekly report standard-nutzungsbedingungen. *Deutsches Institute Fur Wirtschaftsforchung (DIW), 12*. https://doi.org/10.18723/diw_dwr:2022-35-1.

Kröger, M., Neuhoff, K. Richstein, J.C. (2022). Contracts for difference support the expansion of renewable energy sources while reducing electricity price risks .DIW Weekly Report Provided in Cooperation with: German Institute for Economic Research (DIW Berlin) https://www.econstor.eu/bitstream/10419/264921/1/181697188X.pdf

Li, L., Wang, H., Yan, X., & Bian, Y. (2023). Forward contracting and spot trading in electricity markets. *Annals of Operations Research*. https://doi.org/10.1007/s10479-023-05547-4.

Liu, W., Shen, Y., & Razzaq, A. (2023). How renewable energy investment, environmental regulations, and financial development derive renewable energy transition: Evidence from G7 countries. *Renewable Energy, 206*, 1188–1197. https://doi.org/10.1016/j.renene.2023.02.017

Masson-Delmotte, V., Zhai, P., Pörtner, H.-O., Roberts, D., Skea, J., Shukla, P. R., Pirani, A., Moufouma-Okia, W., Péan, C., Pidcock, R., Connors, S., Matthews, J. B. R., Chen, Y., Zhou, X., Gomis, M. I., Lonnoy, E., Maycock, T., Tignor, M., & Waterfield, T. (2018). *Global warming of 1.5°C An IPCC Special Report on the impacts of global warming of 1.5°C above pre-industrial levels and related global greenhouse gas emission pathways, in the context of strengthening the global response to the threat of climate change, sustainable development, and efforts to eradicate poverty*. Summary for Policymakers Edited by Science Officer Science Assistant Graphics Officer Working Group I Technical Support Unit.

McCabe, A., Pojani, D., & van Groenou, A. B. (2018). The application of renewable energy to social housing: A systematic review. *Energy Policy, 114*, 549–557. https://doi.org/10.1016/j.enpol.2017.12.031

Moher, D., Liberati, A., Tetzlaff, J., & Altman, D. G. (2010). Preferred reporting items for systematic reviews and meta-analyses: The PRISMA statement. *International Journal of Surgery, 8*(5), 336–341. https://doi.org/10.1016/j.ijsu.2010.02.007

Nelson, T., & Dodd, T. (2023). Contracts-for-difference: An assessment of social equity considerations in the renewable energy transition. *Energy Policy, 183*, 113829. https://doi.org/10.1016/j.enpol.2023.113829

Nelson, T., Easton, S., Wand, L., Gilmore, J., & Nolan, T. (2024). Electricity contract design and wholesale market outcomes in Australia's National Electricity Market. *Australian Journal of Agricultural and Resource Economics, 68*, 784. https://doi.org/10.1111/1467-8489.12588

Nelson, T., Nolan, T., & Gilmore, J. (2022). What's next for the renewable energy target—resolving Australia's integration of energy and climate change policy?*. *Australian Journal of Agricultural and Resource Economics, 66*(1), 136–163. https://doi.org/10.1111/1467-8489.12457

Neuhoff, K., May, N., & Richstein, J. C. (2022). Financing renewables in the age of falling technology costs. *Resource and Energy Economics, 70*, 101330. https://doi.org/10.1016/j.reseneeco.2022.101330

Neuhoff, K., Richstein, J. C., & Kröger, M. (2023). Reacting to changing paradigms: How and why to reform electricity markets. *Energy Policy, 180*, 113691. https://doi.org/10.1016/j.enpol.2023.113691

Newbery, D. (2023). Efficient renewable electricity support: Designing an incentive-compatible support scheme. *Energy Journal, 44*(3), 1–22. https://doi.org/10.5547/01956574.44.3.dnew

Nicolosi, E., & Corbett, J. B. (2018). Engagement with climate change and the environment: a review of the role of relationships to place. *Local Environment, 23*(1), 77–99. https://doi.org/10.1080/13549839.2017.1385002

Noble, D. R., Grattan, K., & Jeffrey, H. (2024). Assessing the costs of commercialising tidal energy in the UK. *Energies, 17*(9), 2085. https://doi.org/10.3390/en17092085

Onifade, T. T. (2016). Hybrid renewable energy support policy in the power sector: The contracts for difference and capacity market case study. *Energy Policy, 95*, 390–401. https://doi.org/10.1016/j.enpol.2016.05.020

Oak, N., Daniel L., Alan C. (2014) Performance comparison of renewable incentive schemes using optimal control, Energy http://doi.org/10.1016/j.energy.2013.11.038

Page, M. J., McKenzie, J. E., Bossuyt, P. M., Boutron, I., Hoffmann, T. C., Mulrow, C. D., Shamseer, L., Tetzlaff, J. M., Akl, E. A., Brennan, S. E., Chou, R., Glanville, J., Grimshaw, J. M., Hróbjartsson, A., Lalu, M. M., Li, T., Loder, E. W., Mayo-Wilson, E., McDonald, S., et al. (2021). The PRISMA 2020 statement: An updated guideline for reporting systematic reviews. *BMJ, 372*, n71. https://doi.org/10.1136/bmj.n71

Pickering, C., & Byrne, J. (2014). The benefits of publishing systematic quantitative literature reviews for PhD candidates and other early-career researchers. *Higher Education Research and Development, 33*(3), 534–548. https://doi.org/10.1080/07294360.2013.841651

Pickering, C., Grignon, J., Steven, R., Guitart, D., & Byrne, J. (2015). Publishing not perishing: How research students transition from novice to knowledgeable using systematic quantitative literature reviews. *Studies in Higher Education, 40*(10), 1756–1769. https://doi.org/10.1080/03075079.2014.914907

Ross, F. (2014). Climate change and energy working party: Introduction to contracts for differences. *Environmental Law and Management*, 141–143.

Savelli, I., Hardy, J., Hepburn, C., & Morstyn, T. (2022). Putting wind and solar in their place: Internalising congestion and other system-wide costs with enhanced contracts for difference in Great Britain. *Energy Economics, 113*, 106218. https://doi.org/10.1016/j.eneco.2022.106218

Schlecht, I., Maurer, C., & Hirth, L. (2024). Financial contracts for differences: The problems with conventional CfDs in electricity markets and how forward contracts can help solve them. *Energy Policy, 186*, 113981. https://doi.org/10.1016/j.enpol.2024.113981

Sequeira, T. N., & Santos, M. S. (2018). Renewable energy and politics: A systematic review and new evidence. *Journal of Cleaner Production, 192*, 553–568. https://doi.org/10.1016/j.jclepro.2018.04.190

Siddik, A. B., Khan, S., Khan, U., Yong, L., & Murshed, M. (2023). The role of renewable energy finance in achieving low-carbon growth: contextual evidence from leading renewable energy-investing countries. *Energy, 270*, 126864. https://doi.org/10.1016/j.energy.2023.126864

Simshauser, P. (2019). On the stability of energy-only markets with government-initiated contracts-for-differences. *Energies, 12*(13), 2566. https://doi.org/10.3390/en12132566

Simshauser, P. (2020). Merchant renewables and the valuation of peaking plant in energy-only markets. *Energy Economics, 91*, 104888. https://doi.org/10.1016/j.eneco.2020.104888

UNFCCC (2018) Report on the 2018 Forum of the Standing Committee on Finance - Climate finance architecture: enhancing collaboration, seizing opportunities. https://unfccc.int/sites/default/files/resource/2018%20SCF%20Forum%20Report_Final.pdf

Veenstra, A. T., & Mulder, M. (2024). Impact of contracts for differences for non-carbon electricity generation on efficiency of electricity market. *Energy Economics, 136*, 107674. https://doi.org/10.1016/j.eneco.2024.107674

Watson, N., & Bolton, P. (2023). Contracts for difference summary 1 Contracts for difference scheme 2 Stakeholder commentary on AR5 3 Proposed changes to the CfD scheme. In *commonslibrary.parliament.uk Research Briefing* (vol. 12).

Wild, P. (2017). Determining commercially viable two-way and one-way 'Contract-for-Difference' strike prices and revenue receipts. *Energy Policy, 110*, 191–201. https://doi.org/10.1016/j.enpol.2017.08.020

Wilson, P., Rush, R., Hussey, S., Puckering, C., Sim, F., Allely, C. S., Doku, P., McConnachie, A., & Gillberg, C. (2012). How evidence-based is an "evidence-based parenting program"? A PRISMA systematic review and meta-analysis of Triple P. *BMC Medicine, 10*, 130. https://doi.org/10.1186/1741-7015-10-130

Wohlin, C. (2014). Guidelines for snowballing in systematic literature studies and a replication in software engineering. In *ACM international conference proceeding series*. https://doi.org/10.1145/2601248.2601268.

Does Geopolitics Impede or Facilitate the Global Transition to Renewable Energy?

Khalid Khan and Adnan Khurshid

1 Introduction

Energy plays a critical role in achieving sustainable economic growth and ensuring national security. The strategic benefit of controlling energy supplies is present in both peace and war (Su et al., 2021; Khan et al., 2023a). Understanding the geopolitics of renewable energy (REN) is essential to comprehending the transition away from fossil fuels, as it affects energy security (Khan et al., 2023b, 2024), economic stability, climate change mitigation, and global power dynamics. Many nations are investing in local REN sources, such as solar and wind, which not only enhance their energy independence but also bring economic benefits. Nevertheless, the uneven distribution of REN resources can give rise to new dependencies and geopolitical tensions. The geopolitics of energy is influenced by the risks associated with limited and concentrated resources in politically unstable countries (Scholten, 2018; Su et al., 2019). Energy trade plays a vital role in the global economy and continues to shape global geopolitics. However, the rapid development of REN is bringing about changes in geopolitics, impacting the international political system. Trade disputes predict that REN and its products will become as important and widely traded as oil. As global demand for REN increases, countries may compete to protect their renewable resources. REN is becoming a major player in the global economy, surpassing fossil fuels due to concerns over climate change, resource depletion, and diversification (Khan et al., 2021; Khurshid et al., 2024). The rapid growth of REN presents challenges for oil- and gas-producing countries and has the potential to cause disruptions. Climate change has caused global investment in REN, which

K. Khan
International Education Department, Hengxing University of Science and Technology, Qingdao, China

A. Khurshid (✉)
School of Economics and Management, Zhejiang Normal University, Jinhua, China

S. S. Oncel (ed.), *A Green Vision Towards a Renewable Energy Future*,
https://doi.org/10.1007/978-3-031-93760-6_4

is both environmentally friendly and economical in the long run (Vakulchuk et al., 2020). However, the rapid growth of REN also brings about geopolitical risks (GPR) and changes in countries' status in the international system. The shift toward REN in the geopolitical landscape is more complex than that of traditional energy system (Sivaram & Saha, 2018; Duan et al., 2021; Xinyu et al., 2024). The present geopolitical environment may change as a result of rising petroleum prices, making REN more desirable (Xie et al., 2023; Khurshid et al., 2024). Additionally, maritime trade routes may diminish, which reflects the regionalization of the energy market (Khan et al., 2021), and some countries may produce REN domestically, defying international trade rules against protectionism. Some countries are utilizing REN investments to forge stronger international relationships and shape policies, emphasizing the importance of understanding these geopolitical factors.

The production of renewable energy technologies (RETs) may lead to increased competition for strategic minerals, potentially causing conflicts between countries. Rare earth metals, which are crucial for RETs, are often produced and imported across borders. The transit countries can influence energy production and consumption by controlling the flow of energy. The pursuit of RETs could also bring to light long-dormant territorial disputes among states, potentially leading to tensions between producers and consumers (Hatipoglu et al., 2020). In many cases, geopolitical rivalries are closely linked to reliance on fossil fuel imports from politically unstable regions. Countries transitioning to RETs aim to decrease these dependencies and improve their energy security (Khan et al., 2023). However, this shift may create new dependencies on renewable resources or technologies. Countries that heavily rely on fossil fuel exports may resist this transition due to the economic impact it could have, which could potentially hinder global climate agreements or RET initiatives. As the world moves toward RETs, new geopolitical alliances are being formed, with countries promoting their interests through partnerships. This shift also introduces competition for critical materials, such as lithium, cobalt, and rare earth elements, which are essential for technologies like batteries and solar panels.

On the other hand, the transition to REN can significantly impact global energy security by changing geopolitics. While nations may face fewer conflicts related to fossil fuel resources, they may encounter new challenges in controlling and distributing RETs and resources. This, in turn, can affect the global political and economic landscapes. By investing in and adopting RETs, countries can shift their economic influence. Those leading in clean energy technology innovation (TIN), which is percentage of a country's GDP spent on research and development, can gain both economic and political leverage. In contrast, economies dependent on fossil fuels may experience a decline, resulting in changes in geopolitical power dynamics (Khan et al., 2022). Additionally, the energy transition necessitates new strategies for managing resources and energy infrastructure. Countries may prioritize securing supply chains for REN components, which can influence global trade patterns and potentially lead to new geopolitical conflicts or alliances. Furthermore, the energy transition has implications for global climate policies and international agreements. Countries at the forefront of REN may drive the agenda for international climate action, shaping global policies and potentially creating geopolitical divisions between leading and lagging nations.

The transition to REN has the potential to improve interdependence and geopolitical stability by decreasing dependence on traditional energy sources and reducing GHG emissions (Guler et al., 2018). While disputes over REN are currently minimal due to its small share in the energy mix, they may become more frequent in the future. The shift to REN could result in fewer conflicts and less competition, but it may also sever ties with fossil fuels. This complex new system has the potential to reshape interstate relations, becoming more localized and less centralized, which poses a risk of tension (Hache, 2018). Environmental concerns are driving the transition toward sustainable energy systems (Vakulchuk et al., 2020), which are more conducive to international peace compared to traditional energy sources (Overland, 2019). This strategy assures energy security for developing nations while lowering greenhouse gas emissions and encouraging diversity (Scholten et al., 2020). The need for environmental protection, cost-cutting measures, cost-efficient technology improvements, energy security, and favorable policies are what propel the expansion of renewable energy sources (Khan et al., 2022). As of 2020, the global installed capacity for REN was 1708 GW, with China leading the way in clean energy trade. China has made substantial investments in REN and plans to invest $780 million by 2030, potentially leading to increased competition. In the future, REN will foster a more symmetrical relationship between countries, with greater mutual dependency. However, sustainable energy pathways are long and challenging. Therefore, REN cannot be monopolized by competitors or countries leading in renewable development (Vakulchuk et al., 2020).

The primary goal of the research is to evaluate the connection between GPR and REN transition. From 1990 to 2024, this relationship has been influenced by various economic trends, policy developments, and geopolitical events. The Gulf War and its aftermath had a significant impact on global oil markets and energy security concerns. Following the war, there was a notable decrease in oil prices due to increased production and reduced geopolitical tensions. As a result, fossil fuels became more economically appealing compared to REN sources. The Iraq War in 2003 further destabilized the region, leading to disruptions in oil supplies and a growing interest in alternative energy sources. The first significant international agreement to set legally binding goals for cutting GHG emissions was the Kyoto Protocol, which was unveiled in 1997. This pact launched worldwide efforts to combat climate change and pushed countries to switch to cleaner energy sources. The global financial crisis in 2008 caused a sharp decline in oil prices, dropping from $147 per barrel to below $40 by the end of the year. This drop had a profound impact on investment decisions in the energy industry. Similarly, oil prices experienced another significant decline from over $100 per barrel in 2014 to below $30 in early 2016, triggered by oversupply and economic downturns. This decline inspired a significant increase in interest in REN as countries endeavored to build up their resource portfolios and decrease their dependence on conventional sources.

Russia's 2014 invasion of Crimea and subsequent conflict with Ukraine triggered sanctions and cut off energy supplies to Europe. Consequently, there has been an increased emphasis on diversifying energy sources. In response, Obama's government put in place its Clean Power Plan, which seeks to lower carbon dioxide emissions. In addition, China launched a global initiative in 2013 involving substantial investments in REN infrastructure. This initiative has not only bolstered China's geopolitical influence but also significantly impacted global energy markets. The Paris Agreement established ambitious targets for mitigating global warming and required countries to establish and achieve their own emissions reduction goals. This agreement has prompted substantial policy changes worldwide, resulting in increased investments in REN and the development of national climate action plans. Investment in REN has surged from around $50 billion in 2004 to approximately $500 billion in 2021. The release of the IPCC Special Report on Global Warming in 2018 highlighted the urgent requirement for climate action. This report had a significant influence on global policies and hastened the shift toward REN sources. Furthermore, Russia's full invasion of Ukraine caused significant disruptions to European energy supplies. This event further motivated the European Union to strive for energy independence and encourage investments in REN. As a result of these disruptions, oil prices experienced fluctuations, reaching a peak of over $100 per barrel in 2014 and dropping below $30 in 2016.

In 2019, REN investment saw significant growth driven by cost reductions and technological advancements. By 2021, global REN investment had increased, showcasing the rising economic competitiveness of renewables. Notably, investments in solar, wind, and battery storage technologies played a pivotal role in this growth. The global sales of electric vehicles also experienced a remarkable surge, increasing from 100,000 units in 2012 to over ten million units in 2022. Consequently, the demand for lithium and cobalt has risen, as these materials are essential for electric vehicle production. As a result, the price of lithium carbonate has skyrocketed, increasing nearly fivefold between 2015 and 2022. Countries like Australia, which possesses significant lithium reserves, and the Democratic Republic of Congo, known for its cobalt reserves, have gained economic and geopolitical advantages due to this surge in demand. The United States' withdrawal from the Paris Agreement by the Trump administration significantly affected global climate policies. However, the Biden administration reversed this decision and rejoined the Paris Agreement. They also implemented a set of ambitious clean energy policies, including the Inflation Reduction Act of 2022, which provides substantial incentives for REN sources. In 2019, the European Green Deal was introduced to make Europe the first climate-neutral continent by 2050. This initiative includes policies to promote REN, energy efficiency, and carbon neutrality (Mengxuan et al., 2024). As a result, there has been a significant increase in investment in green technologies, and the influence of these policies is evident throughout Europe's energy sector (Su et al., 2022; Khurshid et al., 2024). NATO's expansion into Nordic countries signifies a significant shift in security dynamics that may impact energy policies and investments in RETs within the region. Concurrently, China's New Silk Road initiative seeks to bolster trade and infrastructure, potentially influencing energy supply chains and investments in

REN. Furthermore, geopolitical tensions surrounding natural gas in the Mediterranean introduce uncertainty that could disrupt energy markets, presenting both challenges and opportunities in the transition to REN sources. The role of nuclear power, particularly new technologies like the TOKAMAK fusion reactor, is crucial in understanding the impact of geopolitical risks on renewable energy. While renewables are a key focus in the global energy transition, nuclear power offers stable, large-scale power generation. However, this combination carries risks like safety concerns, waste management, and geopolitical tensions. Integrating renewables and nuclear power into a long-term energy strategy requires navigating both benefits and risks.

The energy transition and its geopolitical implications have received less attention than the oil and gas shortages that have been the subject of most studies in international relations. REN now plays a small part in the energy structure. However, because of political and societal forces, its share could grow and affect geopolitics. Foreign investment in REN could result in conflicts of interest, with some investments surpassing those of fossil fuels. This study brings a distinctive contribution to the existing literature on REN and GPR worldwide. Past research has ignored the importance of GPR by examining this subject exclusively from the standpoint of international relations. Hence, this research aims to explore the intersection between REN and GPR and examine whether it has the potential to redefine geopolitical dynamics. The structure of relationships, collaborations, and conflicts can be impacted by the rise of REN (O'Sullivan et al., 2017). This study examines the deviations in GPR and REN using a rolling window approach, which allows for the identification of time-varying patterns. The existence of structural changes necessitates a time-sensitive standpoint analysis of the causal link between GPR and REN. The findings of this study reveal significant macro correlations and immediate effects across both time and frequency domains. The results suggest that an upsurge in GPR leads to a corresponding increase in REN and vice versa. This emphasizes the crucial role of GPR in driving the growth of REN. Additionally, new powers and GPR expansion are also driven by international GDP growth, growing energy prices, technological advances, and ecological issues (Khurshid et al., 2024; Khan et al., 2023d). Moreover, the study used technological innovation as a mediating factor in the relationship between GPR and REN, and the results confirmed that TIN plays a critical role in the relationship. The transition to REN is advantageous for nations in the context of energy security and supply, as it can resolve ecological conflicts and regional disputes through increased investment. Governments should encourage the private sector through special incentives for REN projects.

We divide the study into seven sections. Section 2 looks at the latest studies. The rolling window technique is outlined in Sect. 3. The data are defined in Sect. 4. The analysis and policy implications are discussed in Sect. 5. The study concludes in Sect. 6.

2 Theoretical Framework and Literature Review

2.1 Theoretical Framework

The study explores the relationship between GPR and REN using theories from international relations, economics, and energy studies. The Realist Theory in International Relations suggests that states prioritize self-interest to maximize power and security within an anarchic international system (Buck & Hosli, 2019). GPR events, like instability in fossil fuel-rich regions or energy supply disruptions, prompt countries to diversify energy through REN, potentially leading to shifts in power dynamics. Liberal Institutionalism emphasizes the role of international institutions in promoting state interaction and addressing global issues (Moravcsik, 1997). The Paris Agreement, for instance, promotes the REN transition by setting global climate targets and encouraging cooperation. However, geopolitical tensions can sometimes lead to collaboration, and global climate agreements can reduce these tensions and create a more stable international energy landscape.

The political economy concerning energy examines how political and economic factors influence energy policies and international relations (Kuzemko et al., 2019). Countries are shifting toward REN sources to address economic vulnerabilities caused by fossil fuel reliance. This shift could reshape global economic and political power dynamics, potentially leading to new geopolitical alliances. According to the resource curse theory, nations rich in natural resources face economic and political consequences, including instability and ineffective governance (Su et al., 2020). This shift could alter geopolitical dynamics and require managing new dependencies or opportunities associated with RETs. Energy security theory emphasizes the impact of GPR and supply disruptions on national security and stability. Countries invest in REN to reduce vulnerability and enhance energy security, thereby reducing reliance on fossil fuels from unstable regions. Technological determinism suggests that technological advancements drive social and political change, and challenges in GPR and energy supply can accelerate innovation. RETs can reshape power dynamics as countries leading in technology development or controlling critical materials gain strategic advantages.

2.2 Literature Review

O'Sullivan et al. (2017) noted that the share of REN in the global supply of energy is growing, which has substantial implications for the world's GPR. This result implied that, in contrast to an oil-dominated world energy system, a REN-dominated system will be more durable and peaceful. Cai and Wu (2020) demonstrated that GPR has a positive effect on REN, meaning that a higher GPR encourages an increase in REN. However, a higher proportion of REN has a negative impact on GPR, making it a useful tool for reducing risks. In the short run, Rasoulinezhad et al. (2020) observed that GPR will be beneficial for energy transition. Yang et al. (2021) confirmed significant spillover effects from GPR to REN stocks. Vakulchuk et al. (2020) investigated the impact of geopolitical instability on the transition to REN. The findings revealed that REN will heighten security threats and geopolitical conflicts, with critical supplies and cybersecurity likely to suffer the most from the energy shift. Su et al. (2021) demonstrated a reciprocal link between GPR and REN. Specifically, GPR played a crucial role in the development of REN. In contrast, the expansion of the global economy, the rise in the price of fossil fuels (Khurshid et al., 2024), and technological advancements all contribute to the substantial influence of REN on GPR. Zheng et al. (2024) focused on the nonlinear effect of GPR on REN technology budgets. Their findings revealed that GPR influences REN technology across different quantile levels in the selected economies.

On the other hand, the existing literature has examined the impact of REN on GPR. For instance, Kostyuk et al. (2012) claimed that the world is at a critical energy crossroads and that REN may reduce geopolitics in international relations. They clarified that as REN expanded, ample energy would become available, which would lessen international political disputes. Khurshid et al. (2025) provided evidence that REN can assist in lowering GPR and carbon footprints. Similar conclusions were drawn by Krewitt et al. (2009), who held that the growth of REN was probably going to allay worries about unequal distributions of economic resources and lessen the possibility of resource-related conflicts. Casertano (2012) emphasized that although renewable energy is not subject to government regulation or market forces, it is not as regionally concentrated as conventional energy sources. According to Scholten et al. (2020), the growing use of REN has changed the structure of the petroleum market. It may affect patterns of international conflict and cooperation. It was anticipated that, overall, REN would stabilize internal politics and international relations despite the possibility that it might increase security risks and GPR (Vakulchuk et al., 2020).

The contemporary literature encompasses the nexus between GPR and TIN. Herrera (2003) noted that TIN is vulnerable to competitiveness in politics and changes in the global structure. Dahlman (2007) posits that GPR had a significant role in the progress of TIN, given that the acquisition of novel technologies and their adoption might stimulate GPR. Khan et al. (2025) found that the increasing utilization of REN has no significant impact on energy security and GPR levels. Fritsch (2011) asserted that TIN has a crucial function in the structure of the global system, being deeply intertwined with global political affairs and having a significant impact on economics and politics. Hulsman and Liedtke (2019) found that GPR escalation causes TIN. Rühlig et al. (2019) found that TIN has evolved into a strategic battlefield in prospective geopolitics and a cause of geopolitical friction amongst the world's superpowers. Triolo (2019) showed that advanced TIN drives the Sino–US relationship and fragments technology due to rivalry. Chin (2019) discovered that developments in defense TIN have considerable effects on disputes, resulting in new strategic shapes and power transfers within governments. According to Rahman's (2020) research, TIN plays a significant role in GPR by safeguarding national interests and implementing geopolitical strategies. Tekir (2020) affirmed that TIN has transformed country interactions, with new geopolitical struggles taking place on global platforms instead of discrete geographical units. Wu (2020) concluded that the Sino–American race to advance TIN has been spurred by the Fourth Industrial revolution, thereby raising GPR.

The existing literature on geopolitics has primarily focused on oil and gas, leaving a gap in research analyzing REN within the context of global political relations and GPR. Although renewable energy sources account for a modest percentage of the world's energy consumption, they have not yet had a major influence on international politics (O'Sullivan et al., 2017). Furthermore, most of the existing literature in this field is qualitative rather than quantitative. Existing work focuses primarily on how TIN contributes to geopolitical tension, ignoring the function of GPR in technological growth. Nonetheless, it is essential to recognize that GPR also plays a significant role in driving technological innovation as powerful nations develop more advanced TEC to exert their dominance in the global political arena. Moreover, many of these studies tend to focus on the qualitative correlation between GPR and TEC rather than providing quantitative analysis. To fill these gaps and possibly obtain precise results, we use a rolling window approach to detect dynamic properties and identify structural shifts caused by external perturbations. Additionally, while previous research has primarily explored the international relations aspect of REN, our research utilizes GPR to analyze global REN and vice versa.

3 Methodology

3.1 *Bootstrap Full-Sample Causality Test*

The time series are not stationary; hence, they do not follow a normal asymptotic distribution. This rules out the use of VAR estimation, according to Sims et al. (1990). As a solution, Toda and Yamamoto (1995) offered using the Wald test to measure the asymptotic distribution and avoid this problem. For medium- and small-sized samples, Shukur and Mantalos (2000) used a Monte Carlo simulation to enhance the Wald test further. The residual-based bootstrap (RB) approach is used to deal with the power and size characteristics. Therefore, this work uses this method to check whether GPR and REN are associated. The stated method is explained as follows:

$$x_t = \theta_0 + \theta_t x_{t-1} + \ldots + \theta_\rho x_{t-\rho} + \mu_t \qquad t = 1, 2\ldots, T \tag{1}$$

where $\varepsilon_t = (\varepsilon_{1t}, \varepsilon_{2t})'$ relates to the zero-mean white-noise procedure. The lag length is selected based on the Schwarz information criterion (SIC). Equation (1) is then divided into two sub-vectors.

$$\begin{bmatrix} GPR_t \\ REN_t \end{bmatrix} = \begin{bmatrix} \theta_{10} \\ \theta_{20} \end{bmatrix} + \begin{bmatrix} \theta_{11}(L)_{12}(L) \\ \theta_{21}(L)_{22}(L) \end{bmatrix} \begin{bmatrix} GPR_t \\ REN_t \\ TIN_t \end{bmatrix} + \begin{bmatrix} \mu_{it} \\ \mu_{2t} \end{bmatrix} \quad (2)$$

where $\theta_{ij}(L)\sum_{k-1}^{p+1}\theta_{ij,k}$, L^k, $i, j = 1, 2$ and L is the lag operator, defined as $L^k\theta_t = \theta_{t-k}$. The control variable in use is TIN. In Eq. (2), we see how GPR affects REN, and REN impacts GPR.

3.2 Parameter Stability Test

The VAR model's parameter is believed to be consistent across the board for the whole sample. Nevertheless, unsteady variables and erroneous outcomes can occur from changes in structures (Balcilar et al., 2010). Multiple researches have demonstrated that variable uncertainty is a complicating factor (Granger, 1969). This problem is tackled by employing tests for the stability of parameters across shorter periods. The parameter stability in the short run can be investigated using the *Mean-F*, *Sup-F*, and *Exp-F* tests, as suggested by Andrews and Ploberger (1994). Hanson (2002) and Nyblom (1989) also recommend implementing a test to evaluate the overall consistency of the variables within a VAR system. This test is predicated on LR stats and enables the identification of structural breaks. The parametric bootstrap method is used to determine the critical and p-values.

3.3 Subsample Rolling Window Causality Test

Balcilar et al. (2010) described the rolling window test as a method for evaluating full-sample causality while also considering structural changes. It is recommended to conduct pretests if the overall sample shows nonstationarity and volatility among subsections. The transformation of a steady rolling window with l full-size evaluations into an order of $T\text{-}l+1$ samples, where each sample corresponds to the window indices $\tau - l + 1, \tau - l+2, \ldots, \tau$ *for* $\tau=l, l+1, \cdots, T$. The subsample causality is estimated using the updated LR test using the RB. The bootstrap p-values of the LR statistics found through $T\text{-}l$ subsamples are used to determine the time-varying properties and the strength of the causal relationship between GPR and REN. The average of all bootstrap estimations, given by $N_b^{-1}\sum_{k-1}^{p}\hat{\theta}^*_{21,k}$, wherever N_b is the number of bootstrap repetitions, is used to compute the outcome from GPR to REN. Likewise, $N_b^{-1}\sum_{k-1}^{p}\hat{\theta}^*_{12,k}$ is the formulation that expresses the causal relationship between REN and GPR. Equation (2) serves as the foundation for assessing the coefficients $\hat{\theta}^*_{21,k}$ and $\hat{\theta}^*_{12,k}$ in the VAR model. The accuracy of parameter estimation and the model's being representative throughout the subsampling period serve distinct purposes.

4 Data

The study evaluates the nexus between global GPR and REN from 1990:01 to 2023:12. The period is crucial due to the shift from international military spending to increased innovation and research as well as a downturn in GPR. During the 2000s, expenditure on renewable energy technologies surged, leading nations with greater investments to emerge as significant actors in global politics (Criekemans, 2018). However, several forces have shifted the gas and oil paradigm to REN, affecting energy, economies, and geopolitics. The depletion of oil fields and the influx of prospective buyers from Asia (India and China) caused a rise in oil prices at the start of the twenty-first century, which in turn sparked competitiveness (Criekemans, 2018). The renewable energy technologies market experienced a remarkable emergence in China in the 2000s, and the unipolar framework is no longer viable; power is now distributed within multiple polarities (Hache, 2018). This transformation process encompasses different nations and traverses boundaries that are particularly susceptible to cyber and terrorist attacks. The involvement of outside groups and the escalating competition for precious metals made from rare earths are now considered threats to the power industry (Overland, 2019; O'Sullivan et al., 2017). The European Union has expanded its capacities in the realm of renewable sectors and demonstrated its commitment to energy advancement in the initial stages of the 2000s. Figure 1 shows the REN and GPR trends during the

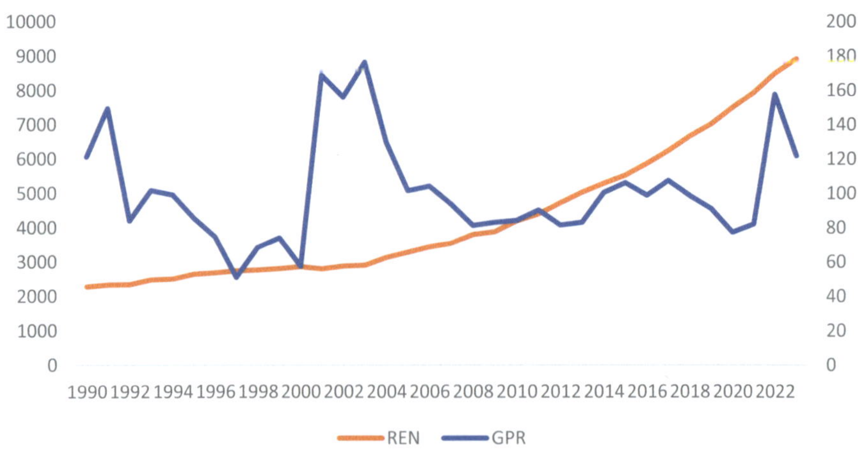

Fig. 1 Trends of REN and GPR

investigation. It illustrates that REN and GPR saw numerous fluctuations over this period due to different policies and incidents.

The GPR index is estimated by Caldara and Iacoviello (2018). It represents the results of war, interstate conflicts, and terrorist attacks, which disrupt the regular course of domestic policies and global relationships. REN is the aggregate of hydropower, solar, and wind power, quantified in TeraWatt hours. It is sourced from World Development Indicators (WDI). The percentage of a country's GDP spent on research and development (R&D) in current and capital forms is referred to as TIN. The WDI is the source of the global R&D expenditures, which serve as the TIN. It encompasses the financial and operations expenses for business firms, government entities, academic institutions, and commercial nonprofit organizations.

The summary data are displayed in Table 1. REN exhibits a larger standard deviation, indicating significant volatility due to multiple factors. The skewness of GPR and TIN is positive, whereas REN exhibits negative skewness. The leptokurtic distribution of the kurtosis of GPR is indicated by values exceeding 3. Nevertheless, REN and TIN are classified as platykurtic distributions. The results of the Jarque–Bera test demonstrate the non-normal distribution of GPR, REN, and TIN.

5 Results Analysis

The stationarity of GPR, REN, and TIN are tested using the Augmented Dickey–Fuller (1981), Phillips–Perron (1988), and the Kwiatkowski–Phillips–Schmidt–Shin (1992) tests. The outcomes presented in Table 2 indicate that GPR, REN, and TIN exhibit stationarity at the first difference. Based on these findings, we proceed to construct a VAR model for conducting Granger causality tests.

We conducted a full sample causality test to determine whether the null hypothesis of GPR causes REN and vice versa is valid. Table 3 presents the findings, which indicate that the underlying premise was not accepted. This can be attributed to the fact that REN shares are currently small in the energy mix. Therefore, they do not have a significant impact on GPR or vice versa. The global shift from a traditional energy framework to REN is a gradual process aimed at lowering

Table 1 Summary statistics

	GPR	REN	TIN
Mean	101.172	352.641	2.053
Std. dev.	50.637	201.008	0.102
Skewness	4.158	−0.008	2.044
Kurtosis	27.757	1.779	7.355
Jarque-Bera	11595.120***	25.356***	428.284***

Note: The 1% is illustrated by ***

Table 2 Unit root outcomes

Variables	ADF	PP	KPSS
GPR	−6.244	−5.804	0.728***
REN	3.241	7.443	1.820***
TIN	−2.484	−1.448	1.824***
ΔGPR	−12.401***	−34.402***	0.710
ΔREN	−7.190***	3.144***	1.428
ΔTIN	−2.443***	−10.449***	0.243

Note: The 1% is illustrated by ***

Table 3 Full-sample causality outcomes

Test *i*	H_0: GPR does not Granger cause REN		H_0: REN does not Granger cause GPR	
Bootstrap	Statistics	*p*-values	Statistics	*p*-values
i*LR* itest	1.819	1.000	0.103	0.900

reliance on petroleum-based fuels and improving energy security. Nevertheless, because of the focus on technological dominance and fiscal disparities, this change also carries with it new issues that can only be resolved by international cooperation.

The sample is entirely made up of one causation and lacks any structural changes. However, because there are changes in the causative relationship between GPR and REN, the findings contradict this assumption. This leads to short-term parameter instability, and the results are therefore not suitable for analysis. This problem is solved by looking at the model's short-term instability using a variety of parameter stability tests, such as *Sup-F, Exp-F,* and *Mean-F*. Table 4 presents the outcomes of these examinations. The underlying hypothesis is rejected for both GPR and REN, indicating that the series exhibits short-term instability. In a similar vein, the *Exp-F* and *Mean-F* tests are employed to explore the possibility of a sharp change. However, we are unable to accept this explanation because it suggests that GPR and REN are changing. We employ the Lc test to analyze the VAR model's overall stability, evaluating variable persistence against random walks. The findings support the full-size VAR model's short-run instability. This confirms that the whole sample is unpredictable and exhibits parametric consistency in the short run as a result of structural changes.

We considered structural changes when analyzing the link between GPR and REN using a rolling window technique. The differences in characteristics across subsamples enabled us to distinguish between traditional and rolling window approaches. We used the VAR model to test the hypothesis that GPR does not cause REN, and we got bootstrap *p*-values. Figure 2 presents the outcomes of the causal link between GPR and REN in various subsamples, including 1998M03–1999M10, 2000M10–2001M09, 2008M04–2010M03, 2014M02–2015M04, and 2022M04–2023M09. The figure demonstrates the significant role of GPR in the development of REN and suggests that, in a world where REN is the primary energy

Table 4 Parameter stability results

	GPR		REN		VAR system	
	Statistics	*p*-values	Statistics	*p*-values	Statistics	*p*-values
Sup-F	164.477***	0.000	32.870	0.1927	217.022***	0.000
Mean-F	21.423***	0.000	16.845	0.1787	65.847**	0.000
Exp-F	77.847***	0.000	13.894	0.1020	103.056***	0.000
Lc					5.207**	0.024

Note: The 1% is illustrated by ***

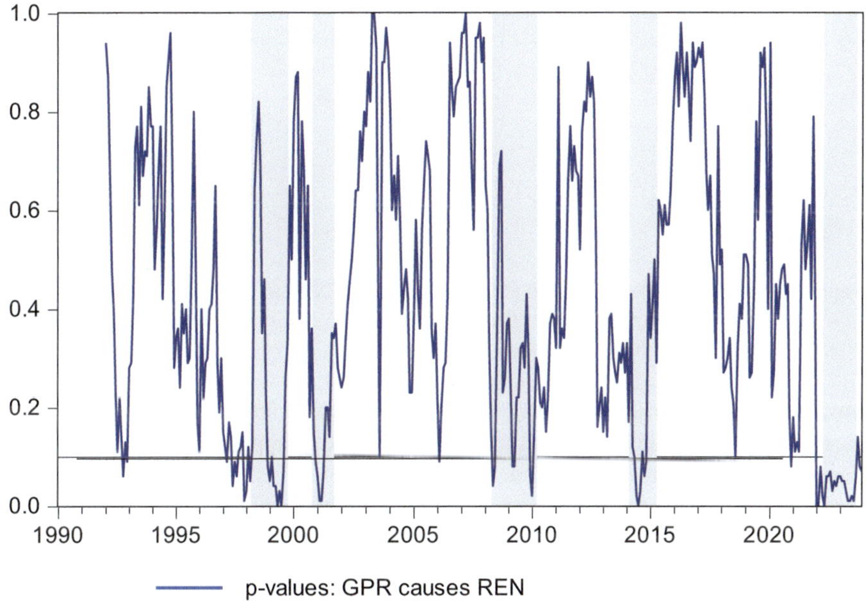

Fig. 2 GPR impact on REN. *Note: The null hypothesis that REN does not cause GPR is tested using a bootstrap p-value of the rolling test statistic. For p-values less than 10%, the null hypothesis cannot be true. The shaded line represents a p-value of 0.1. A line that is less than 0.1 demonstrates causation*

source, investment and technology may become sources of cooperation or protuberances of geopolitical competition (O'Sullivan et al., 2017).

Figure 3 depicts the dynamics of causation, revealing that GPR has positive as well as negative effects on REN. The subsample from 1998:03 to 1999:10 comprises a positive impact of GPR on REN. During this period, the REN sector was significantly impacted by several GPR events. For instance, the Asian financial crisis resulted in a severe economic downturn in many countries, which affected their ability to invest in REN infrastructure. Moreover, the conflict in Kosovo and the resulting geopolitical tensions have implications for global energy markets. This region served as a strategic transit point for oil and natural gas pipelines, causing uncertainty and disruptions in energy supply chains, which in turn affected

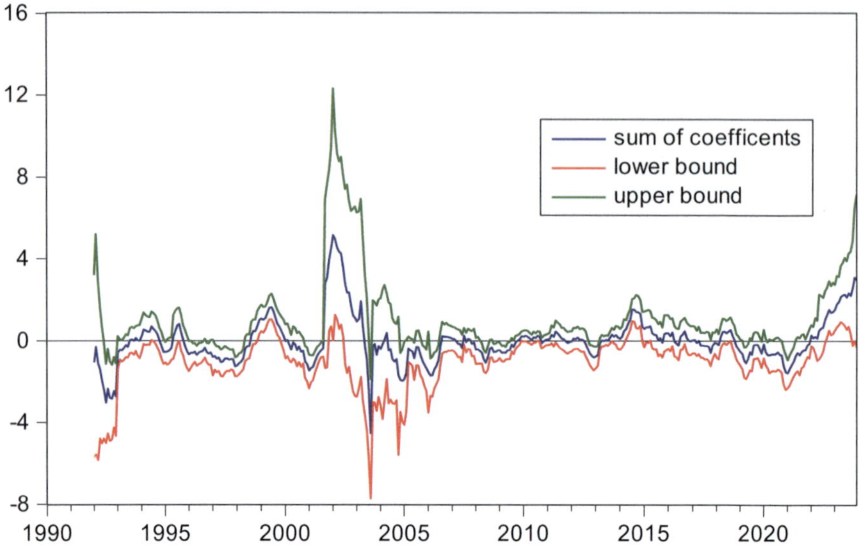

Fig. 3 Bootstrap estimates the causal effect of GPR on REN from rolling window coefficients. *Note: It indicates the magnitude of the effect,* i.e., *positive, adverse, or no causality. The magnitude is represented by the colored line zero*

investments in REN. In the late 1990s, oil prices experienced fluctuations due to tensions in the Middle East, which emphasized the importance of diversifying energy sources. On the other hand, advancements in RETs, such as wind and solar technologies, make them more economically viable. Political instability in oil-producing regions led to global oil supply uncertainties, prompting countries to seek more stable and secure energy sources. The EU has initiated policies to promote REN as part of its overall environmental and energy security strategies. During this period, many countries were in the early stages of developing comprehensive renewable energy policies, leading to geopolitical and regulatory risks associated with investing in REN. While some countries implemented subsidies or incentives, they were often subject to political and budgetary considerations, creating uncertainty for investors. The variability and potential withdrawal of subsidies further increased the perceived risk. The REN industry's vulnerability to geopolitical and macroeconomic factors underscores the need for stable, long-term policy frameworks to support its growth and adoption.

GPR has a negative impact on REN during 2000M10–2001M09. The outbreak of the Second Intifada in September 2000 caused political and economic instability in the Middle East, an important region for global energy supply. This instability affected investor confidence and had a negative effect on REN. The dot-com bubble collapse led to an economic downturn, reducing investment capital for REN initiatives and affecting the industry, which was heavily reliant on public and private investments, resulting in spillover effects. The 2000 US presidential election led to a shift in policy priorities toward traditional fossil fuels. The Organization of

Petroleum Exporting Countries (OPEC) enacted production cuts to keep oil prices stable, which caused the price of oil globally to rise. The Enron scandal, a major player in the REN market, shook investor confidence and raised concerns about corporate governance and transparency, making it harder for REN companies to attract investment. The September 11 terrorist attacks significantly impacted global geopolitical stability and economic confidence, leading to a slowdown in economic activity and increased risk aversion among investors. Post-attacks, policy priorities shifted toward energy security, with traditional fossil fuels often preceding REN. This created uncertainty and reduced the attractiveness of REN investments. At the time, REN technologies were still in their early stages and not cost-competitive with traditional fossil fuels. The combination of geopolitical instability, policy shifts favoring fossil fuels, economic uncertainty, and a focus on short-term energy security negatively impacted the development and investment in REN.

GPR has a significant negative impact on REN during 2008M04–2010M03. The global financial crisis that unfolded in 2008 had a severe impact on the world economy, leading to a decline in energy demand and investment, including in REN. The crisis triggered a "flight to safety" among investors, reducing the availability of capital for REN. Tensions between Russia and Ukraine over natural gas supply and transit caused disruptions in the Europe's gas supply, exposing the vulnerability of the region's energy security. This geopolitical conflict brought attention to the potential role of REN in improving energy independence and diversification. The 2009 United Nations Climate Change Conference in Copenhagen failed to establish a legally binding agreement on climate change, leading to uncertainty for investors in REN. Governments introduced economic stimulus packages, some of which included measures to support the REN sector. However, their implementation and sustainability were influenced by political and budgetary factors, further exacerbating the situation. Several countries have implemented feed-in tariff policies and REN targets to boost sector growth. However, changes in these policies, such as reduced tariff rates or shifts in targets, have raised geopolitical and regulatory risks. During the economic downturn, political and budgetary pressures impacted subsidies and financial incentives for REN, adding further risks for investors.

GPR had a significant impact on REN during 2014M02–2015M04. The Russia annexation of Crimea in 2014 and the conflict in Eastern Ukraine sparked global energy market uncertainty, causing concerns about energy supply security, particularly natural gas. The United States, European Union, and other Western countries imposed economic sanctions on Russia, targeting key sectors, including energy, in response to these actions. Oil prices dropped significantly from over $100 per barrel in mid-2014 to around $50 per barrel by early 2015, making it harder for RETs to compete with cheaper fossil fuels, thereby slowing down REN adoption. Ongoing conflicts and political instability in Middle East, including the rise of the Islamic State and the Yemen crisis, have heightened global GPR and raised concerns about the reliability and security of energy systems, including REN. The Paris Agreement, adopted in 2015, aimed to address climate change, but its implementation led to policy uncertainty in the REN sector due to unclear specific policies and commitments from individual countries. Despite these challenges, significant advancements in RETs, especially solar and wind, led to cost reductions and increased competitiveness with fossil fuels. China became a global leader in REN, investing heavily in solar and wind power to reduce pollution and improve energy security.

GPR has a positive impact on REN during 2022M04–2023M09. The conflict between Russia and Ukraine has disrupted global energy markets, leading to increased demand for REN to address security concerns and reduce reliance on Russian energy. The Ukraine conflict and COVID-19 have disrupted global supply chains, causing shortages of critical raw materials and components in renewable energy (REN) technologies. This has increased costs and timelines, affecting their competitiveness and deployment. The disruption of global energy market has caused price fluctuations in fossil fuels and electricity, creating uncertainty for REN investments due to the potential impact on their cost-competitiveness. The European Union has launched the REPowerEU plan to address the energy crisis caused by the Russia–Ukraine war, aiming to accelerate the transition to REN and lessen reliance on Russian fossil fuels. The United States passed the Inflation Reduction Act, aiming to boost energy security, reduce GHG emissions, and enhance the competitiveness of the REN industry. This policy included significant investments and tax credits for REN technologies, energy efficiency, and electric vehicles. However, geopolitical tensions and energy market disruptions have led to increased scrutiny and potential adjustments to these targets as governments balance energy security, economic concerns, and sustainability goals.

Alternatively, REN had a significant effect on GPR from 2002M10 to 2003M12. The results are illustrated in Figs. 4 and 5. The period was significant for the REN

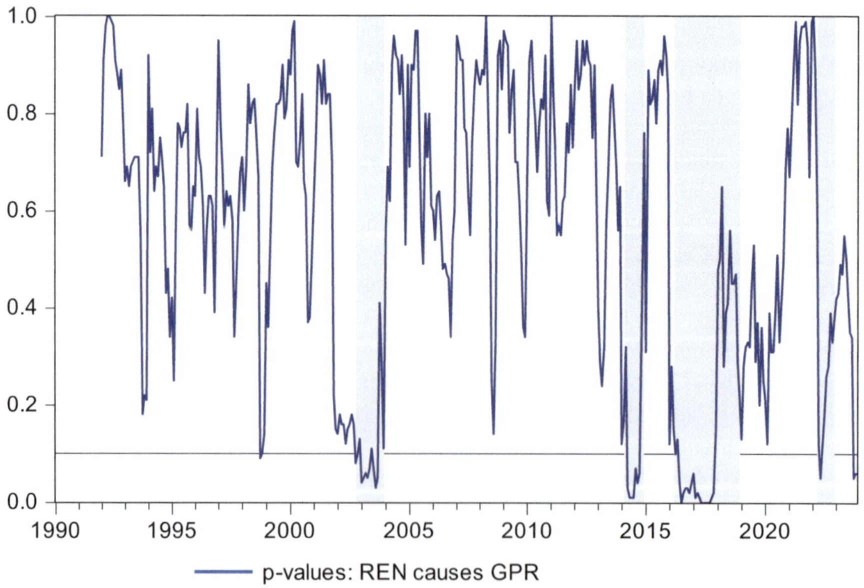

Fig. 4 REN impact on GPR

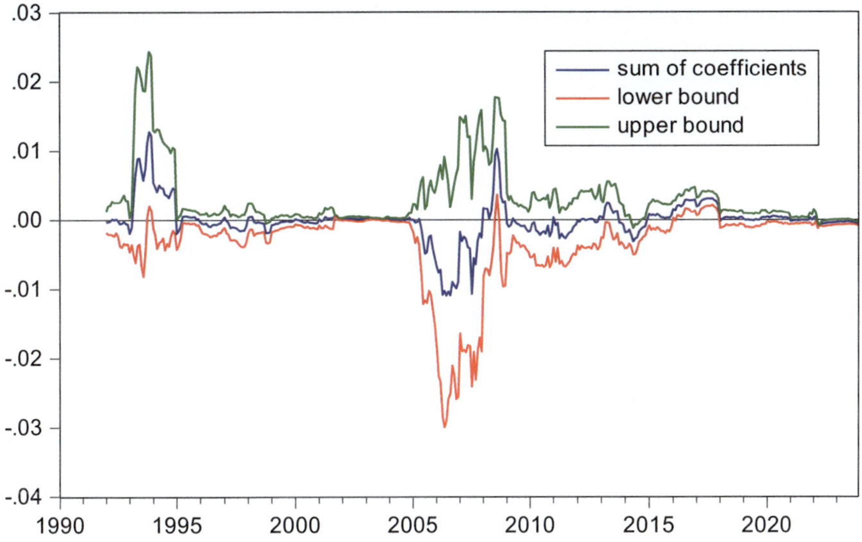

Fig. 5 Bootstrap estimates of the causal effect of REN on GPR from rolling window coefficients

sector, and its impact on GPR was significant for several reasons. The US-led invasion of Iraq significantly impacted global oil markets and energy security, leading to increased concerns about supply disruptions. This situation emphasized the need for diversification of energy sources, including the development of REN. OPEC was managing oil production levels to influence global oil prices. At the same time, uncertainty in the Middle East, especially in Iraq, caused fluctuations in oil prices, promoting GPR associated with fossil fuel reliance, making REN more attractive for reducing dependence on volatile oil markets. The Kyoto Protocol, which came into effect in 2005, was influenced by international discussions and preparations in the early 2000s. Countries began implementing policies to reduce greenhouse gas emissions, including promoting REN sources. This shift influenced national energy policies, driving investments in REN and altering geopolitical dynamics, particularly concerning fossil fuel-exporting nations. The European Union was at the forefront of REN promotion, setting ambitious targets and reducing dependence on external energy sources, aiming to enhance energy security and reduce greenhouse gas prices associated with energy imports. In the early 2000s, technological advancements in REN, such as solar and wind power, made it economically viable, encouraging countries to invest in these technologies. This shift altered the global energy mix, reducing the influence of fossil fuel-rich regions on geopolitics. The focus on REN was driven by enhancing energy security and reducing dependence on fossil fuels, which were subject to global political risks. Countries began diversifying their energy sources, reducing the geopolitical leverage of traditional oil and gas exporters. This diversification reshaped global power dynamics, with REN-producing countries gaining influence. This shift redefined geopolitical risk, with energy and environmental sustainability becoming increasingly interconnected.

REN had a significant negative impact on GPR from 2014M02 to 2014M12. The 2014 Russia annexation of Crimea and the subsequent conflict in Eastern Ukraine heightened European geopolitical tensions. Russia's dominant role as a natural gas supplier exposed European countries' vulnerabilities, highlighting the need for energy diversification and pushing toward renewable energy sources to lessen reliance on Russian gas. In response to the Ukraine crisis, the European Union adopted the European Energy Security Strategy in May 2014. The policy prioritized expanding renewables' proportion of the energy mix to improve energy security and lessen reliance on foreign sources. This geopolitical dimension of energy policy made renewable energy not just an environmental issue but also a strategic tool. In November 2014, OPEC's decision not to restrict oil extraction caused a major decline in the price of crude, affecting countries that rely on oil exports. This challenge to REN investments made fossil fuels cheaper but also highlighted the volatility of oil markets and the strategic value of investing in stable, sustainable energy sources. The 2015 Paris Agreement marked a significant shift in global efforts to combat climate change, with the United Nations Climate Summit in 2014 highlighting the importance of REN adoption. This shift in the geopolitical landscape led to countries increasingly acknowledging the need for climate action. During this period, advancements in RETs led to cost reductions, making them more competitive with traditional energy sources. As renewables became economically viable, they became more prominent in national energy strategies, reducing reliance on geopolitically sensitive fossil fuels. The instability in the Middle East, including ISIS and conflicts in Syria and Iraq, heightened global oil markets and energy security concerns, promoting the interest in REN as a stable and secure energy source.

REN exerted a significant effect on GPR between 2016M03 and 2018M12. The Paris Agreement, implemented in 2015, aimed to reduce carbon emissions globally, leading to increased investments in REN. This shift in focus altered geopolitical dynamics, shifting power from fossil fuel-rich countries to those investing in clean energy technologies. In response to declining oil prices in the mid-2010s, OPEC and non-OPEC countries agreed to production cuts in late 2016. These actions increased oil prices but also highlighted the volatility of fossil fuel markets, emphasizing the need for diversification through renewable energy sources. In June 2017, the United States announced its withdrawal from the Paris Agreement, causing global climate policy uncertainty and geopolitical tensions. This decision led to shifts in international alliances. It increased tensions with major powers like the EU and China, who remained committed to REN transitions. During this period, Europe and China saw a surge in REN investments, filling the leadership void left by the United States. China's Belt and Road Initiative (BRI) focused on REN investments, positioning itself as a global leader in clean energy and contributing to the global shift toward renewables, particularly in developing countries. The EU's Energy Union Strategy, launched in 2015, aimed to diversify energy sources and reduce dependence on Russian gas. By 2018, the EU had integrated renewables into its energy mix, reducing vulnerability to external energy shocks and GPR. Conflicts in the Middle East, particularly in Syria and Yemen, have heightened instability in relying on fossil fuel supplies, prompting countries to accelerate their REN strategies to reduce their dependence on Middle Eastern oil and gas. Advancements in RETs, particularly solar and wind power, have led to cost reductions and increased investments, transforming the global energy landscape and altering traditional geopolitical dynamics tied to fossil fuels.

REN has a significant influence on GPR during 2022M03–2022M12. The Russian intervention in Ukraine in February 2022 had enormous effects on global energy markets, triggering a severe energy crisis in Europe. This prompted a shift toward REN to reduce reliance on Russian fossil fuels, highlighting the link between energy security and geopolitical stability. The EU launched the REPowerEU plan in March 2022 to address the energy crisis caused by the Russia–Ukraine war. The plan aims to cut Europe's dependence on Russian fuels by accelerating renewable energy deployment, increasing efficiency, and diversifying energy supplies. This reduces the geopolitical leverage of energy-rich countries like Russia. The war worsened unpredictable energy sectors, causing increases in fossil fuel prices. This volatility made REN more attractive, influencing global energy strategies and reducing GPR associated with fossil fuel dependencies. The crisis also impacted OPEC+ dynamics, with countries adjusting output to influence global oil prices. Uncertainty pushed countries toward renewables.

The IRA, passed in August 2022, emphasized clean energy technologies and investments in the United States, signaling a commitment to decarbonization and influencing global markets. This policy reduced global dependence on fossil fuels, diminishing the geopolitical influence of fossil fuel-rich states. The 27th UN Climate Change Conference in Egypt in November 2022 saw renewed commitments to REN as a strategy to handle climate change and reduce geopolitical tensions. The focus on renewables underscored a global consensus on energy transition and the link between REN deployment and geopolitical risk mitigation. China's significant investments in REN in 2022, both domestically and through its BRI, strengthened its global leadership in clean energy and shifted geopolitical power away from traditional superpowers. China's control over critical components of the REN supply chain, including solar panels and battery technologies, became a strategic asset, impacting global energy security dynamics. Emerging markets, particularly Asia and Africa, accelerated REN adoption to protect their economies from price shocks.

6 Conclusion

Energy trade is crucial for global economic growth and geopolitics, with control over energy sources providing a geostrategic advantage. Geopolitics significantly influences the REN transition due to security concerns, rare earth metal competition, and trade disputes. Governments should incentivize private sector participation for security. This study investigates the connection between GPR and REN. The findings indicate a reciprocal link between GPR and REN in various subsamples. It demonstrates how GPR influences the development of REN due to issues such as energy security, rivalry for mineral resources, and trade conflicts. These factors act as drivers for the shift towards REN. However, the findings also indicate that REN has a notable impact on GPR. This is due to global growth, rising fuel prices, and technological improvements. The adoption of REN can help emerging nations integrate into the global political framework, making it a valuable tool for promoting world peace and potentially lowering GPR. Thus, governments can profit from switching to REN because of energy security and supply. Governments ought to engage the private sector more actively by offering unique incentives that entice businesses to participate in renewable energy.

The study presents several policy implications. The findings reveal a positive association between GPR and REN. In other words, expanding REN can lead to increased interdependence between economies, which is better for world peace than conventional energies. Furthermore, REN may help to minimize GPR and integrate emerging nations into the global political system. Thus, changing to REN is suitable for nations' energy security and supply. The findings also highlight that GPR has a greater contribution to REN. This emphasizes the importance of prioritizing energy security and focusing on developing REN. Governments should design REN projects with consideration for cybersecurity, energy flow, and location to reduce the risk of supply disruptions. Similarly, REN can be used to reduce the potential GPR stemming from countries that use oil and gas as weapons. Investing more in RETs will help countries become critical participants in the years ahead. It may assist in addressing environmental issues and international conflicts at regional scales. Therefore, global collaboration is necessary to promote the exploration and development of REN. To increase demand for REN, governments must promote the use of renewable-powered electric vehicles and public transportation. Going green with REN is a long and challenging process. However, governments cannot let their rivals handle it because it could affect international relations. Governments should engage the private sector in order to motivate firms to participate in REN initiatives by offering special incentives. This will help countries achieve their transition targets to RE and fulfill their environmental obligations outlined in international agreements.

References

Andrews, D. W., & Ploberger, W. (1994). Optimal tests when a nuisance parameter is present only under the alternative. *Econometrica: Journal of the Econometric Society, 62*(6), 1383–1414.

Balcilar, M., Ozdemir, Z. A., & Arslanturk, Y. (2010). Economic growth and energy consumption causal nexus viewed through a bootstrap rolling window. *Energy Economics, 32*(6), 1398–1410.

Buck, D., & Hosli, M. (2019). *Traditional theories of international relations* (pp. 3–21).

Cai, Y., and Wu, Y. (2020). *Time-varying interactions between geopolitical risks and renewable energy consumption*. ADBI working paper 1089. Asian Development Bank Institute.

Caldara, D., and Iacoviello, M. (2018). *Measuring geopolitical risk*. FRB international finance discussion paper, (vol. 1222, pp. 1).

Casertano, S. (2012). *Risiken neuer Energie—Konflikte durch erneuerbare Energien und Klimaschutz* (Risks of new energy—Risks posed by renewable energy and climate protection). Brandenburg Institute for Society and Security (BIGS), No. 9.
Chin, W. (2019). Technology, war and the state: Past, present and future. *International Affairs, 95*(4), 765–783.
Criekemans, D. (2018). Geopolitics of the renewable energy game and its potential impact upon global power relations. In *The geopolitics of renewables* (pp. 37–73). Springer.
Dahlman, C. (2007). Technology, globalization, and international competitiveness: Challenges for developing countries. In *Industrial development for the 21st century: Sustainable development perspectives* (pp. 29–83).
Dickey, D. A., & Fuller, W. A. (1981). Likelihood ratio statistics for autoregressive time series with a unit root. *Econometrica, 49*, 1057–1072.
Duan, W., Khurshid, A., Rauf, A., Khan, K., & Calin, A. C. (2021). How geopolitical risk drives exchange rate and oil prices? A wavelet-based analysis. *Energy Sources, Part B: Economics, Planning, and Policy, 16*(9), 861–877.
Fritsch, S. (2011). Technology and global affairs. *International Studies Perspectives, 12*(1), 27–45.
Granger, C. W. (1969). Investigating causal relations by econometric models and cross-spectral methods. Econometrica: journal of the Econometric Society, 424-438.
Guler, B., Çelebi, E., & Nathwani, J. (2018). A 'regional energy hub' for achieving a low-carbon energy transition. *Energy Policy, 113*, 376–385.
Hache, E. (2018). Do renewable energies improve energy security in the long run? *International Economics, 156*, 127–135.
Hanson, B. E. (2002). Tests for parameter instability in regressions with I (1) processes. *Journal of Business and Economic Statistics, 20*(1), 45–59.
Hatipoglu, E., Al Muhanna, S., & Efird, B. (2020). Renewables and the future of geopolitics: Revisiting main concepts of international relations from the lens of renewables. *Russian Journal of Economics, 6*, 358–373.
Herrera, G. L. (2003). Technology and international systems. *Millennium, 32*(3), 559–593.
Hulsman, J. C., & Liedtke, B. N. (2019). Digital geopolitics and political risk. Aspenia online, International Analysis and Commentary.
Khan, K., Khurshid, A., & Cifuentes-Faura, J. (2023a). Energy security analysis in a geopolitically volatile world: A causal study. *Resources Policy, 83*, 103673.
Khan, K., Khurshid, A., & Cifuentes-Faura, J. (2023b). Investigating the relationship between geopolitical risks and economic security: Empirical evidence from central and eastern European countries. *Resources Policy, 85*, 103872.
Khan, K., Khurshid, A., & Cifuentes-Faura, J. (2023d). Is geopolitics a new risk to environmental policy in the European union? *Journal of Environmental Management, 345*, 118868.
Khan, K., Khurshid, A., Cifuentes-Faura, J., & Xianjun, D. (2024). Does renewable energy development enhance energy security? *Utilities Policy, 87*, 101725.
Khan, K., Su, C. W., Khurshid, A., & Qin, M. (2023). Does energy security improve renewable energy? A geopolitical perspective. *Energy, 282*, 128824.
Khan, K., Su, C. W., Rehman, A. U., & Ullah, R. (2022). Is technological innovation a driver of renewable energy? *Technology in Society, 70*, 102044.
Khan, K., Su, C. W., & Tao, R. (2021). Does oil prices cause financial liquidity crunch? Perspective from geopolitical risk. *Defence and Peace Economics, 32*(3), 312–324.
Khan, K., Su, C. W., Tao, R., & Umar, M. (2021). How do geopolitical risks affect oil prices and freight rates? *Ocean & Coastal Management, 215*, 105955.
Khan, K., Su, C. W., Umar, M., & Zhang, W. (2022). Geopolitics of technology: A new battleground? *Technological and Economic Development of Economy, 28*(2), 442–462.
Khan, K., Khurshid, A., & Cifuentes-Faura, J. (2025). Russia-Ukraine and Israel-Palestine wars on the asymmetric multifractals of defense stocks: a novel A-MFDFA method. *Defence and Peace Economics*, 1–20. https://doi.org/10.1080/10242694.2025.2495253

Khurshid, A., Huang, Y., Cifuentes-Faura, J., & Khan, K. (2024). Beyond borders: Assessing the transboundary effects of environmental regulation on technological development in Europe. *Technological Forecasting and Social Change, 200*, 123212.

Khurshid, A., Huang, Y., Khan, K., & Cifuentes-Faura, J. (2024). Innovation, institutions, and sustainability: Evaluating drivers of household green technology adoption and environmental sustainability of Africa. *Gondwana Research, 132*, 88–102.

Khurshid, A., Khan, K., Cifuentes-Faura, J., & Chen, Y. (2024). Asymmetric multifractality: Comparative efficiency analysis of global technological and renewable energy prices using MFDFA and A-MFDFA approaches. *Energy, 289*, 130106.

Khurshid, A., Khan, K., Rauf, A., & Cifuentes-Faura, J. (2024). Effect of geopolitical risk on resources prices in the global and Russian-Ukrainian context: A novel Bayesian structural model. *Resources Policy, 88*, 104536.

Khurshid, A., Khan, K., Cifuentes-Faura, J., & Rauf, A. (2025). Examining the nexus of clean energy access, adaptive governance and resilient futures for Africa: A novel empirical approach. Renewable Energy, 239, 122100.

Kostyuk, V., Makarov, A., & Tatyana M. (2012). Энергетика и геополитика (Energy and geopolitics). *Energoacademy, 1*(44), 46–59.

Krewitt, W., Nitsch, J., & Kristina, N. (2009). The importance of renewable energy and energy efficiency in various global energy scenarios. Forschen für globale Märkte erneuerbarer Energien. *Annual meeting of the ForschungsVerbunds Erneuerbare Energien (FVEE) in cooperation with the Agency for Renewable Energy (AEE)*, 24–25 November 2009, Umweltforum Berlin, pp. 18–23.

Kuzemko, C., Lawrence, A., & Watson, M. (2019). New directions in the international political economy of energy. *Review of International Political Economy, 26*, 1–24.

Kwiatkowski, D., Phillips, P. C. B., Schmidt, P., & Shin, Y. (1992). Testing the null hypothesis of stationarity against the alternative of a unit root: How sure are we that economic time series have a unit root. *Journal of Econometrics, 54*, 159–178.

Mengxuan, T., Khan, K., Cifuentes-Faura, J., & Sukumaran, S. (2024). Technological innovation and energy efficiency in central eastern European countries. *Utilities Policy, 88*, 101761.

Moravcsik, A. (1997). Taking preferences seriously: A Liberal theory of international politics. *International Organization, 51*, 513–553.

Nyblom, J. (1989). Testing for the constancy of parameters over time. *Journal of the American Statistical Association, 84*(405), 223–230.

O'Sullivan, M., Overland, I., & Sandalow, D. (2017). *The geopolitics of renewable energy*. Center on Global Energy Policy Columbia University| SIPA.

Overland, I. (2019). The geopolitics of renewable energy: Debunking four emerging myths. *Energy Research & Social Science, 49*, 36–40.

Phillips, P. C. B., & Perron, P. (1988). Testing for a unit root in time series regression. *Biometrika, 75*, 335–346.

Rahman, M. F. B. A. (2020). *The intersection of emergent technologies and geopolitics*. RSIS working paper, Nanyang Technological University Singapore.

Rasoulinezhad, E., Taghizadeh-Hesary, F., Sung, J., & Panthamit, N. (2020). Geopolitical risk and energy transition in Russia: Evidence from ARDL bounds testing method. *Sustainability, 12*(7), 2689.

Rühlig, T., Seaman, J., & Voelsen, D. (2019). *5G and the U.S.-China tech rivalry test for Europe's future in the digital age: How can Europe shift back from back foot to front foot?*

Scholten, D. (2018). The geopolitics of renewables—An introduction and expectations. In *The geopolitics of renewables* (pp. 1–33). Springer.

Scholten, D., Bazilian, M., Overland, I., & Westphal, K. (2020). The geopolitics of renewables: New board, new game. *Energy Policy, 138*, 111059.

Shukur, G., & Mantalos, P. (2000). A simple investigation of the Granger-causality test in integrated-cointegrated VAR systems. *Journal of Applied Statistics, 27*(8), 1021–1031.

Sims, C. A., Stock, J. H., & Watson, M. W. (1990). Inference in linear time series models with some unit roots. *Econometrica, 58*(1), 113–144.
Sivaram, V., & Saha, S. (2018). The geopolitical implications of a clean energy future from the perspective of the United States. In *The geopolitics of renewables* (pp. 125–162). Springer.
Su, C. W., Khan, K., Tao, R., & Nicoleta-Claudia, M. (2019). Does geopolitical risk strengthen or depress oil prices and financial liquidity? Evidence from Saudi Arabia. *Energy, 187*, 116003.
Su, C. W., Khan, K., Tao, R., & Umar, M. (2020). A review of resource curse burden on inflation in Venezuela. *Energy, 204*, 117925.
Su, C. W., Khan, K., Umar, M., & Chang, T. (2022). Renewable energy in prism of technological innovation and economic uncertainty. *Renewable Energy, 189*, 467–478.
Su, C. W., Khan, K., Umar, M., & Zhang, W. (2021). Does renewable energy redefine geopolitical risks? *Energy Policy, 158*, 112566.
Tekir, G. (2020). Huawei, 5G network and digital geopolitics. *International Journal of Politics and Security*, 2 (4 (Çin Özel Sayısı)), 113–135.
Toda, H. Y., & Yamamoto, T. (1995). Statistical inference in vector auto-regression with possibly integrated processes. *Journal of Econometrics, 66*(1–2), 225–250.
Triolo, P. (2019). Geopolitics and technology—U.S.-China competition: The coming decoupling? *RSIS Commentaries* (pp. 214–19).
Vakulchuk, R., Overland, I., & Scholten, D. (2020). Renewable energy and geopolitics: A review. *Renewable and Sustainable Energy Reviews, 122*, 109547.
Wu, X. (2020). Technology, power, and uncontrolled great power strategic competition between China and the United States. *China International Strategy Review, 2*(1), 99–119.
Xie, H., Khurshid, A., Rauf, A., Khan, K., & Calin, A. C. (2023). Is geopolitical turmoil driving petroleum prices and financial liquidity relationship? Wavelet-based evidence from Middle East. *Defence and Peace Economics, 34*(6), 810–826.
Xinyu, W., Haoran, L., & Khan, K. (2024). Innovation in technology: A game changer for renewable energy in the European Union?. In *Natural resources forum*. Blackwell Publishing
Yang, K., Wei, Y., Li, S., & He, J. (2021). Geopolitical risk and renewable energy stock markets: An insight from multiscale dynamic risk spillover. *Journal of Cleaner Production, 279*, 123429.
Zheng, X., Li, C., Ali, S., & Adebayo, T. S. (2024). *Global chessboard: Analyzing how geopolitical risk shapes renewable energy technology investments* (Vol. 44, p. 2429). Risk Analysis.

Renewable Energy: Key Aspects of Patent Rights to Promote Investment in Innovation

Luiz Otavio Pimentel and Salete Oro Boff

1 The Renewable Energy: Concept

According to the United Nations (UN) "Renewable energy is energy derived from natural sources that are replenished at a higher rate than they are consumed. Sunlight and wind, for example, are such sources that are constantly being replenished. Renewable energy sources are plentiful and all around us. Fossil fuels—coal, oil, and gas—on the other hand, are non-renewable resources that take hundreds of millions of years to form. Fossil fuels, when burned to produce energy, cause harmful greenhouse gas emissions, such as carbon dioxide. Generating renewable energy creates far lower emissions than burning fossil fuels. Transitioning from fossil fuels, which currently account for the lion's share of emissions, to renewable energy is key to addressing the climate crisis. Renewables are now cheaper in most countries and generate three times more jobs than fossil fuels."[1]

For more information on renewable energy sources, among others, visit the websites of the following international organizations and programs:

- International Renewable Energy Agency
  https://www.irena.org.
- International Energy Agency (IEA)
  https://www.iea.org/energy-system/renewables.
- Intergovernmental Panel on Climate Change

[1] UN: *What is renewable energy?* Available online at https://www.un.org/en/climatechange/what-is-renewable-energy (accessed in 2024).

L. O. Pimentel (✉) · S. O. Boff
Postgraduate Program, School of Law, Atitus Education,
Passo Fundo, State of Rio Grande do Sul, Brazil
e-mail: salete.boff@atitus.edu.br

S. S. Oncel (ed.), *A Green Vision Towards a Renewable Energy Future*,
https://doi.org/10.1007/978-3-031-93760-6_5

- https://www.ipcc.ch/report/renewable-energy-sources-and-climate-change-mitigation/.
- UN Environment Program
  https://www.unep.org/interactive/sectoral-solution-climate-change/.
- Sustainable Energy for All
  https://www.seforall.org/goal-7-targets/renewable-energy.

According to Graca-Aranha (2023), WIPO's Special Representative to the United Nations Sustainable Development Goals (UNSDGs), "the #climatecrisis is an undeniable fact. In a historic gesture, at #COP28 in Dubai, 116 countries signed an agreement to triple the capacity for the production of clean energy. One could discuss the origins of #GlobalWarming, but it is a fact that the Planet is getting warmer and warmer. The highest temperatures ever have been reached in the last 12 months. Extreme temperatures and more frequent natural disasters are taking place in all corners of the Planet. As the UN Secretary-General #AntónioGuterres (#UNSG) stressed, the world is going toward a "#ClimateCatastrophe." The sad reality is that natural disasters are happening increasingly and resulting in the loss of millions of lives and assets."[2]

Graca-Aranha asks and brings the answer: "What is the main instrument for addressing and, hopefully, revert, even partially, the forecasted disaster? #Innovation will be the key (SDG9). In fact, innovation has been the key to solve all problems since centuries ago. Innovation allied with #IPRs, such as #patents, is the way to advance society, to achieve SDGs and what will come after the UN 2030 Agenda. New forms of energy are being developed. Heavy investment is being made on Research and Development (#R&D). The potential of several areas of the world to identify and diversify new sources of energy for the badly needed emergency transition will be the answer to the current climate disaster that is threatening our lives. And again, this innovation will come thanks to IPRs in general, and the patent system in particular, which have served to improve living conditions since the first patent-type legislation [...] Developing countries have an enormous potential to contribute to achieving the SDGs. Small or big ones. [...]"

2 Key Aspects of Patent Rights

Ten years ago, the World Intellectual Property Organization (WIPO) published a report on the increase in the number of renewable energy patents (Helm et al., 2014): "Climate change is one of the greatest challenges of our times. Fortunately, much can be done to address this challenge through technological innovation, including renewable sources of energy which help reduce carbon emissions.

[2] Graca-Aranha, J. *Brief reflection on SDGs, climate crisis and IP*. 2023. Available online at https://www.linkedin.com/pulse/brief-reflection-sdgs-climate-crisis-ip-jose-graca-aranha-dw7te/?trackingId=CD0KFAI9TMu7aNyLmtzhXg%3D%3D (accessed in 2024).

Governments throughout the world have embraced these technologies as a big part of the solution to climate change and have implemented regulatory frameworks and incentives to encourage their development and uptake."[3]

Green-tech patent surge (Helm et al., 2014): "The top line: more patents have been filed in the last 5 years than in the previous 30 years across key CCMTs [Climate Change Mitigation Technologies] examined by the report (biofuels, solar thermal, solar photovoltaics [PV] and wind energy). While the average global rate of patent filing grew by 6% between 2006 and 2011, these CCMTs have experienced a combined growth rate of 24% over the same period–four times the rate of growth."

Patenting activity is taking off in several green-tech sectors. More on patent landscape reports by WIPO.[4]

Also notable (Helm et al., 2014) is the shift in innovation from West to East. China and the Republic of Korea have filed the most patents in recent years across all four technology areas, while in solar PV. The top ten technology owners are all based in Asia.

Regarding globalization and business competition, "Patent concentrations have also decreased across three of the four patent landscapes (wind being the exception), reflecting greater globalization as well as higher competition between players from more countries. WIPO is doing other important work in the battle against climate change. […] WIPO GREEN, an interactive marketplace that promotes innovation and diffusion of green technologies by connecting technology and service providers with those seeking innovative solutions. WIPO GREEN consists of an online database and network that brings together a wide range of players in the green technology innovation value chain and connects owners of new technologies with individuals or companies looking to commercialize, license or otherwise access or distribute a green technology" (Helm et al., 2014).

2.1 *Patent and Patent Rights*

Patent is not a new product; patent is a document of a private intangible property right that guarantees ownership (property) and exclusive use of an invention of a product or process, that is, of a technology that, when applied, solves a technical problem, in this case, related to the use of energy.

The holder or owner of a patent is a person, whether natural or legal, who has this title. One can say something based on a title; someone who has a patent. Therefore,

---

[3] Helm, S., Q. Tannock and I., Iliev, 2014: *Renewable Energy Technology: Evolution and Policy Implications—Evidence from Patent Literature.* Global Challenges Report, WIPO. Available online at https://www.wipo.int/edocs/pubdocs/en/wipo_pub_gc_3.pdf (accessed in 2024).

[4] WIPO: *WIPO GREEN—The Marketplace for Sustainable Technology.* Available online at https://www3.wipo.int/wipogreen/en/ (accessed in 2024).

ownership is the quality of holder, of owner of something that is defined in the title, in the patent document.

Patent protection is territorial and does not extend to other countries unless there is also protection in their territories, which is an effect of the rule of territoriality of this type of rights.

Therefore, when protecting the technology of an invention by means of a patent, it is necessary to take into account the countries where the market will be relevant for commercializing the products or processes of a company, university, or even the independent inventor.

The nullity of a patent is a case of extinction of the title of ownership and consequently of the respective rights. It can be annulled whenever it has been granted contrary to the provisions of the law of the country in which the application was filed. The declaration of nullity takes effect from the date of filing of the patent application.

With regard to the nullity procedure, it can be instituted administratively by the national office that granted it or requested by any interested party. In the laws that deal with patent protection in the respective countries, there is a section that deals with nullity, which can be the subject of an administrative process in the office that granted it or through the courts.

Patent protection produces important effects, the main one being to generate a temporary right of exclusivity for its holder, also known as a monopoly, that extends for the period of protection of the patent.

Thus, whoever has a good technology, incremental or radical, protected by a patent, can guarantee, with support from the law, that competitors do not copy it or use it without their authorization.

This invention does not necessarily need to be revolutionary, nor is a prototype required for filing a patent application. It is enough that the invention meets the legal requirements set forth in the law of the country where the application was filed. The requirements are novelty, inventive step, industrial application, and sufficient descriptiveness of the invention.

Another important effect of patent rights is exhaustion since the right is generally exhausted with the sale of the product or a product containing the patented process. The holder cannot prevent the circulation of a product that was legitimately acquired based on his right. For example, if a company sells the patented product cheaper in one part of the territory of a country and someone legitimately acquires it to then resell it in the same market in another part of the same territory at a better price, in competition with the holder himself, the latter cannot do anything, because the patent rights have been exhausted in relation to this third party in the same territory, the same being true in the context of the European patent with protection in several countries.

2.2 *Administrative Procedure for Obtaining Patent Protection*

To obtain patent protection in the territory of a country or a group of countries, such as the European patent granted by the European Patent Office (EPO), the application document must describe the invention and go through the administrative procedure and, at the end, be granted.

The application will be filed and receive its own number. The patent application filing form generally has a model established by the office of each country or the EPO.

The patent application will be considered filed after the preliminary formal examination.

According to the Patent Cooperation Treaty (PCT) and the Paris Convention for the Protection of Industrial Property, Stockholm revision of 1967 (Article 4), there is a period of 12 months, counted from the date of filing in the country, for the holder to decide in which countries he wishes to obtain international protection, which is called the priority period.

From the filing date, the patent application will be kept confidential for a certain period and after this period it will be published, and then examined by the office's technicians to verify compliance with the basic and intrinsic requirements, as well as whether the holder has made the payments for the deposit, publication, examination, annual fees and other legal fees, all governed by administrative deadlines. We can summarize the steps to obtain patent protection at a national office or at the EPO in six steps:

(i) The first is to search for prior art or prior search, essential for drafting a patent application and verifying the novelty that must be worldwide. This search can be carried out in specialized public databases such as those available on the WIPO, EPO, or national offices websites, or private databases such as Derwent World Patents Index™ and Clarivate™.
(ii) The second is to draft the patent application, which is not a simple task, requiring technical knowledge of both the invention and the legislation that regulates its application. The drafter must be familiar with the invention, its technological area, materials used, methods and their results, so that he or she can sufficiently describe the invention and claim it correctly.
(iii) The third, as part of the administrative procedure of an office, is to file the application at the national priority office.
(iv) The fourth is to request for the examination within 36 months of the filing date.
(v) The fifth is to monitor the processing of the patent application.
(vi) The sixth is to issue the patent title, if granted, and to maintain the annual fees. If the patent application is denied, an administrative appeal may be filed with the national office or the EPO; if the denial is upheld, the invention will fall into the public domain.

2.3 Patent Classification

The CPC[5] is an extension of the International Patent Classification (IPC)[6] and is jointly managed by the EPO and the US Patent and Trademark Office; there are approximately 250,000 classification entries. CPC is divided into nine sections, A–H and Y, which in turn are subdivided into classes, subclasses, groups, and subgroups.[7]

The IPC, established by the Strasbourg Agreement 1971, from the WIPO, provides for a hierarchical system of language independent symbols for the classification of patents and utility models according to the different areas of technology to which they pertain. However, the IPC divides technology fields into eight sections (A–H) with approximately 75,000 subdivisions, each represented by a language independent symbol consisting of Latin alphabet characters and Arabic numerals. For this reason, the CPC is more accurate for searching for renewable energy patent documents.

Relevant documents in the EPO's World Patent Statistical Database[8] make it easier to retrieve patent documents that cover technologies, such can be found in section Y, specially in Y02 classes, and subclasses: Y02B, Y02C, Y02D, Y02E, Y02P, Y02T, and Y04S.

2.4 Patenting Trends in Renewable Energy

By protecting intellectual property rights (Nurton, 2020),[9] patents serve the purpose of protecting the economic interests of inventors and encouraging them to innovate. The innovations they are designed to protect are a driving force behind technological and social change. Many inventions that have had profound impacts on society can be traced back to patent documents applied for or granted by patent offices. Not all inventions have a high level of economic success, but they are indicative of some value to those who patented them. Given the expense and effort required to innovate, it follows that most patent applicants expect to be compensated for the ideas they seek to protect. This may be through exclusive sale of a new product,

[5] CPC: *EPO and USPTO launched the Cooperative Patent Classification System*. Available online at https://www.cooperativepatentclassification.org/home (accessed in 2024). https://www.cooperativepatentclassification.org/sites/default/files/cpc/scheme/Y/scheme-Y.pdf

[6] WIPO: IPC. Available online at https://www.wipo.int/classifications/ipc/en/ and https://www.wipo.int/edocs/pubdocs/en/wipo-pub-rn2022-7-en-international-patent-classification-ipc.pdf (accessed in 2024).

[7] CPC: *EPO and USPTO launched the Cooperative Patent Classification System*.

[8] CPC: *Section Y, Classes Y02, and Y04, and their subclasses*. Available online at https://www.cooperativepatentclassification.org/sites/default/files/cpc/scheme/Y/scheme-Y.pdf (accessed in 2024).

[9] Nurton, J., 2020, March: *Patenting trends in renewable energy*. WIPO Magazine, 50–56.

competitive advantages gained from improved operational processes, or royalties from a breakthrough technology. All things considered, patents provide a signal of future economic value by creating a competitive edge. "From 2002 to 2012, a decade in which innovation was being promoted in the renewables sector, the number of patent applications published under the Patent Cooperation Treaty for renewables increased by 547%. Although this figure has declined, in 2019, it was still 3.5 times higher than in 2002."

According to Tang, WIPO director general, the UNSDGs are hanging in the balance. To deliver this, we need to harness the innovation and creative potential of humankind, with IP critical to making this happen. IP incentivizes innovation, rewards creativity, and brings new technologies, ideas, and concepts to the market. All these can help us address common global challenges like climate change.[10]

There is a challenge in understanding the pathways of innovation. Technology data is captured in patent data, most of which is publicly available. This creates an opportunity through patent analytics to transform patent data into actionable insights that highlight the trajectory of technological advancement, as well as measure and track technological progress in areas of interest. Patent analytics also provides us with a sharper sense of the gaps that exist and points to where more resources need to be directed into R&D and translation.[11]

It is hoped that its insights will also serve as catalysts inspiring stakeholders across industries, governments, and academia to use PI to make a positive difference and truly leave no one behind.[12]

3 To Promote Investments in Innovation

3.1 Innovation and the Oslo Manual[13]

According to the Oslo Manual, regarding the Economy of Innovation, it was stated that: "The work of Joseph Schumpeter has greatly influenced theories of innovation. He argued that economic development is driven by innovation through a dynamic process in which new technologies replace the old, a process he labeled "creative destruction." In Schumpeter's view, "radical" innovations create major disruptive changes, whereas "incremental" innovations continuously advance the process of

[10] Tang, D., 2024: *Foreword to Mapping Innovations Patents and the Sustainable Development Goals*. WIPO. Available online at https://www.wipo.int/edocs/pubdocs/en/wipo-pub-rn2024-18-en-mapping-innovations.pdf (accessed in 2024).

[11] Tang, D. (2024).

[12] Tang, D. (2024).

[13] OECD, Oslo Manual, 2005: *Guidelines for collecting and interpreting innovation data—The Measurement of Scientific and Technological Activities.* Third edition. Available online at https://www.oecd-ilibrary.org/science-and-technology/oslo-manual_9789264013100-en (accessed in 2024).

change. Schumpeter (1934) proposed a list of five types of innovations: (i) introduction of new products, (ii) introduction of new methods of production, (iii) opening of new markets, (iv) development of new sources of supply for raw materials or other inputs, and (v) creation of new market structures in an industry."

It is essential to know why technological change occurs so that companies innovate. The reason given in Schumpeter's work is that they are looking for profits: a new technological device brings some advantage to the innovator. In the case of a process that increases productivity, the company obtains a cost advantage over its competitors, an advantage that allows it to obtain a higher margin at current market prices or, depending on the specificity of demand, use a combination of lower price and higher margin than its competitors to gain market share and obtain even more profits. In the case of product innovation, the company obtains a "monopoly" position due to either a patent (known as legal monopoly) or the time it takes competitors to imitate it. This monopoly position allows the company to set a higher price than would be possible in a competitive market, thus making a profit.

A patent is a property right over an invention, granted by national patent offices (institutes). A patent gives its holder an exclusive (limited) right (known as legal monopoly) to exploit the patented invention in return for disclosure, which is intended to enable a wider social use of the invention. Increasingly, patent statistics are used in various ways by students of technology as indicators of the performance of invention activities. The number of patents granted to a given company or country can reflect its technological dynamism; examining the growth of patent classes can give some indication of the direction of technological change.

Examining the technologies described in the patent document also provides a guide to the direction of technological change and serves as a basis for improvements and new technological advances.

The problems of using patents as indicators are well known: many innovations do not cover the patented inventions; many patents are available for inventions of almost no technological and economic value; many of them have quite significant value, while many others have never resulted in innovation.[14]

3.2 *The Appropriability of Innovations*

In the Oslo Manual, when dealing with questions about the appropriability of innovations, it was observed that:

- "The ability of enterprises to appropriate the gains from their innovation activities is an important factor affecting innovation. If, for example, enterprises are unable to protect their innovations from imitations by competitors, they will have

[14] OCED, Patent Manual, 1994: *The Measurement of Scientific and Technological Activities Using Patent Data as Science and Technology Indicators*. Available online at https://doi.org/10.1787/9789264065574-en (accessed in 2024).

less incentive to innovate. On the other hand, if an industry functions well without formal protection methods, promoting these may slow the flow of knowledge and technology and lead to higher prices for goods and services."[15]

- "Policy plays a central role in the design of legal methods of protecting innovations. Data on which types of methods are used and their relative importance can help inform policies to maximize the economic and social benefits from IP rights."[16]
- "Patent data, both applications and grants, function as an intermediate output indicator for innovation activity and provide information on the innovative capabilities of the enterprise. For example, an enterprise that has applied for patents can be assumed to be capable of developing innovations that are new to the world (occasionally only new to the market, depending on the patent strategies of other firms). Data on whether or not enterprises have applied for or been granted a patent can thus provide useful data for innovation surveys and be used in specialized surveys on IP rights. Note that patent data should refer to the country where the patent was developed and not the country where the patent application was made. Patent data is discussed in greater detail in the Patent Manual (OECD, 1994)."[17]

According to Nurton (2020), when analyzing patenting trends in renewable energy, the development of renewable energy sources (such as solar, wind, and tidal) is essential to tackling the climate crisis. What can we learn about them by looking at patent data? The past decade has seen unprecedented investment in renewable energy, as well as significant development of new technologies. Evidence of this can be seen in the proliferation of solar cells and wind turbines dotting the landscape. But it can also be measured by observing trends in the number of patent applications published. Patents are widely used as an indicator of how much innovation is taking place, where, and in which areas. A closer look at the data can therefore provide a range of insights into innovation in this sector.

3.3 *Overall Trends*

As shown, the total number of international applications filed and published under the PCT for renewable technologies increased each year from 2002 (831 applications) to 2012 (4541 applications), when it peaked at 4541. Since then, the number of applications declined each year from 2013 (4308 applications) to 2018 (2689 applications), although the numbers increased slightly in 2019 (2863 applications).[18]

[15] OECD, Oslo Manual (2005), paragraph 414.
[16] OECD, Oslo Manual (2005), paragraph 415.
[17] OECD, Oslo Manual (2005), paragraph 417.
[18] Nurton, J. (2020).

3.4 Improving Technologies

According to Nurton (2020), "increasing the use of renewable energy is key to limiting global warming to 1.5 °C above pre-industrial levels, one of the goals set by the Paris Agreement. [...] The evidence from the publication of patent data supports this finding and suggests that innovation in the renewables sector took off in the decade up to 2012, particularly in solar technologies. Over the next few years, we will see how that innovation helps to tackle global warming in practice."[19]

3.5 Investment

According to IEA (2024a, b, c): "Global spending on clean energy technologies and infrastructure on track to hit USD 2 trillion in 2024 even as higher financing costs hinder new projects, notably in emerging and developing economies."[20]

Despite pressures on financing, global investment in clean energy is set to reach almost double the amount going to fossil fuels in 2024, helped by improving supply chains and lower costs for clean technologies, according to a new IEA report.

Total energy investment worldwide is expected to exceed USD 3 trillion in 2024 for the first time, with some USD 2 trillion set to go toward clean technologies—including renewables, electric vehicles, nuclear power, grids, storage, low-emissions fuels, efficiency improvements, and heat pumps—according to the latest edition of the IEA's annual World Energy Investment report. The remainder, slightly over USD 1 trillion, is going to coal, gas, and oil. In 2023, combined investment in renewable power and grids overtook the amount spent on fossil fuels for the first time.[21]

The new report warns, however, that there are still major imbalances and shortfalls in energy investment flows in many parts of the world. It highlights the low level of clean energy spending in emerging and developing economies (outside China), which is set to exceed USD 300 billion for the first time—led by India and Brazil. Yet, this accounts for only about 15% of global clean energy investment, far below what is required to meet the growing energy demand in many of these countries, where the high cost of capital is holding back the development of new projects.[22]

"Clean energy investment is setting new records even in challenging economic conditions, highlighting the momentum behind the new global energy economy. For

[19] Nurton, J. (2020).

[20] IEA, 2024 (b): *Investment in clean energy this year is set to be twice the amount going to fossil fuels*. Available online at https://www.iea.org/news/investment-in-clean-energy-this-year-is-set-to-be-twice-the-amount-going-to-fossil-fuels (accessed in 2024).

[21] IEA (2024b).

[22] IEA (2024b).

every dollar going to fossil fuels today, almost two dollars are invested in clean energy," said IEA Executive Director Fatih Birol. "The rise in clean energy spending is underpinned by strong economics, by continued cost reductions and by considerations of energy security. But there is a strong element of industrial policy, too, as major economies compete for advantage in new clean energy supply chains. More must be done to ensure that investment reaches the places where it is needed most, in particular the developing economies where access to affordable, sustainable and secure energy is severely lacking today."[23]

When the Paris Agreement was reached in 2015, the combined investment in renewables and nuclear for electricity generation was twice the amount going to fossil fuel-fired power. In 2024, this is set to rise to ten times as much, the report highlights, with solar PV leading the transformation of the power sector. More money is now going into solar PV than all other electricity generation technologies combined. In 2024, investment in solar PV is set to grow to USD 500 billion as falling module prices spur new investments.[24]

China is set to account for the largest share of clean energy investment in 2024, reaching an estimated USD 675 billion. This results from strong domestic demand across three industries in particular—solar, lithium batteries, and electric vehicles. Europe and the United States follow, with clean energy investment of USD 370 billion and USD 315 billion, respectively. These three major economies alone make up more than two-thirds of global clean energy investment, underlining the disparities in international capital flows into energy.[25]

Global upstream oil and gas investment is expected to increase by 7% in 2024 to reach USD 570 billion, following a similar rise in 2023. The growth in spending in 2023 and 2024 is predominantly by national oil companies in the Middle East and Asia. The report finds that oil and gas investment in 2024 is broadly aligned with the demand levels implied in 2030 by today's policy settings, but far higher than projected in scenarios that hit national or global climate goals. Clean energy investment by oil and gas companies reached USD 30 billion in 2023, accounting for only 4% of the industry's overall capital spending, according to the report. Meanwhile, coal investment continues to rise, with more than 50 GW of unabated coal-fired power approved in 2023, the highest since 2015.[26]

In addition to economic challenges, grids and electricity storage have been a significant constraint on clean energy transitions. But spending on grids is rising and is set to reach USD 400 billion in 2024, having been stuck at around USD 300 billion annually between 2015 and 2021. The increase is largely due to new policy initiatives and funding in Europe, the United States, China, and some countries in Latin America. Meanwhile, investments in battery storage are taking off and set to reach USD 54 billion in 2024 as costs fall further. Yet again, this spending is highly

---

[23] IEA (2024b).

[24] IEA (2024b).

[25] IEA (2024b).

[26] IEA (2024b).

concentrated. For every dollar invested in battery storage in advanced economies and China, only one cent was invested in other emerging and developing economies.[27]

Inventions generally result from R&D, which in turn seeks to solve technical problems using basic scientific research, applied research, and technology development.

Improving productivity rates, reducing costs, eliminating damage and restoring the environment, and finding alternatives for fuels and energy generation are some of the problems that will continue to challenge scientists and technologists who invent.

Investment in research has increased in recent years, contributing to the advancement of technological innovations in countries. Different segments of society participate in this process, such as the government and its agencies, universities, independent research institutes, industrial companies, cooperatives, rural producer organizations, among others—often forming strategic alliances to withstand national and global competition and create wealth.

According to the data available on the website "Energy Technology Patents Data Explorer," from the IEA, on clean energy patents from 44 countries, updated on 08/05/2024, it was found that, in general, patents provide early constraints on technological developments with the potential to transform the economy and to promote the energy transition.

How is the clean energy transition reflected in energy patents? Energy patents can be either fossil fuel-specific patents or clean energy patents.

The EPO classification taxonomy for climate mitigation technologies was used to classify the patents as clean energy patents. The EPO also classifies the patents within different fossil fuel categories. Note that each patent can belong to several technology categories. For example, some patents related to electric vehicles belong to both storage and e-mobility. Therefore, the sum of subcategory patents is not equal to the sum of patents of the corresponding category family.

The EPO classification scheme for climate mitigation technologies was used to classify the patents as clean energy patents. A collaboration with the EPO was undertaken to perform text searches on abstracts to classify patents within different fossil fuel categories.

Each patent may belong to more than two technology categories. For example, some patents related to electric vehicles belong to both storage and electric mobility. Therefore, the sum of the patents of the subcategories is not equal to the sum of the patents of the corresponding parent category.

Which energy technology has the most patents? The patent tree map provides a broad overview of how patents are distributed among different energy technologies. A total of 529,898 energy-related patents were counted worldwide between the years 2000 and 2023.

Which clean energy technologies have the most patents?

[27] IEA (2024b).

The IEA (2024a, b, c) Energy Technology Patents Data Explorer,[28] on clean energy and fossil fuel patents in 44 countries, is very important because patents provide early indications of technological developments that may transform the economy and drive the energy transition. Energy patents can be classified as fossil fuel patents or clean energy patents, and the graphs show trends in each type of patent since 2000. Classification[29] to follow, in alphabetical order: agriculture energy efficiency, air, rail, and marine, bioenergy, building energy efficiency, carbon capture and storage, e-mobility, energy efficiency, grid, hydrogen and fuel cells, industry energy efficiency or substitution, nuclear, renewable energy integration in buildings, solar, storage (not e-mobility), wind, and other renewables.

According to the IEA (2024a, b, c) methodological notes for patent counts: The relevant patent documents are identified based on a classification scheme (the Y02 scheme) developed by the EPO for technical attributes of technologies that are related to CCMTs. The Y02 subclasses include areas related to specific clean energy technologies, namely Y02B (CCMTs related to buildings), Y02C (greenhouse gases—capture and storage/sequestration or disposal), Y02E (greenhouse gases—emissions reduction technologies related to energy generation, transmission, or distribution), Y02T (CCMTs related to transport), etc. In addition, the ancillary Y04S scheme identifies CCMTs related to smart grids.

According to Harrison (2024) the "Climate Action" is driven by technologies that aim to reduce greenhouse gas emissions. "Affordable and Clean Energy" (SDG7), the focus of our study, benefits from advances in renewable energy such as solar and wind power. This shows a slightly stronger upward trend than most other SDGs, reflecting a growing awareness among suppliers and consumers of cleaner alternatives. The upward trends in SDG-related patents, particularly for renewable energy and emissions reductions, reflect a growing focus on sustainable technologies.[30]

So, looking at patents in renewable energy gives us a snapshot of the development of renewable energy sources, and we learn about them by looking at patent data and documents.

We see that there is unprecedented investment in renewable energy, as well as significant development of new technologies. Evidence of this can be measured by looking at trends in the number of patent applications published.

Patents are widely used as an indicator of how much innovation is happening, where, and in which areas. A closer look at the data can therefore provide a wealth of insights into innovation in this sector.

[28] IEA, 2024 (a): *Energy Technology Patents Data Explorer*. Available online at https://www.iea.org/data-and-statistics/data-tools/energy-technology-patents-data-explorer (accessed in 2024).

[29] IEA, 2024 (c): *Methodological notes for patent counts* (is part of the "IEA's Clean Energy Transitions in Emerging Economies programme". Available online at https://iea.blob.core.windows.net/assets/bb70e652-6862-4f15-9ba7-f3249d1551ea/Methodologicalnotesforpatentcounts.pdf (accessed in 2024).

[30] Harrison, C., 2024, February*: Patent Data Show One-Third of Inventions Relate to the SDGs*. WIPO Magazine, 27–32.

Patents is an important indicator for guiding investment in technologies, knowing which sectors are concentrating on R&D and its results, and defining public policies for less developed sectors.

References

Graca-Aranha, J. (2023). Brief reflection on SDGs, climate crisis and IP. Available online at https://www.linkedin.com/pulse/brief-reflection-sdgs-climate-crisis-ip-jose-graca-aranha-dw7te/?trackingId=CD0KFAI9TMu7aNyLmtzhXg%3D%3D (accessed in 2024).

Harrison, C. (2024, February). Patent data show one-third of inventions relate to the SDGs. *WIPO Magazine*, 27–32. Retrieved from https://www.wipo.int/wipo_magazine/en/2024/01/article_0007.html (accessed in 2024)

Helm, S., Tannock, Q., & Iliev, I. (2014). *Renewable energy technology: Evolution and policy implications-evidence from patent literature*. Global Challenges Report, WIPO. Retrieved from https://www.wipo.int/edocs/pubdocs/en/wipo_pub_gc_3.pdf (accessed in 2024)

IEA (2024a). *Energy technology patents data explorer*. Retrieved from https://www.iea.org/data-and-statistics/data-tools/energy-technology-patents-data-explorer (accessed in 2024).

IEA (2024b). *Investment in clean energy this year is set to be twice the amount going to fossil fuels*. Retrieved from https://www.iea.org/news/investment-in-clean-energy-this-year-is-set-to-be-twice-the-amount-going-to-fossil-fuels (accessed in 2024).

IEA (2024c). *Methodological notes for patent counts* (is part of the "IEA's Clean Energy Transitions in Emerging Economies programme). Retrieved from https://iea.blob.core.windows.net/assets/bb70e652-6862-4f15-9ba7-f3249d1551ea/Methodologicalnotesforpatentcounts.pdf (accessed in 2024).

Nurton, J. (2020, March). Patenting trends in renewable energy. *WIPO Magazine*, 50–56. Retrieved from https://www.wipo.int/edocs/pubdocs/en/wipo_pub_121_2020_01.pdf (accessed in 2024)

OCED, Patent Manual. (1994). *The measurement of scientific and technological activities using patent data as science and technology indicators*. https://doi.org/10.1787/9789264065574-en. (accessed in 2024).

OECD, Oslo Manual. (2005). *Guidelines for collecting and interpreting innovation data—The measurement of scientific and technological activities* (3rd ed.) Retrieved from https://www.oecd-ilibrary.org/science-and-technology/oslo-manual_9789264013100-en (accessed in 2024)

Schumpeter, J. A. (1934). *The theory of economic development: An inquiry into profits, capital, credit, interest, and the business cycle*. Harvard University Press.

Renewable Energy and New Computational Intelligence

Cosimo Magazzino

1 Introduction

Over the past few years, climate change has had an increasingly severe impact, prompting countries to transition to more sustainable economic models and practices. Countries worldwide face significant challenges in their efforts to build a sustainable future, driven by the pressing issues of climate change and global warming. Environmental degradation has long been a major concern, but climate change has become an equally critical issue for nations (Magazzino, 2023b). The widespread dissemination of sustainable policies and the shift toward a green economy are essential to achieving a future that balances economic development, social welfare, and environmental standards (Munasinghe, 2004).

The United Nations (UN) introduced the 2030 Agenda to guide member countries in transitioning to a sustainable development approach in the long run. As outlined in the 2030 Sustainable Development Goals (SDGs), transitioning to accessible and renewable energy is crucial for both advanced and emerging nations. However, a close examination reveals that countries predominantly rely on fossil fuels for energy consumption. The widespread use of fossil fuels has led to extensive emissions of greenhouse gases (GHG) and carbon dioxide (CO_2) into the atmosphere (Mohsin et al., 2023). This reliance on fossil-fuel-based energy, which has been intensively used throughout countries' economic development, threatens the future.

Increasing production has led to a corresponding increase in energy consumption, which in turn has caused harmful emissions such as CO_2 to rise. This results in

C. Magazzino (✉)
Department of Political Science, Roma Tre University, Rome, Italy

Economic Research Center, Western Caspian University, Baku, Azerbaijan
e-mail: cosimo.magazzino@uniroma3.it

S. S. Oncel (ed.), *A Green Vision Towards a Renewable Energy Future*,
https://doi.org/10.1007/978-3-031-93760-6_6

the deterioration of environmental quality and irreversible global climate changes. Developed and developing countries alike struggle to minimize these effects, with the Paris Agreement representing a significant step in this direction. The Paris Agreement aims to reduce the global average temperature. However, sustainable policy transitions aimed at improving environmental quality can negatively affect state budgets. Therefore, adverse environmental quality developments often result in a difficult choice between increasing production and protecting the environment. Energy availability is crucial to development, and energy consumption is a fundamental need in every aspect of life. The rising energy demand has led to the depletion of natural resources and environmental degradation resulting from fossil fuel consumption. Additionally, reliance on fossil fuels has revealed significant energy security issues. Consequently, countries have turned to alternative and sustainable energy sources to address both environmental sustainability and economic energy security. One of the optimal ways to mitigate the climate crisis in the twenty-first century is by using cleaner electricity generation sources across all sectors.

Renewable energy sources, such as hydroelectricity and biomass, are pivotal in the transition to a sustainable energy future. Hydroelectric energy, obtained from flowing water, is one of the most significant renewable sources. However, hydroelectric power production is contingent on specific water conditions and exhibits substantial variability. For example, the drought experienced in Turkey in 2014 resulted in a decrease in hydroelectricity production (Acaroğlu et al., 2023). Despite its potential, hydroelectric energy sources can have mixed impacts on environmental quality and local ecosystems. Hydroelectric power is favored for its affordability and minimal environmental impact (Mohsin et al., 2023). In 2004, the UN Beijing Declaration on Hydropower and Sustainable Development emphasized the importance of hydropower for economies (Bildirici & Gökmenoğlu, 2017). However, the environmental benefits of hydroelectric power depend on the share of this energy source in a country's overall energy mix. Studies have shown that hydroelectric power plants can affect air temperature, negatively impact river ecosystems, and influence overall ecosystem health. Additionally, some chemicals and toxic substances used in hydroelectric power plants can contaminate water resources. Consequently, hydroelectric power plants can pollute both water and the atmosphere, adversely affecting nearby homes, lands close to dams, and the structure of natural habitats (Pata et al., 2023). For instance, Alsaleh and Abdul-Rahim (2023) examined the impact of increasing hydroelectric energy usage on fish supply in the EU region from 1990 to 2019, finding that fish supply declines as hydroelectric production expands.

Biomass energy, distinct from other alternatives, is the only renewable organic petroleum substitute that can be converted into liquid fuel. Biomass energy meets various needs, including home heating, electricity generation, vehicle fueling, and providing processed heat for industrial factories. Biomass is derived from organic matter produced from forests, living organisms, seaweed, and waste from agricultural and forestry processes, and it is a significant contributor to the shift toward renewable energy sources due to its abundance, low emission levels, and high application potential (Yasmeen et al., 2022). Beyond boosting renewable energy production, biomass energy also contributes to economic growth by creating more job opportunities (Mehmood, 2022).

The links between environmental quality and economic growth are frequently studied using the Environmental Kuznets Curve (EKC) hypothesis (Magazzino et al., 2023). During the initial phases of economic growth, ecological awareness is low, and technology does not support energy efficiency or renewable energy production. Therefore, resource use and harmful emissions increase as the economy grows (Aydin et al., 2023; Golpîra et al., 2023; Magazzino, 2023a). As ecological awareness increases, this process begins to reverse. Waste management policies, the adoption of renewable energy sources, and energy efficiency technologies become crucial. Renewable energy sources are more sustainable than conventional alternatives.

2 The Status Quo of Research on Renewable Energy Sources

Numerous studies have investigated the influence of renewable energy sources, including hydroelectricity and biomass, on pollution, with many reporting favorable results for environmental quality. For example, Alnour et al. (2022) examined the connection among hydroelectric energy consumption, population growth, natural resource rent, and environmental pollution in Sudan from 1990 to 2018, controlling for structural breaks. The literature revealed that hydroelectric energy usage and economic growth substantially lower environmental pollution over the long run, whereas short-term reductions in environmental pollution were observed with urban population growth. Mohsin et al. (2023) investigated the nonlinear effects of hydroenergy consumption on carbon emissions in Switzerland, Austria, Norway, Italy, Spain, Germany, Sweden, France, Finland, and Portugal from 1991 to 2019. They found that the use of hydroelectric energy significantly reduced carbon emissions. Magazzino (2023a) explored the relationship among ecological footprint, electricity consumption, and GDP in China. Quantile Regression estimates indicate that electricity consumption and real GDP increase environmental degradation, while trade and urbanization reduce it. Yasmeen et al. (2022) focused on the potential of biomass energy to promote technological innovations in energy efficiency through foreign direct investments. Their study highlighted negative connections between biomass energy consumption and the ecological footprint. Adebayo et al. (2023) examined the effects of structural change on the ecological footprint of hydropower and coal consumption in India, finding that structural changes reduce the ecological footprint in both the short and long run. Additionally, hydropower consumption was found to reduce the ecological footprint in the long run. Magazzino (2012) assessed the relationship between disaggregated energy production and real aggregate income in Italy, highlighting a bidirectional flow in the long run.

Adebayo et al. (2022) investigated the relationship between renewable and nonrenewable energies and the ecological footprint in newly industrialized countries. They considered hydro and geothermal energy for renewable sources and coal, gas, and oil for nonrenewable sources. Their findings indicated that economic growth increases the ecological footprint, as expected, while the consumption of oil, coal, and gas exacerbates environmental pollution. Conversely, hydro and geothermal energies improve environmental quality. Wang (2023a, b) examined the relationship between oil substitute energies and economic growth, concluding that hydro, wind, and solar energies support economic growth and have a reducing effect on carbon emissions. Bello et al. (2018) investigated the effects of hydropower consumption on the environment in emerging economies, considering different environmental degradation criteria. Their results showed that hydropower consumption improves environmental quality.

Some studies, however, have reported mixed or negligible effects of renewable energy sources on environmental degradation. For instance, Acaroğlu et al. (2023) examined the relationships among environmental pollution, coal consumption, energy consumption, economic growth, trade openness, and hydroelectric consumption in Turkey. Their empirical findings indicated that while hydroelectric energy reduces carbon emissions, it does not significantly impact overall environmental quality. Pata et al. (2023) investigated the relationship between hydroelectric energy consumption and life expectancy in Turkey, including variables such as income, trade openness, and globalization in the context of the EKC hypothesis. Their study showed that hydroelectric energy consumption does not affect environmental pollution. Pata and Aydin (2020) examined the relationship among hydroelectric energy consumption, economic growth, and the ecological footprint, finding no causal link between hydroelectric consumption and the ecological footprint.

Bildirici and Gökmenoğlu (2017) explored the causal link between hydroelectric energy consumption and carbon emissions in different regimes of business cycles in G7 countries. They found evidence that carbon dioxide emissions cause hydroelectric energy consumption in the first, second, and third regimes, while hydroelectric energy consumption causes carbon dioxide emissions. Wang et al. (2023a) investigated the role of hydroelectric, nuclear, and biomass energy on the ecological footprint in China, revealing strong causal links but no significant effect in the upper quantiles of the distribution. Saadaoui et al. (2024) examined the link among hydroelectric energy production, geopolitical risk, income, foreign direct investments, financial development, and FDIs in Turkey, finding a unidirectional causality from hydroelectricity to carbon emissions at medium and low frequencies. Perone (2024) analyzed the long-term relationship between renewable energy production and carbon dioxide emissions in 27 OECD countries, employing PVAR and causality analysis. They determined that geothermal, biomass, solar, and hydroelectric energy production are the most effective energy sources in reducing carbon emissions.

3 Computational Intelligence and Machine Learning in Renewable Energy Research

The ongoing shift toward renewable energy sources presents both challenges and opportunities that require innovative solutions. In this context, the role of Computational Intelligence (CI) and Machine Learning (ML), integral components of the broader field of Artificial Intelligence (AI), cannot be overstated. They have emerged as powerful tools for addressing the complexities inherent in renewable energy systems, offering advanced methodologies for optimizing energy production, enhancing grid stability, and improving predictive capabilities, particularly when coupled with the vast amounts of data generated in the energy sector (Magazzino, 2023a).

Big Data (BD) is not just a buzzword, but a crucial player in the successful deployment of CI and ML in renewable energy research. The proliferation of sensors, smart grids, and IoT devices has resulted in an unprecedented volume of data across various aspects of energy generation, distribution, and consumption. This data encompasses weather patterns, energy output from renewable sources, grid performance metrics, and consumer usage patterns, among others. Leveraging this data through CI and ML allows for the development of more accurate models, which can predict energy generation and consumption with greater precision.

In particular, BD facilitates the training of ML models, which require large datasets to identify patterns, correlations, and anomalies that might be overlooked by traditional statistical methods. For instance, in wind energy research, data on wind speed, direction, and atmospheric pressure collected over the years can be used to train ML models that accurately forecast wind power generation. Similarly, solar energy forecasting benefits from BD by incorporating factors such as cloud cover, solar irradiance, and temperature into predictive models.

The application of ML in renewable energy is primarily focused on forecasting and optimization. Accurate energy production forecasting from renewable sources is essential for maintaining grid stability and ensuring efficient energy management. ML techniques, such as Support Vector Machines (SVMs), Artificial Neural Networks (ANNs), and ensemble learning methods, have been widely adopted for these purposes (Magazzino, 2024).

SVMs are supervised learning models that are particularly effective in classification and regression tasks. In renewable energy research, SVMs are used to predict energy output based on historical data. Their ability to handle high-dimensional spaces makes them suitable for modeling complex relationships between variables, such as the nonlinear effects of weather conditions on energy production.

ANNs mimic the human brain's structure and are highly effective in recognizing patterns and making predictions. They are especially useful in renewable energy forecasting, where the input data is complex and nonlinear. For example, ANNs can predict solar power output by processing multiple input variables, including solar radiation, temperature, and humidity.

Ensemble Methods involve combining multiple models to improve prediction accuracy. Techniques such as Random Forests and Gradient Boosting Machines are commonly used in renewable energy research to aggregate predictions from different models, thereby reducing the variance and improving the robustness of the forecasts.

Beyond forecasting, CI and ML are also instrumental in optimizing energy production and distribution. Renewable energy sources like wind and solar are inherently variable, and their integration into the grid poses challenges in maintaining a balance between supply and demand. CI techniques, such as genetic algorithms and swarm intelligence, are employed to optimize these processes:

- Genetic Algorithms: These are inspired by natural selection and are used to find optimal solutions in complex environments. In the context of renewable energy, genetic algorithms can optimize the placement of wind turbines or solar panels to maximize energy output while minimizing costs and environmental impacts.
- Swarm Intelligence: This technique, modeled after the collective behavior of social organisms, is used to solve optimization problems related to power flow and grid stability. For instance, swarm intelligence can be applied to coordinate the actions of multiple distributed energy resources, ensuring that energy is efficiently distributed across the grid.

While the integration of CI and ML in renewable energy research has shown considerable promise, several challenges remain. One of the primary challenges is the quality of data. The accuracy of ML models is highly dependent on the quality and completeness of the data used to train them. Issues such as data sparsity, noise, and missing values can significantly affect model performance. Therefore, developing robust data-preprocessing techniques is crucial to mitigate these issues.

Another challenge lies in the scalability of CI and ML models. As renewable energy systems become more complex, the computational demands of modeling these systems increase. There is a growing need for scalable algorithms that can handle large datasets and perform real-time analysis. Deep Learning (DL), a subset of ML that involves training models with multiple layers of ANNs, offers a potential solution. By capturing hierarchical patterns in the data, DL models can improve the accuracy and efficiency of renewable energy predictions.

- Integration with emerging technologies: The future of CI and ML in renewable energy will likely involve their integration with other emerging technologies, such as the IoT and blockchain. IoT devices can enhance data collection by providing real-time monitoring of energy systems, while blockchain can improve data security and transparency. The combination of these technologies with CI and ML can lead to more resilient and efficient energy systems.
- Ethical and policy considerations: The deployment of CI and ML in the energy sector also raises important ethical and policy considerations. Issues such as data privacy, cybersecurity, and the potential for job displacement due to automation must be addressed. Policymakers will need to create frameworks that encourage the adoption of CI technologies while ensuring that their benefits are equitably distributed and that potential risks are mitigated.

CI and ML represent powerful tools in the advancement of renewable energy research. By harnessing the power of BD, these technologies can enhance the accuracy of energy production forecasts, optimize energy distribution, and contribute to the development of more sustainable energy systems. As the field continues to evolve, addressing the challenges of data quality, scalability, and integration with emerging technologies will be critical to realizing the full potential of CI and ML in renewable energy.

The increasing deployment of renewable energy sources has created a significant demand for advanced computational techniques to optimize energy production, distribution, and consumption (Hong et al., 2016). CI encompasses a range of methodologies and technologies, including ML, ANNs, fuzzy logic, and genetic algorithms, that are employed to solve complex problems and enhance decision-making processes (Goodfellow et al., 2016). ML, a subset of CI, has shown remarkable success in developing predictive models that can analyze vast amounts of historical data to forecast energy production and demand (Hong et al., 2016). ANNs mimic the human brain's structure and functioning and are particularly adept at recognizing patterns and making accurate predictions based on complex datasets. These capabilities are vital for managing the intermittent nature of renewable energy sources, such as wind and solar power, which are highly variable and weather dependent (Raza et al., 2016).

Accurate forecasting of renewable energy production is essential for efficient energy management and grid stability. Traditional forecasting methods often struggle with the variability and unpredictability inherent in renewable energy sources like solar and wind power. CI, particularly ML, offers robust solutions to these challenges by leveraging large datasets and sophisticated algorithms to improve prediction accuracy.

ML algorithms, such as support vector machines, decision trees, and ensemble methods, have been widely adopted for energy forecasting (Hong et al., 2016). These algorithms can process vast amounts of historical and real-time data to identify patterns and trends that traditional statistical methods might miss (Raza et al., 2016). For instance, ANNs, with their ability to learn from data, have been particularly successful in capturing the nonlinear relationships and temporal dependencies that characterize renewable energy production.

One notable application of ANNs in renewable energy forecasting is predicting solar power generation. Solar energy production is influenced by various factors, including sunlight intensity, cloud cover, and temperature, which can change rapidly and unpredictably. ANNs can process these complex inputs to generate accurate short-term and long-term forecasts, thereby helping grid operators to manage supply and demand more effectively. Similarly, wind power forecasting has benefited from the application of CI. Wind energy production is highly dependent on wind speed and direction, which are influenced by atmospheric conditions (Lv et al., 2023).

Despite these advances, several challenges remain in renewable energy forecasting. The accuracy of ML models depends heavily on the quality and quantity of available data (Goodfellow et al., 2016). Data sparsity, noise, and missing values can significantly affect model performance, highlighting the need for robust data preprocessing techniques and advanced model architectures to handle these issues.

Future research should focus on developing more sophisticated algorithms to improve forecasting accuracy and robustness. DL, a subset of ML that involves multiple layers of ANNs, holds significant promise in this regard (Goodfellow et al., 2016). By learning hierarchical representations of data, DL models can capture complex patterns and relationships that simpler models might miss. Additionally, integrating CI techniques with other emerging technologies, such as the IoT and blockchain, can enhance data collection, processing, and security, further improving the reliability of renewable energy forecasts (Sikorski et al., 2017).

4 Applications of Computational Intelligence in Renewable Energy

One of the primary applications of CI in energy storage management is optimizing charging and discharging schedules. Magazzino (2024) presented an overview of the recent developments in CI techniques for renewable energy production. Genetic algorithms, inspired by the principles of natural selection, can be used to determine the optimal times for charging and discharging batteries, thereby minimizing energy losses and maximizing overall efficiency. These algorithms can evaluate multiple factors, including energy demand forecasts, electricity prices, and battery health, to find the best strategies for energy storage utilization (Fathima & Palanisamy, 2015). ML models also play a crucial role in predicting energy demand and supply fluctuations, enabling better planning and utilization of energy storage capacities (Raza et al., 2016). These models can analyze historical and real-time data to forecast short-term and long-term energy needs, helping operators to make informed decisions about when to store or release energy.

Another significant application of CI in energy storage is enhancing the performance and lifespan of energy supplies. Predictive maintenance, powered by ML, can identify potential issues in energy storage systems before they lead to failures (Hong et al., 2016). By analyzing data from sensors embedded in batteries, ML models can detect anomalies and predict maintenance needs, reducing downtime and extending the operational life of energy storage systems. Despite these advancements, there are challenges to be addressed in the application of CI to energy storage management. One key challenge is the integration of CI techniques with existing energy infrastructure, which requires substantial investments in technology and training (Sikorski et al., 2017). Additionally, the development of accurate and reliable CI models depends on the availability of high-quality data, which can be difficult to obtain for some renewable energy sources and storage systems.

Another potential application is the integration of renewable energy into power grids, which presents significant operational challenges, including maintaining grid stability, balancing supply and demand, and minimizing transmission losses. CI might provide advanced tools for optimizing grid operations, addressing these challenges through sophisticated algorithms and real-time data analytics. Inspired by the collective behavior of social organisms such as ants and bees, swarm intelligence algorithms have been used to solve complex optimization problems related to power flow and grid stability. These algorithms can coordinate the actions of numerous distributed energy resources, ensuring efficient energy distribution and minimizing losses. For example, advanced software tools can find optimal paths for energy flow in a power grid, reducing transmission losses and enhancing overall efficiency (Fathima & Palanisamy, 2015).

Magazzino et al. (2021) analyzed the causal relationship among renewable energy technologies, biomass energy consumption, per capita GDP, and CO_2 emissions in Germany. The empirical findings support the powerful role of biomass energy in reducing carbon dioxide emissions. Magazzino et al. (2020) inspected the relationship between nuclear energy consumption and economic growth in Switzerland. The results suggest carefully analyzing the nuclear energy phase-out process to avoid adverse effects on economic growth. Soytas et al. (2022) highlighted the necessity to manage the process of nuclear phase-out in Belgium in a timely manner, along with a progressive deployment of installed renewable energy capacity.

ML also might play a crucial role in optimizing grid operations by providing real-time data analytics and predictive capabilities. By identifying potential issues before they escalate, these models enable proactive maintenance and dynamic adjustments to maintain optimal grid performance. With the ability to learn from data, NN has been particularly effective in enhancing grid stability and reliability. These models can predict fluctuations in energy supply and demand, allowing grid operators to make informed decisions about energy distribution and load balancing. For instance, ANNs can forecast peak demand periods and recommend strategies to shift or reduce loads, ensuring a stable and reliable power supply (Raza et al., 2016).

Future research should focus on developing more robust and scalable CI algorithms that can handle the increasing complexity and scale of renewable energy systems. Deep learning and reinforcement learning, which involve training models through trial and error, offer promising avenues for improving grid operations. By learning from interactions with the environment, these models can develop optimal strategies for energy distribution and load balancing, even under highly dynamic and uncertain conditions.

5 An Example of Empirical Application in Resource Energy Research

The dataset utilized in this empirical application spans from 1970 to 2022, encompassing over five decades of data to provide a comprehensive analysis of the impacts of hydroelectric and biomass energy consumption on the ecological footprint. This rich dataset includes a variety of variables that are crucial for understanding the intricate relationships between energy consumption and environmental sustainability. Each variable offers a unique perspective on different aspects of energy use, economic development, and urbanization, allowing for a multifaceted analysis.

1. **Ecological Footprint (*efp*):** The ecological footprint is a metric that quantifies the environmental impact of human activities in terms of the number of Earths

required to sustain the current level of consumption. It is a holistic indicator that reflects the demand placed on the planet's ecosystems by a given population. The data for this variable is sourced from the Global Footprint Network, which provides robust and internationally recognized measurements of ecological impact.

2. **Hydropower Consumption (*hydro*):** This variable represents the primary energy consumption from hydropower sources. Hydropower is one of the most widely used forms of renewable energy, and its consumption levels are indicative of its role in the overall energy mix. The data for hydropower consumption is obtained from Our World in Data, a comprehensive source that compiles global statistics on energy production and consumption.
3. **Biomass Consumption (*bio*):** Biomass consumption refers to the domestic extraction of biomass, which includes organic materials such as wood, agricultural residues, and other biological matter used for energy production. This variable highlights the reliance on biomass as an energy source and its potential environmental impacts. The data for biomass consumption is sourced from the Global Material Flows database, which tracks the extraction, trade, and use of biomass resources globally.
4. **GDP per Capita (*gdpc*):** GDP per capita is a measure of economic output per person, expressed in current US dollars. It serves as an important indicator of economic development and prosperity. Higher GDP per capita often correlates with greater energy consumption and higher ecological footprints. The data for GDP per capita is sourced from OECD Stat, which provides detailed economic statistics for member countries.
5. **Energy Consumption per Capita (*en*):** This variable measures the average energy consumption per person within a given population. It provides a per capita perspective on energy usage, which is crucial for understanding individual energy demands and consumption patterns over time. The data for energy consumption per capita is sourced from the World Development Indicators (WDI), a reputable database maintained by the World Bank.
6. **Trade Openness (*trade*):** Trade openness is defined as the share of exports and imports in a country's GDP. It reflects the degree of economic integration with the global market and can influence energy consumption patterns through trade-related activities. The data for trade openness is sourced from WDI, which tracks international trade statistics and economic indicators.
7. **Urbanization (*urban*):** This variable measures the population living in the largest city as a percentage of the total urban population. It serves as an indicator of urban concentration and development trends. Due to the higher energy demands of densely populated urban areas, urbanization can have significant implications for energy consumption and ecological footprints. The data for urbanization is also sourced from WDI.

This empirical example employs both traditional statistical methods and advanced ML techniques to analyze the impacts of hydroelectric and biomass energy consumption on the ecological footprint. The objective is to illustrate the effectiveness of these methods in predicting the ecological footprint and understanding the relationships between various energy consumption and socioeconomic variables.

The traditional method used in this empirical example is linear regression. Linear regression models are a fundamental tool in statistical analysis, known for their simplicity and ease of interpretation. In this context, linear regression is used to establish relationships between the dependent variable, the ecological footprint (*efp*), and the independent variables, which include hydropower consumption (*hydro*), biomass consumption (*bio*), GDP per capita (*gdpc*), energy consumption per capita (*en*), trade openness (*trade*), and urbanization (*urban*). Linear regression helps to quantify the strength and direction of the relationships between these variables. For example, it can reveal how an increase in GDP per capita or energy consumption per capita correlates with changes in the ecological footprint. This method provides a clear and straightforward way to understand the potential impacts of each independent variable on the ecological footprint, making it an invaluable tool for initial explorations and establishing baseline relationships.

While linear regression models offer valuable insights, they have limitations, especially when dealing with complex, nonlinear relationships. To address these limitations and improve predictive accuracy, this empirical example also employs advanced ML techniques, specifically SVM and Bagging.

1. **SVM:** SVM is a powerful supervised learning model commonly used for both classification and regression tasks. In this example, SVM is employed to predict the ecological footprint based on the independent variables. The strength of SVM lies in its ability to handle high-dimensional data and its effectiveness in finding the optimal hyperplane that separates the data into different classes or regression outcomes. By transforming the data into a higher dimension, SVM can identify complex patterns and relationships that might be missed by linear models. This capability makes SVM particularly useful in capturing the nuanced interactions between variables such as energy consumption, economic indicators, and urbanization, leading to more accurate and reliable predictions of the ecological footprint.
2. **Bagging (Bootstrap Aggregating):** Bagging is an ensemble learning technique designed to enhance the stability and accuracy of ML models. This method works by generating multiple versions of a model using different subsets of the training data, and then averaging their predictions to produce a final result. The primary advantage of bagging is its ability to reduce variance and mitigate overfitting, which is a common issue in complex models. By combining the strengths of multiple models, bagging can provide more robust and generalizable predictions. In this example, bagging is applied to create an ensemble of regression models, each trained on different subsets of the data. The aggregated predictions from these models offer improved accuracy in estimating the ecological footprint, as they effectively capture the variability and underlying patterns in the data.

The use of SVM and Bagging in this empirical example is driven by the need to overcome the limitations of traditional linear regression models. While linear regression provides a good starting point for understanding basic relationships, it often falls short in capturing nonlinear interactions and handling large, complex datasets. ML techniques, on the other hand, excel in these areas. SVM's ability to handle nonlinear relationships and high-dimensional spaces makes it particularly suitable for this example, where the interactions among variables such as hydropower consumption, biomass consumption, and economic indicators are likely to be complex. By finding the optimal hyperplane in a transformed feature space, SVM can deliver more precise predictions of the ecological footprint. In addition, the ensemble approach of bagging helps in creating a more resilient model by averaging out the biases and variances of individual models. This is particularly beneficial in environmental studies where the data can be noisy and heterogeneous. Bagging enhances the overall predictive performance, making it a crucial tool for reliable forecasting.

Figure 1 illustrates the performance metrics for different models, including Mean Squared Error (MSE), R-squared (R2), and Root Mean Squared Error (RMSE). As shown, ML techniques show superior performance compared to traditional methods in several key areas.

From Fig. 1, it is evident that ML models, particularly SVM, outperform traditional methods such as linear regression. The SVM model shows significantly better performance in terms of R^2 and RMSE, indicating higher accuracy and better predictive power. This aligns with the findings of Lv et al. (2023).

Given the superior performance of the SVM, feature importance scores are reported for this model in Fig. 2. A feature importance analysis helps identify which variables have the most significant impact on the ecological footprint, guiding policy decisions and research priorities.

The SVM model's feature importance scores indicate that energy consumption per capita (*en*), biomass consumption (bio), and GDP per capita (*gdpc*) are the most critical variables influencing the ecological footprint. This insight can be valuable for understanding which factors should be prioritized in efforts to reduce the ecological footprint and promote sustainable energy practices.

SVM offers several advantages over other ML algorithms. Firstly, this method aims to directly maximize generalizability (i.e., accuracy). Because SVM is fundamentally a convex optimization problem, it always finds a global optimum, unlike Deep Neural Networks (DNNs). Additionally, by adjusting the margin with a cost parameter, SVM maintains robustness against outliers. Finally, the use of kernel functions allows SVM to flexibly handle complex nonlinear decision boundaries, effectively modeling nonlinear relationships.

In conclusion, this empirical example demonstrates the effectiveness of ML techniques in analyzing the impacts of hydroelectric and biomass energy consumption on the ecological footprint. The superior performance of ML models, particularly SVM, in predicting the ecological footprint underscores the potential of these advanced techniques in energy research. Policymakers can leverage these insights to develop targeted strategies for reducing the ecological footprint. For example, prioritizing energy efficiency measures and promoting the use of biomass and hydropower can significantly contribute to environmental sustainability.

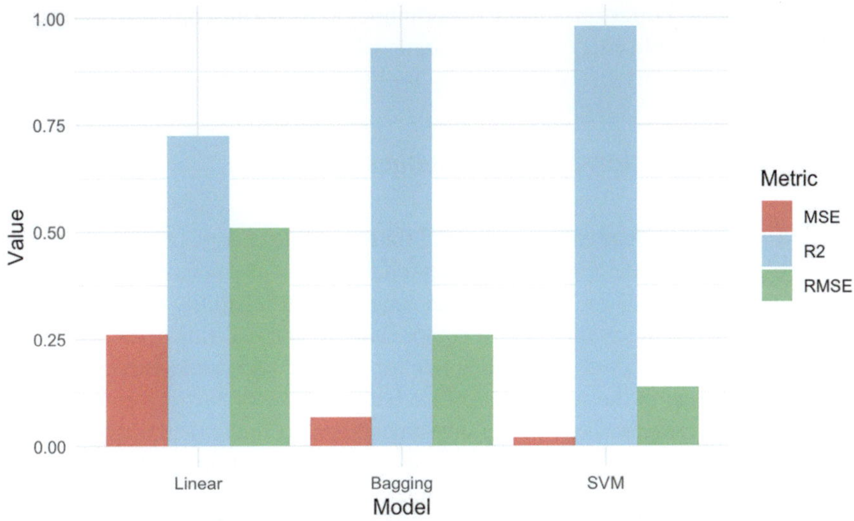

Fig. 1 Performance metrics for different models

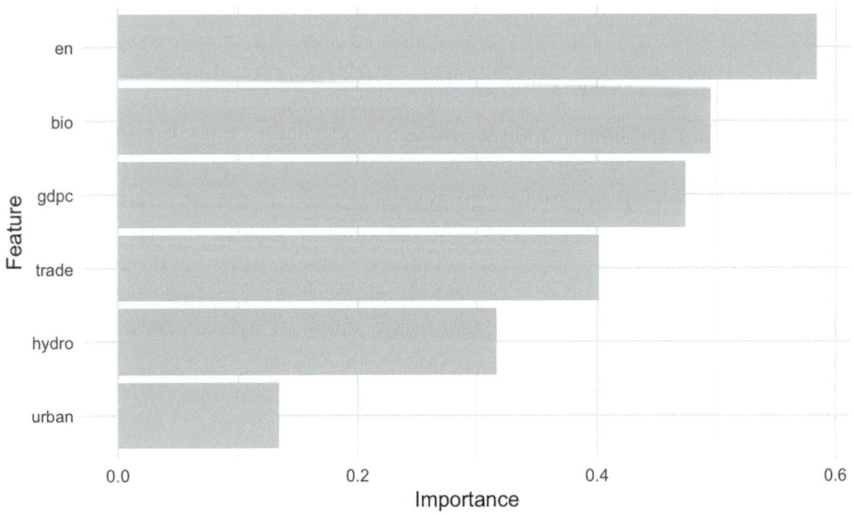

Fig. 2 Feature importance scores from the SVM model

Furthermore, understanding the importance of economic and urbanization factors can help in designing comprehensive policies that address the multifaceted nature of environmental challenges. Moreover, by integrating ML techniques into energy research, researchers can achieve more accurate predictions, deeper insights, and, ultimately, more effective policies for sustainable development. The ongoing advancement in CI offers promising opportunities to tackle the complex issues associated with renewable energy and environmental sustainability.

6 Future Directions and Challenges

Despite the significant advancements in CI for renewable energy, several challenges and opportunities remain. One of the key challenges is the need for high-quality data to train and validate CI models. The accuracy and reliability of these models depend heavily on the availability of comprehensive and real-time data, which can be difficult to obtain in certain regions or for specific renewable energy sources. Another challenge is the integration of CI techniques with existing energy infrastructure, which requires substantial investments and technological upgrades. This includes the deployment of advanced sensors and data collection systems, as well as the development of robust communication networks to facilitate real-time data exchange. Additionally, the integration process may face regulatory and policy hurdles, as governments and regulatory bodies need to create supportive frameworks that encourage the development and deployment of CI technologies.

Future research should focus on developing more sophisticated CI algorithms to handle the increasing complexity and scale of renewable energy systems. DL, a subset of ML that involves multiple layers of ANNs, holds significant promise in this regard (Goodfellow et al., 2016). By learning hierarchical representations of data, DL models can capture complex patterns and relationships that simpler models might miss.

Reinforcement learning, another promising area of CI, involves training models through trial and error to develop optimal strategies for energy management. These models can learn from interactions with the environment, allowing them to adapt to changing conditions and make more informed decisions. For instance, reinforcement learning can be used to optimize the operation of hybrid renewable energy systems, ensuring efficient energy distribution and storage.

The integration of CI with emerging technologies such as the Internet of Things (IoT) and blockchain can create more resilient and transparent energy systems (Sikorski et al., 2017). By combining these technologies with CI, it is possible to enhance the efficiency, reliability, and security of renewable energy systems.

7 Policy and Regulatory Implications

The adoption of CI in renewable energy systems also has significant policy and regulatory implications. Governments and regulatory bodies need to create supportive frameworks that encourage the development and deployment of CI technologies. This includes investing in research and development, providing incentives for innovation, and establishing standards for data sharing and interoperability. Policies should also address potential ethical and social issues related to the use of CI, such as data privacy, cybersecurity, and the impact on employment in the energy sector.

- *Data Privacy and Cybersecurity*
  The widespread use of CI technologies in the energy sector raises concerns about data security and the potential for cyberattacks on critical energy infrastructure. CI systems rely heavily on vast amounts of data to function effectively, which makes them susceptible to breaches and misuse. Ensuring robust cybersecurity measures and data protection standards is essential to mitigate these risks. Governments and regulatory bodies must implement stringent cybersecurity protocols to protect sensitive data and critical infrastructure. This can include mandatory security audits, the establishment of cybersecurity response teams, and the promotion of best practices for data handling and storage. Additionally, policies should encourage the development of advanced cybersecurity technologies that can keep pace with the evolving threat landscape. This might involve investing in research on new encryption methods, anomaly detection systems, and AI-driven cybersecurity tools. By staying ahead of potential threats, policymakers can ensure that the adoption of CI in the energy sector does not compromise national security or consumer privacy.
- *Impact on Employment and Workforce Development*
  The deployment of CI technologies in the energy sector may lead to significant changes in the workforce. As some tasks become automated, there will be a shift in the types of skills required for energy sector jobs. ML models, for example, have been shown to outperform traditional methods in various applications within the energy sector, such as forecasting and optimization (Lv et al., 2023). This superior performance can result in the need for fewer manual interventions and more reliance on automated systems. Policymakers need to consider the potential impact on employment and provide support for workforce retraining and education to ensure a smooth transition to a more technologically advanced energy sector. This can include funding for training programs that teach workers new skills related to CI and renewable energy technologies, as well as incentives for companies that invest in workforce development. By proactively addressing the implications for the workforce, policymakers can help mitigate the negative effects of automation and ensure that the benefits of CI adoption are broadly shared.
- *Advantages of CI Over Traditional Methods*
  CI systems, including ML methods, offer vast improvements in performance over traditional methods in several key areas. Traditional methods often rely on static models and linear assumptions, which can be limiting in the dynamic and complex environment of renewable energy systems. In contrast, CI methods can adapt to changing conditions and learn from new data, leading to more accurate and reliable predictions and optimizations.
  Accurate forecasting is critical for the effective management and integration of renewable energy sources. Traditional forecasting methods, such as time series analysis and regression models, have limitations in handling the variability and intermittency of renewable energy sources like wind and solar power. ML mod-

els, on the other hand, can leverage large datasets to identify patterns and make more accurate predictions. For example, hybrid neural network models have been shown to significantly improve the accuracy of solar irradiance and wind speed forecasts, which are essential for optimizing the operation of renewable energy systems (Lv et al., 2023). CI methods also excel in optimization tasks, which are crucial for the efficient operation of renewable energy systems. Traditional optimization methods often struggle with the complexity and nonlinearity of energy systems, leading to suboptimal solutions. In contrast, CI techniques such as genetic algorithms, particle swarm optimization, and DL can handle complex optimization problems more effectively. These methods can optimize the placement and operation of renewable energy assets, minimize energy losses, and improve the overall efficiency of the energy system.

Effective management and integration of renewable energy sources into the power grid is another area where CI methods outperform traditional approaches. Traditional grid management systems are often not equipped to handle the variability and uncertainty of renewable energy. CI systems can use real-time data to dynamically adjust grid operations, ensuring stability and reliability. For example, ML algorithms can predict energy demand and supply fluctuations, allowing grid operators to make proactive adjustments and avoid blackouts or energy wastage.

– *Supportive Policy Frameworks*

To fully realize the benefits of CI in renewable energy, it is essential for governments and regulatory bodies to establish supportive policy frameworks. These frameworks should promote the adoption of CI technologies through a combination of incentives, standards, and regulations. Governments can encourage innovation in CI technologies by providing financial incentives such as grants, tax credits, and subsidies for research and development. These incentives can help reduce the financial risk for companies and researchers developing new CI applications for renewable energy. Additionally, governments can support innovation hubs and collaborative research initiatives that bring together industry, academia, and government agencies to advance CI technologies.

Data is a critical resource for CI systems, and effective data sharing is essential for their success. Governments can establish standards for data sharing and interoperability to ensure that data from different sources can be easily integrated and used by CI systems. These standards can include guidelines for data formats, communication protocols, and data governance practices. By promoting data sharing and interoperability, policymakers can facilitate the development of more advanced and effective CI applications. Moreover, as CI technologies become more prevalent in the energy sector, it is important to address potential ethical and social issues. Policies should ensure that the deployment of CI technologies does not exacerbate existing inequalities or create new ones. For example, access to CI technologies and the benefits they provide should be equitable, and measures should be taken to prevent discrimination or bias in CI algorithms. Furthermore, the potential impact on employment should be carefully managed. Policymakers can implement social safety nets and support programs for workers who may be displaced by automation. This can include unemployment benefits, retraining programs, and job placement services. By addressing these ethical and social considerations, policymakers can ensure that the transition to CI-driven renewable energy systems is just and inclusive.

CI represents a powerful tool for addressing the complex challenges associated with renewable energy systems. By leveraging advanced algorithms and data-driven approaches, CI can enhance the forecasting, management, and integration of renewable energy sources, leading to more efficient and sustainable energy systems. However, realizing the full potential of CI in renewable energy requires ongoing research, substantial investments, and supportive policy frameworks. As the demand for clean energy continues to grow, the synergy between renewable energy and CI will play a crucial role in shaping a sustainable and resilient energy future. Governments and regulatory bodies have a critical role to play in this transition. By implementing supportive policies and addressing potential challenges, they can ensure that the benefits of CI technologies are maximized while minimizing any negative impacts. This includes investing in cybersecurity, supporting workforce development, promoting data sharing and interoperability, and addressing ethical and social issues. With the right policy frameworks in place, CI can help unlock the full potential of renewable energy, leading to a cleaner, more efficient, and more sustainable energy system for all.

References

Acaroğlu, H., Kartal, H. M., & García Márquez, F. P. (2023). Testing the environmental Kuznets curve hypothesis in terms of ecological footprint and CO_2 emissions through energy diversification for Turkey. *Environmental Science and Pollution Research, 30*(22), 63289–63304.

Adebayo, T. S., Akadiri, S. S., Adamu, Y., & Olasehinde-Williams, G. O. (2022). Disaggregated energy consumption and ecological footprint: Proposing an SDG framework for newly industrialised countries. *International Journal of Oil, Gas and Coal Technology, 32*(4), 417–439.

Adebayo, T. S., Akadiri, S. S., Altuntaş, M., & Awosusi, A. A. (2023). Environmental effects of structural change, hydro and coal energy consumption on ecological footprint in India: Insights from the novel dynamic ARDL simulation. *Environment, Development and Sustainability, 25*, 14309–14332.

Alnour, M., Ali, M., Abdalla, A., Abdelrahman, R., & Khalil, H. (2022). How do urban population growth, hydropower consumption and natural resources rent shape environmental quality in Sudan? *World Development Sustainability, 1*, 100029.

Alsaleh, M., & Abdul-Rahim, A. S. (2023). Does hydropower growth threaten fish security under the pathway of sustainable development? Evidence from European Union economies. *Energy & Environment, 34*(1), 78–98.

Aydin, M., Koc, P., & Sahpaz, K. I. (2023). Investigating the EKC hypothesis with nanotechnology, renewable energy consumption, economic growth and ecological footprint in G7 countries: Panel data analyses with structural breaks. *Energy Sources, Part B: Economics, Planning, and Policy, 18*(1), 2163724.

Bello, M. O., Solarin, S. A., & Yen, Y. Y. (2018). The impact of electricity consumption on CO_2 emission, carbon footprint, water footprint and ecological footprint: The role of hydropower in an emerging economy. *Journal of Environmental Management, 219*, 218–230.

Bildirici, M. E., & Gökmenoğlu, S. M. (2017). Environmental pollution, hydropower energy consumption and economic growth: Evidence from G7 countries. *Renewable and Sustainable Energy Reviews, 75*, 68–85.

Fathima, A. H., & Palanisamy, K. (2015). Optimization in microgrids with hybrid energy systems—A review. *Renewable and Sustainable Energy Reviews, 45*, 431–446.

Golpîra, H., Sadeghi, H., & Magazzino, C. (2023). Examining the energy-environmental Kuznets curve in OECD countries considering their population. *Environmental Science and Pollution Research, 30*, 94515–94536.

Goodfellow, I., Bengio, Y., & Courville, A. (2016). *Deep learning*. MIT Press.

Hong, T., Pinson, P., Fan, S., Zareipour, H., Troccoli, A., & Hyndman, R. J. (2016). Probabilistic energy forecasting: Global energy forecasting competition 2014 and beyond. *International Journal of Forecasting, 32*(3), 896–913.

Lv, S., Wang, L., & Wang, S. (2023). A hybrid neural network model for short-term wind speed forecasting. *Energies, 16*(4), 1841.

Magazzino, C. (2012). On the relationship between disaggregated energy production and GDP in Italy. *Energy & Environment, 23*(8), 1191–1207.

Magazzino, C. (2023a). *Ecological footprint, electricity consumption, and economic growth in China: Geopolitical risk and natural resources governance*. Empirical Economics.

Magazzino, C. (2023b). The inextricable link among climate change, pandemic, migrations, and geopolitics: How artificial intelligence can help. In J. A. Benítez-Andrades, P. García-Llamas, A. Taboada, L. Estévez-Mauriz, & R. Baelo (Eds.), *Global challenges for a sustainable society. EURECA-PRO 2022. Springer proceedings in earth and environmental sciences* (pp. 216–222). Springer.

Magazzino, C. (2024). *AI for Green energy production, reference module in chemistry, molecular sciences and chemical engineering*. Elsevier.

Magazzino, C., Mele, M., Schneider, N., & Vallet, G. (2020). The relationship between nuclear energy consumption and economic growth: Evidence from Switzerland. *Environmental Research Letters, 15*.

Magazzino, C., Mele, M., Schneider, N., & Shahbaz, M. (2021). Can biomass energy curtail environmental pollution? A quantum model approach to Germany. *Journal of Environmental Management, 287*, 112293.

Magazzino, C., Gallegati, M., & Giri, F. (2023). The environmental Kuznets curve in a long-term perspective: Parametric vs semi-parametric models. *Environmental Impact Assessment Review, 98*, 106973.

Mehmood, U. (2022). Biomass energy consumption and its impacts on ecological footprints: Analyzing the role of globalization and natural resources in the framework of EKC in SAARC countries. *Environmental Science and Pollution Research, 29*(12), 17513–17519.

Mohsin, M., Orynbassarov, D., Anser, M. K., & Oskenbayev, Y. (2023). Does hydropower energy help to reduce CO_2 emissions in European Union countries? *Environmental Development, 45*, 100794.

Munasinghe, M. (2004). *Sustainomics: A transdisciplinary framework for making development more sustainable. International Society for Ecological Economics*. Internet Encyclopaedia of Ecological Economics.

Pata, U. K., & Aydin, M. (2020). Testing the EKC hypothesis for the top six hydropower energy-consuming countries: Evidence from Fourier Bootstrap ARDL procedure. *Journal of Cleaner Production, 264*, 121699.

Pata, U. K., Yurtkuran, S., Ahmed, Z., & Kartal, M. T. (2023). Do life expectancy and hydropower consumption affect ecological footprint? Evidence from novel augmented and dynamic ARDL approaches. *Heliyon, 9*(9).

Perone, G. (2024). The relationship between renewable energy production and CO_2 emissions in 27 OECD countries: A panel cointegration and Granger non-causality approach. *Journal of Cleaner Production, 434*, 139655.

Raza, M. Q., Khosravi, A., & Lee, H. W. (2016). A review on artificial intelligence-based load demand forecasting techniques for smart grid and buildings. *Renewable and Sustainable Energy Reviews, 50*, 1352–1372.

Saadaoui, H., Dogan, M., & Omri, E. (2024). The impacts of hydroelectricity generation, financial development, geopolitical risk, income, and foreign direct investment on carbon emissions in Turkey. *Environmental Economics and Policy Studies, 26*(2), 239–261.

Sikorski, J. J., Haughton, J., & Kraft, M. (2017). Blockchain technology in the chemical industry: Machine-to-machine electricity market. *Applied Energy, 195*, 234–246.

Soytas, U., Magazzino, C., Mele, M., & Schneider, N. (2022). Economic and environmental implications of the nuclear power phase-out in Belgium: Insights from time-series models and a partial differential equations algorithm. *Structural Change and Economic Dynamics, 63*, 241–256.

Wang, C., Raza, S. A., Adebayo, T. S., Yi, S., & Shah, M. I. (2023a). The roles of hydro, nuclear and biomass energy towards carbon neutrality target in China: A policy-based analysis. *Energy, 262*, 125303.

Wang, Q., Yang, T., & Li, R. (2023b). Does income inequality reshape the environmental Kuznets curve (EKC) hypothesis? A nonlinear panel data analysis. *Environmental Research, 216*, 114575.

Yasmeen, R., Zhaohui, C., Shah, W. U. H., Kamal, M. A., & Khan, A. (2022). Exploring the role of biomass energy consumption, ecological footprint through FDI and technological innovation in B&R economies: A simultaneous equation approach. *Energy, 244*, 122703.

Renewable Energy Through a New Education Vision

Shoukat Iqbal Khattak

1 Introduction

The world's energy scene is changing dramatically, fueled by the critical push for climate action and sustainability. Cleaner, renewable energy sources are increasingly replacing traditional fossil fuel-based energy systems; it is critical to reimagine and reformulate educational paradigms to support this transition. Education fosters awareness and innovation, and adopting renewable energy technologies cannot be overstated. This chapter delves into the intersection of education and renewable energy, exploring how educational initiatives can catalyze a sustainable energy future.

Historically, energy education has primarily focused on extracting, producing, and consuming fossil fuels. This approach was reflective of the industrial demands of the twentieth century, which prioritized economic growth and energy security. The late twentieth and early twenty-first centuries have seen increasing environmental challenges. Notably, those related to climate change push the necessity to shift toward sustainable practices. Renewable energies like solar, wind, hydro, and bioenergy are now crucial in reducing the environmental footprint of traditional fossil fuels. Incorporating lessons on renewable energy into educational programs is essential for empowering the next generation to lead the shift toward cleaner energy solutions. United Nations Sustainable Development Goal 7.2 targets a greater reliance on renewable energy worldwide, a goal that demands technological innovation and education that promotes renewable technologies. The section discovers the growth of energy education, underlining the significance of renewables today and the educational methods needed to advance a sustainable energy future. The goal is to lay the groundwork for a well-informed, creative, and environmentally mindful

S. I. Khattak (✉)
College of Business Administration, Jimei University, Xiamen, China
e-mail: 201961000012@jmu.edu.edu.cn

S. S. Oncel (ed.), *A Green Vision Towards a Renewable Energy Future*,
https://doi.org/10.1007/978-3-031-93760-6_7

society, focusing on renewable energy technologies and how education can facilitate their broader acceptance.

1.1 Background

1.1.1 Overview of the Global Energy and Educational Landscape

The historical evolution of energy education is meticulously traced from its early focus on traditional energy systems, such as coal and oil, to its contemporary emphasis on sustainability and renewable energy. In the mid-twentieth century, energy education predominantly concentrated on meeting the burgeoning energy demands of industrialized societies by using fossil fuels. However, with the advent of the environmental movement in the 1970s and the subsequent recognition of climate change, the focus shifted decisively toward renewable energy and sustainable practices. This paradigm shift has profoundly impacted societal attitudes and behaviors, fostering heightened responsiveness to the environmental consequences of energy consumption and the imperative for sustainable solutions (Cirella et al., 2021).

It is vital to shift toward renewable energy sources to combat climate change and move toward a sustainable future. This direction supports the United Nations' Sustainable Development Goal 7.2, which aims to rely more on renewables by 2030. Utilizing these resources diminishes dependence on fossil fuels, mitigates climate change's impacts, and supports energy security, economic advancement, and environmental resilience (De Rosa et al., 2024).

1.1.2 The Need for a New Vision in Education to Address the Challenges of Renewable Energy Adoption

Emerging technologies within the physical environment emphasize energy demand, supply, storage, and integration, contributing significantly to the global energy landscape. The authors underscore the importance of energy education in fostering the integration of these advancements to create sustainable urban environments. By presenting notable progress in innovative grid technologies, energy-efficient building designs, and renewable energy sources, the authors highlight how education is the practical application of these innovations. A case study from the Netherlands illustrates how informed approaches can effectively address energy challenges in urban areas. The historical context of energy education is discussed, tracing its evolution from focusing on fossil fuels and traditional energy systems to the contemporary emphasis on renewable energy and sustainability. This perspective underscores the transformative influence of energy education on societal attitudes and behaviors regarding energy consumption. Raising awareness about renewable energy benefits and challenges has effectively reduced energy use and fostered sustainability (Lampropoulos et al., 2020).

Cirella et al. (2021) underscore the critical interplay among urban morphology, landscape ecology, and the global energy landscape. Metropolitan areas, identified as the primary energy consumers, account for more than 60% of global energy usage. Incorporating renewable energy like solar and wind into city infrastructures is key to reducing greenhouse gases. Also, smart grids are crucial in making energy use and storage in cities more efficient.

Experts argue that the current challenges in the adoption of renewable energy can be mitigated through several strategies. For instance, Cirella et al. (2021) advocate for a transformative approach to energy education that integrates technical, ecological, and socioeconomic perspectives. They argue for developing interdisciplinary curricula that extend beyond traditional engineering and technical training, incorporating subjects such as urban planning, landscape ecology, and the social sciences. This innovative educational vision aspires to cultivate critical thinking, innovation, and collaborative problem-solving skills among future energy professionals, equipping them to design and implement effective and sustainable energy solutions in complex urban environments.

1.2 Objectives and Scope

This chapter aims to discuss education's crucial function in supporting the worldwide shift toward renewable energy. It offers an in-depth look at how educational endeavors can boost consciousness, pioneering developments, and the practical application of clean energy solutions. By investigating the evolution of energy-focused education, assessing renewable energy's contribution to vindicating climate change, and examining the inclusion of innovative technologies in infrastructure, this text advocates for an updated educational model that champions ecological sustainability and responsible stewardship. Through detailed case studies and an analysis of multidisciplinary strategies, the chapter seeks to demonstrate the critical capacity of energy education to foster a future that is both sustainable and resilient.

Structured to begin with a summary of the current global energy and educational contexts, the narrative then shifts to emphasize the status of renewable energy and the pivotal role that education must play in its expansion. It covers a range of renewable energy technologies, including solar, wind, and other alternative sources, alongside educational practices that facilitate their wider acceptance. Moreover, the chapter calls for an integrated curriculum that blends technical knowledge with environmental and socioeconomic insights. Featuring examples of effective educational schemes and projects, this segment intends to show how educational efforts can nurture a culture geared toward sustainability. Concluding with a recap of the main arguments and contemplation on the overarching outcome of renewable energy education on the broader society and environment, the chapter aims to leave a lasting impression on the significance of informed and proactive educational policies in renewable energy.

2 The Importance of Education in Renewable Energy

Education is essential for accelerating the shift toward renewable energy sources. It helps to raise awareness, offers necessary training, and readies stakeholders to support and champion the cause of renewable energy. This form of education encompasses various methods, including structured learning in academic institutions, vocational training programs, informal community education efforts, and campaigns to increase public knowledge and involvement (Daoudi, 2024).

Education is critical in transitioning toward sustainable energy and achieving various Sustainable Development Goals (SDGs). By mitigating environmental degradation, education fosters greater awareness and innovation at the grassroots level, leading to the development of cleaner technologies. This, in turn, generates numerous vocational opportunities, promoting social inclusion and fostering partnerships among the public, private, and civil sectors. Specifically, the goal of quality education (SDG 4) is intricately linked to partnerships for the goals (SDG 17), sustainable vocational prospects (SDG 8), industry, innovation, and infrastructure (SDG 9), and climate action (SDG 13). As education levels rise, there is a notable increase in renewable energy consumption, indicating a growing preference for green energy solutions among the educated populace. The economic growth pattern also supports this shift, emphasizing the necessity of a phased shift from fossil fuels to renewable energy sources. Policymakers should, therefore, consider this gradual shift to ensure sustainable economic growth supported by collaborative efforts among various stakeholders (Zafar et al., 2020).

2.1 Understanding Renewable Energy

2.1.1 Basic Concepts and Types of Renewable Energy

The shift toward new energy models offers a substantial opportunity to decrease greenhouse gas emissions and promote creative strategies in energy generation and use (Huijben et al., 2022).

2.1.2 Renewable Energy Systems

The proposed energy education programs are expected to form the core of a dynamic platform that will evolve alongside advancements in relevant fields. The section investigates the basic energy infrastructures and proposes a strategic plan to move toward a future where energy is used sustainably for the environment.

2.1.2.1 Electrochemical Energy

Electrochemical energy covers a wide array of processes, including systems for energy conversion such as fuel cells, devices for electrolysis, various types of batteries, sensors for detecting chemical gases, and electrochemical pumps and separators. For example, batteries play a role in transforming renewable electrical energy into chemical energy and back again. This transformation involves charging, powered by an external voltage source, and discharging, which occurs naturally due to a decrease in Gibbs free energy. Such battery systems are crucial for storing energy derived from solar or wind sources. Ongoing research aims to craft new materials that boast enhanced ionic conductivity and greater charge storage capacity. Meanwhile, fuel cells are at the forefront of converting chemical energy from various fuels directly into electrical energy through a process energized by lowering Gibbs free energy during redox reactions, which occur at the fuel cell's anode and cathode. The energy harnessed from this process is due to the flow of electrons through an external circuit, which is blocked within the cell by solid electrolytes. An example of this technology includes solid oxide fuel cells (SOFCs), which utilize yttria-stabilized zirconia as their electrolyte component (Nowotny et al., 2018).

A significant drawback of Solid Oxide Fuel Cells (SOFCs) is their requirement for high operating temperatures. It is essential to develop solid electrolytes that have improved ionic conductivity at lower temperatures while maintaining low electronic conductivity to overcome this. One promising approach is the creation of low-dimensional interface structures (LDISs), which play a crucial role in advancing electrode reactions. Advancements in technology depend heavily on the use of electrochemical devices. These devices require materials that are efficient in catalyzing electrode reactions and possess high electrical conductivity. A critical task in this field is the innovation of materials that offer better performance, longer lifespan, and cost efficiency, all while functioning effectively at reduced temperatures (Nowotny et al., 2018).

2.1.2.2 Photoelectrochemical Energy

The primary focus of research in photoelectrochemical energy conversion centers on using light to catalyze the separation of water into hydrogen and oxygen. This technique, known as light-induced water oxidation, can achieve complete water splitting to produce hydrogen and also facilitates the partial oxidation needed to break down organic compounds. The success of this method relies on the use of oxide semiconductors, known for their durability in aquatic settings. In this process, oxygen is produced at the anode, while hydrogen gas is generated at the cathode, offering a method to harness solar energy in hydrogen fuel. Enhancing the light absorption ability of oxide semiconductors, such as TiO_2, through electronic structure modifications is a critical study area. The efficiency of photoelectrochemical systems is significantly influenced by the characteristics of the interface layer, including the gradient of defect concentrations, the required electric fields for efficient charge separation, and the abundance of active sites on the surface. To advance this technology, surface engineering is crucial for creating interface layers with better performance, necessitating a deeper understanding of how these properties influence charge transfer during water oxidation. Considerable efforts are being made to innovate in the field of solar energy, particularly through the development of new oxide and oxy-nitride semiconductor materials. These advancements aim to enable the transformation of solar energy into viable fuels, notably hydrogen. Adopting vehicles powered by hydrogen fuel is crucial for minimizing CO_2 emissions. Nonetheless, further advancements are needed across various hydrogen technologies, including solar cells designed for hydrogen production, fuel cells for hydrogen, and electric motors (Nowotny et al., 2018).

2.1.2.3 Solar Energy

Solar power is predominantly utilized for heating purposes. Additionally, solar photons can be captured to create charge carriers in semiconductor junctions, leading to the direct generation of electricity or to the facilitation of hydrogen production from water through photochemical or photoelectrochemical processes. Silicon-based photovoltaic cells are the most widespread technology for transforming solar energy into electrical power. Significant advancements in the technology behind these silicon-based cells have dramatically decreased the costs associated with solar photovoltaics, rendering it a cost-competitive alternative to traditional fossil fuel-generated power (Nowotny et al., 2018).

2.1.2.4 Alternative Renewable Energy

Interest in renewable energy sources such as wind, tidal, and geothermal is rapidly growing, with significant advancements in their development. The expansion of wave and tidal energy technologies is receiving special attention for its potential, while geothermal energy is acknowledged for its constant electricity production capabilities. Furthermore, a novel approach that utilizes the difference in chemical concentration when freshwater rivers and oceans converge is being investigated. Applying the Nernst equation, this method transforms salt concentration differences into electrical energy using cutting-edge membrane technologies (Nowotny et al., 2018).

Eze et al. (2024) provide a comprehensive analysis of advancements in energy efficiency technologies for thermal systems, emphasizing the critical role of advanced materials and control strategies. They highlight that those thermal systems account for approximately 50% of global energy consumption, underscoring the need to enhance efficiency. The authors discuss innovations such as phase change materials (PCMs) and nanomaterials, which significantly improve the thermal performance of energy systems. Additionally, they explore sophisticated control techniques, including model predictive control (MPC) and artificial intelligence (AI), which optimize energy utilization and minimize wastage in thermal applications. These technological advancements are crucial for meeting rising energy demands sustainably while mitigating environmental impacts.

2.1.2.5 Hydrogen Energy

The execution of a hydrogen-based economy necessitates the development of a variety of new technologies. A hydrogen economy's most significant environmental impact is anticipated in the transportation sector, particularly with the advent of affordable hydrogen-powered vehicles. For sustainability, hydrogen production must employ environmentally benign technologies and utilize carbon-free renewable energy sources, fission energy, or, eventually, nuclear fusion (Nowotny et al., 2018).

2.1.2.6 Nuclear Energy

Power generation from nuclear energy depends on the process of nuclear fission. This method involves splitting heavy atomic nuclei into lighter ones, which releases energy. Although nuclear fission can be considered a potent short-term substitute for fossil fuels, it produces radioactive waste that is both long-lasting and hazardous. To tackle these issues, there is a need for improvements in the technology behind nuclear reactors and the creation of effective systems for managing radioactive waste. Following the incident at the Fukushima Daiichi Nuclear Power Station, there have been major strides toward enhancing the safety protocols of nuclear power stations. Today's fission reactors include sophisticated safety features designed to prevent accidents. Efforts are also being made to solve the problem of nuclear waste through the use of transmutation reactions. These reactions can convert radioactive materials that persist for long periods into stable and harmless elements. Moreover, research is ongoing in the pursuit of nuclear reactors that are inherently safe by design. The potential of nuclear fusion is viewed with great optimism as it could provide a supremely eco-friendly method for generating electricity sustainably and without producing greenhouse gases. Although it remains a technology in development, the successful deployment of nuclear fusion could revolutionize energy production by providing a vast amount of clean energy (Nowotny et al., 2018).

The Role of Education in Promoting Renewable Energy Awareness

The increasing awareness of climate change's potential for lasting harm to our planet highlights the urgent need for swift and significant measures to protect our global environment for the long term. This realization points to the vital necessity of implementing wide-ranging plans to safeguard the health of coming generations by ensuring access to pure water and air. Achieving a sustainable future requires a deep understanding of the interconnections among energy, the environment, food, water, and climate. A key achievement in the global fight against climate change is the Paris Agreement, which has the backing of 195 countries. This treaty aims to limit global warming to well below 2 °C, preferably to 1.5°, compared to levels before the industrial era. It represents a collective commitment to decrease dependence on fossil fuels and enhance the use of renewable energy resources. The agreement's broad endorsement underscores the serious worries regarding climate change's effects, such as the loss of ice in polar regions, the rise in sea levels, and the increased frequency of severe weather phenomena like droughts and floods, along with their health consequences (Nowotny et al., 2018).

Efforts to combat climate change necessitate the development of personnel proficient in implementing innovative energy solutions, including nuclear, hydrogen, and renewables. The United Nations initiative, Future Earth, focuses on comprehending global environmental shifts, establishing sustainable development objectives, and providing the global community with the insights needed to meet these goals. The initiative seeks to prevent detrimental environmental changes by fostering an inclusive approach that merges various fields of study and promotes a globally adaptable innovation framework. The engagement and backing of communities are fundamental to the success of this initiative. Recognizing the substantial influence of energy production on climate change, integrating discussions on energy into contemporary educational syllabuses is essential. The global effort predominantly aims to reduce the negative outcomes of climate change through eco-friendly energy production solutions. In response, the Sustainable Energy Network (SEN) is working toward creating a Sustainable Energy curriculum for universities and, possibly, high schools. Initiating this ambitious plan involves drafting a comprehensive textbook on Sustainable Energy that will form the basis of a continually evolving knowledge platform. This platform will combine the latest energy production breakthroughs with modern educational strategies, equipping the next generation for a seamless transition toward global sustainability. This summary reflects global initiatives aimed at addressing climate change, highlighting the key roles and goals of the Future Earth initiative (Nowotny et al., 2018).

Education is crucial in advancing the sustainable energy shift and decarbonization efforts. A study conducted at a Chilean university demonstrates that educational programs significantly impact students' awareness and attitudes toward renewable energy and sustainability. By incorporating energy transition, environmental impact, and sustainable practices into university curricula, educational institutions can effectively prepare the next generation of leaders and professionals to address the complexities of contemporary energy challenges. The findings suggest that a detailed education in sustainable energy leads to greater support among students for initiatives to lower carbon emissions and their increased involvement in projects focused on renewable energy. The study emphasizes the significance of merging diverse fields of study, like engineering and the social sciences, to give students a wide-ranging grasp of the dynamics involved in shifting toward sustainable energy. Equipped with this thorough understanding, students are more capable of contributing significantly to the creation of sustainable energy solutions (Véliz et al., 2023).

The Connection Between Education and Sustainable Development

The concept of sustainable development was first introduced in the influential "Our Common Future" report, also known as the Brundtland Report, published by the World Commission on Environment and Development in 1987. This report defined sustainable development as the capacity to "meet the needs of the present without compromising the ability of future generations to meet their own needs" (Brundtland, 1987).

Nousheen et al. (2020) highlight that sustainable development has become a guiding principle for a wide range of government bodies and civil society organizations worldwide. The notable Earth Summit in 1992, officially known as the United Nations Conference on Environment and Development (UNCED) held in Rio de Janeiro, marked the introduction of Agenda 21. This detailed strategic plan aimed at balancing socioeconomic development with the protection of natural resources to prevent environmental degradation. Furthermore, Agenda 21 emphasized the critical role of education at all levels of schools and universities and the professional development of teachers, both before and during their careers, as vital in equipping people with the skills necessary to address challenges related to sustainability.

Nousheen et al. (2020) advocate for an integrated teaching approach and recommend including key sustainable development aspects of environmental, economic, and social factors in the curricula for both new and current educators. The significance of education in promoting socioeconomic and environmental advancements for sustainable growth was highlighted during the 2002 World Summit on Sustainable Development. Although education by itself cannot accomplish sustainable development, it plays an essential role. In recognition of education's pivotal role in fostering sustainable development, the United Nations declared the period from 2004 to 2015 as the Decade of Education for Sustainable Development (DESD). This initiative aimed to weave sustainable development themes throughout all educational opportunities to shape the perspectives and actions of both individuals and the wider community.

De Rosa et al. (2024) assert that Education is the cornerstone for understanding, articulating, and solving societal challenges. The emphasis on education is notably reflected in the Sustainable Development Goals (SDGs), with a special focus on SDG 4, dedicated to promoting Quality Education. This objective strives to guarantee that everyone has access to equitable and comprehensive education, while also encouraging continuous learning opportunities for all. Recognizing education as a vital component for achieving sustainable development, it aims to enrich human resources by imparting vital knowledge and skills and heightens consciousness regarding the shift toward sustainable practices. Fulfilling these objectives demands an appreciation for the interdisciplinary approach required for sustainable development, which moves beyond conventional disciplinary and sectoral divides in tertiary education.

Leal Filho et al. (2019) posit that the Sustainable Development Goals (SDGs) represent a significant global commitment to sustainability, aiming to secure a legacy for future generations through collective actions by various stakeholders worldwide. Integrating SDGs into education can catalyze a paradigm shift in teaching, learning, and understanding sustainability, benefiting both students and institutions. This shift enhances the role of educators and extends human capital, ultimately increasing sustainable actions among individuals.

Wojuola and Alant (2019) highlight that the integral relationship between renewable energy and the goal of sustainable development prompts countries around the globe to embrace renewable energy technologies as a path to sustainability. The success of these technologies and securing community support are heavily dependent on education about renewable energy. As outlined in Chapter 36 of Agenda 21 by the United Nations Commission on Environment and Development, both formal and informal educational efforts are essential for changing societal attitudes toward sustainable development challenges. Education plays a key role in nurturing environmental consciousness, ethical values, and skills that are in harmony with sustainable practices, while also encouraging public participation in the decision-making process. Ensuring that all stakeholders are well educated is essential for the success of any sustainable energy program.

2.2 Environmental and Social Benefits

2.2.1 Educating About the Environmental Impacts of Fossil Fuels

Photosynthesis has been key in absorbing atmospheric carbon dioxide, leading to the formation of fossil deposits. Today, burning these fossils releases extensive CO_2, intensifying the greenhouse effect and accelerating climate change, which poses serious environmental threats. Human actions further compound these issues through pollution and degradation of natural resources, underscoring the urgency to combat human-driven climate change. The United Nations emphasizes this urgency through global conferences, advocating for practical solutions and international cooperation to curb greenhouse gas emissions. Most of these emissions stem from using fossil fuels in energy production, transport, and heating. Although reducing emissions is critical for climate protection, it poses financial challenges, especially for densely populated developing economies. Balancing the economic aspects of emission reduction is thus crucial. As awareness of climate change grows, there is a shift toward eco-friendly solutions, including green buildings, solar farms, and hydrogen-powered transport. Embracing renewable energy is central to these efforts, offering a path to mitigate environmental impacts significantly (Nowotny et al., 2018).

2.2.2 Social Benefits of Renewable Energy Education

Wüstenhagen et al. (2019) underscore the essential role of social acceptance in successfully implementing renewable energy technologies. Despite the technical and economic feasibility of such projects, including wind energy in Europe, they highlight those delays often arise due to local opposition, thereby emphasizing the critical importance of securing public support. Traditionally, energy education has concentrated on technical aspects; however, there is an increasing acknowledgment of the necessity to integrate social and environmental dimensions to manage public opposition effectively. The authors advocate for an interdisciplinary educational framework that combines technical training with insights from economics, policy, and social sciences. They demonstrate that educational programs incorporating public engagement and policy analysis better equip students to address the complex challenges associated with renewable energy adoption. Graduates from such programs are more proficient in designing socially acceptable projects, facilitating smoother project approvals and higher adoption rates.

2.2.3 Empowering Communities Through Knowledge

Véliz et al. (2023) highlight social factors' significant influence on public acceptance of renewable energy sources, particularly education. Communities with advanced educational attainments often favor renewable energy sources more favorably. The key to increasing environmental consciousness lies in education, often propelled by efforts from local community groups and neighborhood associations through information sharing and educational initiatives. Such increased awareness drives changes in the behaviors of these communities, encouraging a more urgent move toward adopting renewable forms of energy. As a result, communities with higher education levels are more likely to support environmentally friendly policies and practices. The current climate emergency has further mobilized youth worldwide, who are increasingly engaged in advocacy for intergenerational justice and demand swift, sustainable climate action. This activism underscores the importance of understanding today's university students' perceptions, views, and knowledge levels. "The upcoming generation of students is poised to be at the forefront of shaping policies and innovations in renewable energy and the wider efforts to reduce carbon emissions. Acknowledging their future influence as discerning consumers, key stakeholders, and leaders is essential. Investing in their development and understanding of social dynamics will empower them to make strategic choices vital for driving the energy sector's evolution."

Mahalik et al. (2021) emphasize the pivotal role of human capital in improving environmental quality and driving economic progress, particularly in developing countries. They caution that neglecting human capital can lead to a "race-to-the-bottom" scenario, where economic growth is pursued at the cost of environmental degradation. The core question raised in their research is: To what extent can human capital enhance environmental quality? The answer to this question hinges on both the level of education within a society and the awareness and knowledge regarding energy security and efficient resource use. Their studies present contrasting effects of education on environmental quality in different contexts. Studies indicate that in nations like Pakistan and Australia, lower educational attainment is linked with higher consumption of nonrenewable resources and increased carbon footprint, largely due to insufficient understanding of environmental issues. In contrast, individuals with higher levels of education demonstrate a greater capacity for adopting eco-friendly technologies and adhering to green regulations, reflecting the pivotal role of education in enhancing energy sustainability and minimizing carbon emissions. Individuals knowledgeable about the environmental impact of their choices tend to move away from fossil fuels and support the adoption of cleaner, alternative energy solutions. Therefore, investing in education is essential for fostering economic development while improving environmental standards.

2.3 Economic Implications

2.3.1 Economic Advantages of a Well-Educated Workforce in Renewable Energy

A well-educated workforce in the renewable energy sector is vital for driving economic growth, fostering innovation, and ensuring sustainable development. Switching to renewable energy is not only vital for tackling environmental issues but also opens up vast economic benefits. As the appetite for these technologies grows, so does the necessity for skilled professionals to develop, implement, and upkeep them, fueling job growth and skill enhancement in the process. The renewable energy field is globally recognized as a key contributor to generating employment. According to the International Renewable Energy Agency (IRENA), approximately 12 million people were employed in the sector in 2020, with projections suggesting a significant increase in the coming years. The transition to renewable energy creates employment across various domains, including manufacturing, installation, maintenance, and research and development. These roles span various skill levels and disciplines, from engineering and technical positions to project management and policy development (IRENA, 2019; IRENA, 2022). This diverse array of opportunities highlights the sector's potential to contribute to economic growth and sustainable development.

2.3.2 Job Creation and Skill Development

Job creation within the renewable energy sector frequently fosters local economic development. Renewable energy projects are commonly implemented across diverse regions, contributing to economic diversification and generating employment opportunities in rural and underserved areas. This regional development is essential for promoting balanced economic growth and mitigating socioeconomic inequalities. Additionally, the sector provides opportunities for upskilling and reskilling workers from declining industries, thereby supporting broader economic stability and growth (IRENA, 2022).

2.3.3 The Role of Education in Driving Innovation and Economic Growth

Education fosters innovation and economic growth within the renewable energy sector. A well-educated workforce is adept at developing and implementing new technologies, optimizing processes, and tackling the complex challenges of the energy transition. Educational institutions, research centers, and industry partnerships are vital for advancing innovation and supporting the renewable energy agenda. In particular, higher education institutions are leading in researching and developing renewable energy technologies. These institutions provide essential training and engage in pioneering research that propels technological advancements (IRENA, 2023).

The integration of education into the discourse on carbon emissions is primarily because increased carbon emissions, associated with rising economic activities, are primarily driven by human actions. Economic growth and education are intrinsically linked, with economists widely acknowledging the critical role of human capital in fostering long-term economic development. Education, skill development, and research are crucial components of human capital, significantly impacting productivity. An examination of human capital within growth frameworks reveals that advanced economies have transitioned from labor-intensive models to those that emphasize human capital and skill-intensive approaches. This shift has enabled many advanced countries to achieve sustainable economic growth while minimizing adverse environmental impacts. Developing countries can benefit from this model by enhancing their human capital and skilled workforce. Such an approach is pivotal for attaining higher economic growth and crucial for the sustainable management of environmental quality (IRENA, 2019, 2022).

Collaborative efforts among universities, research centers, and the private sector are integral to developing innovative solutions that enhance renewable energy systems' efficiency, cost-effectiveness, and scalability. Education in the renewable energy sector also fosters entrepreneurship and the establishment of new business ventures. Graduates with specialized knowledge and skills are more inclined to launch their enterprises, thus driving economic growth and job creation. These emerging businesses often concentrate on developing novel products and services for the renewable energy market, further stimulating economic activity and promoting innovation. Government policies and incentives significantly support education and innovation within the renewable energy sector. Investments in educational infrastructure, research and development funding, and tax incentives for renewable energy projects create a favorable environment for growth and innovation. Such policies advance the renewable energy sector and contribute to broader economic development objectives (IRENA, 2019).

Zafar et al. (2020) emphasize that enhancing the educational curriculum and fostering technological advancements in renewable energy domestically can substantially decrease reliance on technology imports through Foreign Direct Investment (FDI). This approach revitalizes domestic industries, stimulates research and development, creates new vocational opportunities, and mitigates CO_2 emissions. By reducing dependence on FDI, nations can more effectively manage their natural resources and preserve the natural habitats of flora and fauna. Therefore, effective policy design should concentrate on fortifying public property rights and incentivizing the adoption of cleaner technologies in current and future production processes. Aligning the goals of Quality Education (SDG 4) with Responsible Consumption and Production (SDG 12), Life Below Water (SDG 14), and Life on Land (SDG 15) ensures a holistic approach to sustainable development. Ultimately, education is crucial for implementing and sustaining cleaner technologies, contributing to environmental preservation and sustainable economic advancement.

3 Current State of Renewable Energy Education

3.1 Traditional Education Models

3.1.1 Overview of Existing Educational Curricula

Husanu et al. (2018) highlight that renewable energy education has become increasingly essential as the global focus shifts toward sustainable energy solutions. However, traditional educational models have relatively slowly adapted to these emerging needs. A comprehensive study conducted by the World Economic Forum reveals that, despite the growing demand for green skills, universities worldwide continue to produce more graduates oriented toward fossil fuel industries rather than renewable energy sectors. Specifically, approximately 68% of energy-related degrees are concentrated on fossil fuels, leaving only 32% dedicated to renewable energy. This disparity is particularly pronounced in regions outside the Asia-Pacific, where countries such as China and India have made notable advancements in supporting clean energy education. Conversely, many developing countries in Central and South America, Africa, and the Middle East continue to fall behind in offering degrees focused on renewable energy.

Geurts et al. (2024) note that the curriculum within many engineering programs has been slow to incorporate renewable energy topics comprehensively. The

American Society for Engineering Education reports that, although capstone projects related to green and renewable energy have contributed to some curriculum development, these projects are frequently supplementary rather than central components of the educational framework. Traditional engineering principles continue to dominate, with insufficient emphasis on the specific skills required for the renewable energy sector. Historically, educational research has predominantly focused on investigating students from an external perspective, treating them as passive recipients of predeveloped curricula. In the late 1970s and early 1980s, it became apparent that student perspectives were lacking in academic research and curriculum development. Since this realization, there has been notable advancement in recognizing the advantages and effects of direct student involvement in these areas. Various strategies have been developed to demonstrate student participation levels, ranging from minimal involvement to initiatives led by students. These strategies typically indicate a tiered approach, implying that engagement deepens at each successive level. Addressing earlier critiques, some frameworks now categorize student engagement into six specific roles: being a subject of research; providing information; advising, reviewing, collaborating as a coresearcher; or taking the lead, thereby highlighting the equal importance of these diverse contributions. This approach advocates for organizing research projects that allow students to choose roles aligned with their abilities, interests, and needs, thereby acknowledging the equal importance of all phases and roles in the research process.

3.1.2 Limitations and Gaps in Current Renewable Energy Education

The International Energy Agency (2023) identifies a significant limitation in current renewable energy education: the entrenched "carbon lock-in" within universities. Significant financial support from governments and fossil fuel companies has led to a profoundly ingrained infrastructure prioritizing fossil fuel education for decades. This entrenched infrastructure includes physical resources such as laboratories, equipment, and human capital, including faculty expertise, course materials, and textbooks predominantly oriented toward fossil fuels. Furthermore, traditional educational models exhibit a notable deficiency in interdisciplinary approaches. The challenges associated with renewable energy are inherently multidisciplinary, necessitating engineering, environmental science, economics, and policy expertise. Despite this, most educational programs remain compartmentalized, restricting students' capacity to comprehensively understand and address the multifaceted nature of renewable energy systems.

3.1.3 Examples of Successful Programs and Their Impacts

According to Pennsylvania State University (2021), despite existing challenges, there are notable examples of successful renewable energy education programs that have achieved substantial impacts. The Renewable Energy and Sustainability

Systems (RESS) program at Pennsylvania State University exemplifies excellence in integrating renewable energy education within its curriculum. The RESS program provides a comprehensive online master's degree that encompasses a range of renewable energy topics, including solar, wind, and bioenergy technologies, as well as sustainability management and policy. This program has successfully produced graduates proficiently prepared to address the complexities and demands of the renewable energy sector.

According to the Masdar Institute (2022), another exemplary case is the Masdar Institute of Science and Technology in the United Arab Emirates. This institution provides specialized programs in sustainable engineering and renewable energy, offering students practical experience in advanced research facilities. Graduates of the Masdar Institute have assumed influential roles in both the public and private sectors, contributing significantly to innovation and policy development within the renewable energy field.

To effectively facilitate the transition to a green and sustainable civilization, educational systems must embed sustainability throughout their curricula, emphasizing renewable energy and environmental stewardship. Schools are progressively incorporating sustainability topics into various subjects, ensuring students comprehensively understand green technologies and practices. For example, Stanford University's Energy Resources Engineering program combines theoretical instruction with practical field projects and industry collaborations, equipping students for real-world applications in the green economy (Stanford University, 2023).

Moreover, emphasizing practical skills and experiential learning is crucial for aligning education with industry requirements. Programs like the MSc in Wind Energy at the Technical University of Denmark provide students with hands-on experience and access to advanced research, essential for progress in wind technology and offshore projects (DTU, 2023). Similarly, the Master's in Sustainable Energy at the University of Copenhagen focuses on real-world applications through internships and research opportunities, ensuring that educational outcomes meet industry needs (University of Copenhagen, 2023).

Table 1 provides an overview of renewable energy education programs and their enrollment growth in various countries between 2010 and 2023.

Table 1 Renewable energy education in selected countries (2023)

Country	Type of program	Enrolment growth (2010–2023)	Key features of curriculum
Germany	Bachelor's in renewable energy systems	+45%	Strong emphasis on engineering and technology
United States	Master's in sustainable energy	+60%	Focus on policy, economics, and technology integration
China	Renewable energy engineering degree	+80%	Large-scale national initiatives for energy transition
Denmark	Certificate programs	+30%	Short-term programs focusing on wind energy

Source: Global Education Monitoring Report (2023)

3.2 *Policy and Institutional Support*

3.2.1 National and International Policies Promoting Renewable Energy Education

The International Renewable Energy Agency (IRENA, 2018) underlines the crucial role of national and international policies in advancing renewable energy education. Such policies aim to prepare future generations with the skills and knowledge to address climate change and enable the transition to sustainable energy systems. Numerous countries have enacted policies to integrate renewable energy topics into curricula and to support research and development in this domain. IRENA, for example, has launched several initiatives, including the Renewable Energy Learning Partnership (IRELP), which provides resources and training opportunities to enhance global renewable energy education.

The U.S. Department of Energy has implemented various funding programs to advance renewable energy education, focusing on minority-serving institutions and community colleges. These initiatives aim to cultivate a diverse workforce with the skills necessary for the clean energy sector. Similarly, the European Union has integrated renewable energy education into its overarching climate and energy policies, urging member states to incorporate renewable energy topics into their national education systems and vocational training programs (European Commission, 2020).

3.2.2 Role of Educational Institutions and Government Bodies

Educational institutions and government bodies are instrumental in advancing renewable energy education and aligning curricula with industry requirements. Higher education institutions (HEIs) are central to this effort, providing specialized renewable energy and sustainability programs. These programs offer theoretical instruction and emphasize practical skills through laboratory work, internships, and field projects. For instance, universities in Europe and the United States have established comprehensive curricula encompassing courses on renewable energy technologies, policy, and management. Government agencies contribute by facilitating renewable energy education through funding, policy frameworks, and industry partnerships. In India, the Ministry of New and Renewable Energy (MNRE) has formed collaborations with educational institutions to advance renewable energy research and training, resulting in the development of specialized courses and research centers dedicated to renewable energy technologies. Similarly, the U.S. Department of Energy (2022) supports educational institutions with grants and funding programs to enhance their ability to deliver high-quality renewable energy education.

International organizations, such as the International Renewable Energy Agency (IRENA), significantly contribute to advancing renewable energy education by providing a global platform for knowledge exchange and capacity building. IRENA's initiatives, including the Global Renewable Energy Education and Training

(GREET) program, are designed to recover the quality and accessibility of renewable energy education internationally (Irena & Ren, 2018).

3.2.3 Analysis of Funding and Resource Allocation

Funding and resource allocation are fundamental to successfully implementing renewable energy education programs. Governments and international organizations have acknowledged the need for significant investment to equip educational institutions with the resources to provide high-quality instruction. For example, the European Union has designated substantial financial support through its Horizon 2020 program to advance research and innovation in renewable energy, encompassing educational initiatives as a critical component (European Commission, 2020).

The Department of Energy (DOE) has allocated substantial funding to universities and community colleges to advance renewable energy education in the United States. These financial resources are dedicated to developing and enhancing renewable energy curricula, including creating new courses, upgrading laboratory facilities, and supporting research projects that provide practical experience to students (U.S. Department of Energy, 2022). Additionally, the DOE's Solar Training and Education for Professionals (STEP) funding program specifically addresses deficiencies in solar education by supporting the development of targeted training programs for the solar workforce (U.S. Department of Energy, 2022).

4 A New Vision for Renewable Energy Education

4.1 Integrative Curriculum Design

4.1.1 Developing Interdisciplinary Curricula

Huijben et al. (2022) discuss the importance of interdisciplinarity in engineering education, emphasizing the integration of diverse knowledge bases and practices to address real-world problems. Unlike multidisciplinary interactions, which may lack integrative processes, interdisciplinarity combines various disciplines' theories, concepts, and methods to tackle complex challenges. Furthermore, the transdisciplinary extends this integrative approach by applying knowledge across disciplines in collaboration with external stakeholders.

Networked learning is highlighted as a pivotal approach to promoting interdisciplinary education. This method encourages students to seek expertise beyond their immediate disciplinary boundaries and utilize these connections for problem-solving. Networked learning fosters collaboration among learners, tutors, and external resources, creating a shared language and facilitating co-creation. This approach equips learners with the necessary abilities to innovate and solve critical problems, focusing especially on issues related to sustainability, such as the shift toward renewable energy. Though its impact is promising, research on how connected learning influences sustainability changes is still scarce. Nonetheless, there is a clear call within the connected learning community for a greater emphasis on tackling significant issues and pursuing sustainable approaches, highlighting the potential of connected learning in advancing the shift to sustainable energy sources (Huijben et al., 2022).

Nowotny et al. (2018) emphasize the critical importance of integrating diverse scientific disciplines to understand energy conversion processes thoroughly and identify the functional materials crucial for optimizing system performance. For example, developing advanced solar cells requires a comprehensive knowledge base spanning multiple fields, including electrochemistry, surface chemistry, catalysis, solid-state science, and materials science. Although significant advancements in energy technologies have been made over the past sixty years, this progress has led to increasingly specialized fields such as solid-state chemistry, solid-state electrochemistry, materials interface science, and nanotechnology. The conceptual and theoretical gaps between these disciplines are substantial, underscoring the need for interdisciplinary collaboration to drive innovation in energy technology.

The increasing specialization of fields in energy technology, Nowotny et al. (2018) propose establishing a comprehensive and cohesive educational program designed to minimize disciplinary barriers. This program would achieve this by employing standardized terminology and consistent notation across various fields. The proposed program would be inherently multidisciplinary, with individual courses structured in an interdisciplinary manner to ensure the seamless integration of fundamental concepts from various scientific disciplines.

Furthermore, Nowotny et al. (2018) recommend that these courses are designed to be delivered through a collaborative teaching method, involving a team of faculty members who each contribute their distinct expertise. This educational strategy serves to widen students' understanding and fosters an appreciation for the complexity of energy sciences, along with their connection to vital areas such as water, food, the environment, and climate change. Creating an integrated learning atmosphere equips students more effectively to address the complex issues surrounding contemporary energy systems and their role in sustainable development.

When determining which disciplines to include in particular courses, the selection should be customized to fit the specific needs of the university's program, but it should generally cover essential subjects like materials science and engineering, electrochemistry, solid-state science, and surface chemistry. For degree programs with a focus on sustainable energy technologies, incorporating cutting-edge fields such as materials interface science, solid-state electrochemistry, nanotechnology, and principles of sustainability is critical (Nowotny et al., 2018).

Furthermore, Daoudi (2024) emphasizes that education on renewable energy should be all-encompassing, covering the spectrum from elementary through to tertiary levels, ensuring a thorough grasp of renewable energy concepts across all stages of learning. This broad approach aligns with insights from authoritative institutional reports, which stress the importance of embedding sustainability concepts throughout the educational journey. Horizontally, renewable energy education should also include the entire active population, ensuring that people from various sectors and backgrounds have the knowledge and skills essential to contribute to the renewable energy transition. Ensuring widespread involvement is key to encouraging a collective dedication to sustainable practices and the broad implementation of renewable energy technologies.

4.1.2 Vertical Level

Remarkable advancements have been made in the sector of higher education and vocational training concerning renewable energy. Numerous academic institutions now present specialized programs in renewable energy engineering, energy management, and other related fields. These courses aim to furnish students with the practical skills needed, alongside a thorough appreciation for the economic, social, and environmental impacts of renewable energy. Research and innovation are integral, with universities and research centers actively advancing solar, wind, and hydropower technologies. Vocational training complements this by equipping professionals with practical skills for implementing and maintaining renewable energy systems. However, renewable energy education is largely absent from primary and secondary curricula. Integrating renewable energy concepts across all educational levels, from primary school to higher education, fosters environmental responsibility and prepares future generations to contribute to a sustainable energy transition (Daoudi, 2024).

4.1.3 Horizontal Level

Daoudi (2024) emphasizes that to effectively promote renewable energy and eco-friendly habits, it is essential to launch widespread education and awareness initiatives. These should target a range of groups, including communities, companies, and nonprofits, aiming to highlight the perks of using renewable energy—like diminishing reliance on fossil fuels, fighting climate change, and bettering air quality. The messaging needs to stress the finite nature of fossil fuels and the environmental harm from greenhouse gases, underscoring the shift toward renewables as critical. These initiatives should also motivate personal actions toward energy saving and renewable adoption, leveraging platforms like social media and local events for broader reach. Such efforts are key to fostering an informed society ready to actively support a sustainable energy shift.

4.1.3.1 Incorporating Renewable Energy Topics into Science, Technology, Engineering, and Mathematics (STEM) Education

Huijben et al. (2022) emphasize that interdisciplinary teams enable individuals to learn from diverse disciplinary perspectives, producing integrative work that would not be possible in a single-disciplinary setting. This approach requires incorporating multiple disciplines, fostering interdisciplinarity by seeking a common methodology and theoretical foundation as a synthesis of the participating disciplines, ultimately leading to a shared language among participants. In interdisciplinary engineering education, students collaborate within learning networks—flexible, borderless, and innovative spaces that connect ideas, share problems, and integrate new knowledge through dialogue and social interactions. These networks, enriched by online and in-person interactions, allow students to develop flexibly alongside their peers, focusing on shared learning needs while remaining within their expertise. Combining interdisciplinary approaches with networked learning is an effective strategy for training change agents in energy transitions, as it encourages value creation across personal, community, and societal dimensions.

4.1.3.2 Case Studies of Innovative Curricula

Huijben et al. (2022) conducted a case study at a Dutch University of Technology, exploring the integration of networked learning in interdisciplinary education for sustainable technology development. The study engaged a diverse team of around 20 students from various engineering disciplines, who voluntarily participated in developing a sustainable bus prototype powered by formic acid. This initiative aimed to demonstrate technological feasibility and consider broader societal adoption and systemic change, aligning with insights from sustainability transition studies. Key findings from the study highlighted the effectiveness of networked learning in broadening students' perspectives beyond technical aspects. Workshops and lectures co-created with students, focusing on disciplines such as finance, marketing, and sustainability transitions, fostered a collaborative environment. External experts further enriched the learning experience by validating theoretical concepts and providing practical insights, enhancing the relevance of the educational content. The 'Open Space Studio' workshop, structured around co-creation principles, was pivotal in establishing a shared vision and organizational structure within the student team. Additionally, the integration of Lego Serious Play in the final workshop exemplified innovative pedagogical methods to enhance co-creation and equal participation among participants. This holistic educational approach deepened students' understanding of sustainability transitions and equipped them with essential teamwork, problem-solving, and interdisciplinary collaboration skills. Overall, the case study underscored the value of integrating networked learning methodologies in engineering education to prepare students for addressing complex societal challenges associated with sustainability transitions. By bridging disciplinary divides and fostering collaborative learning environments, such initiatives are crucial in nurturing future professionals capable of driving meaningful change in sustainable technology development.

Stanford University's "Energy Resources Engineering" program provides an interdisciplinary curriculum centered around renewable energy technologies such as solar, wind, and geothermal. Students participate in hands-on experiences, including simulations and fieldwork, to apply their knowledge in practical settings. The program collaborates with industry leaders to offer internships and career opportunities, resulting in a 50% growth in student enrollment between 2010 and 2023. With 85% of graduates securing employment in renewable energy within 6 months, many find roles in technology development and policy (Stanford University, 2023; IEA, 2023).

The "MSc in Wind Energy" at the Technical University of Denmark (DTU) stands out as a premier global program in wind energy education. It emphasizes turbine technology, wind farm management, and grid integration, providing students with opportunities for hands-on projects and access to advanced research. Through collaborations with international partners and industry experts, the program equips students for careers in both research and industry. Between 2010 and 2023, the program experienced a 35% increase in enrollment, with graduates making significant contributions to turbine innovation and the development of offshore wind farms (DTU, 2023; GWEC, 2023).

Tsinghua University's "Renewable Energy and Smart Grid Program" focuses on training students in integrating renewable energy technologies with smart grid systems. Backed by robust government support, the program provides hands-on experience with major national energy projects like wind and solar farms. Between 2010 and 2023, enrollment surged by 80%, reflecting China's ambitious renewable energy targets. Graduates play a key role in advancing China's carbon neutrality goals for 2060, engaging in energy management, and significant national initiatives (Tsinghua University, 2023; CNREC, 2023).

The "Master's in Sustainable Energy" program at the University of Copenhagen provides an interdisciplinary education in renewable energy, including solar, wind, and biomass technologies, as well as sustainability and energy policy. Emphasizing industry collaboration, the program offers students practical experience through internships and research projects in offshore wind and energy storage. Enrollment grew by 60% from 2010 to 2023, with graduates finding positions in policy, environmental NGOs, and renewable energy firms (University of Copenhagen, 2023; EREC, 2023).

4.2 *Experiential Learning and Hands-On Training*

4.2.1 Importance of Practical Experiences and Laboratory Work

Broman and Robèrt (2017) emphasize the indispensable role of practical experiences and laboratory work in renewable energy education. These experiences connect theoretical knowledge with practical application, enhancing understanding of renewable energy technologies. Hands-on training in laboratory settings enables students to experiment with and comprehend the operational intricacies of renewable energy systems, such as solar panels, wind turbines, and bioenergy plants. The hands-on work helps solidify theoretical ideas and improves critical thinking and problem-solving skills, which are essential for progress and innovation in the renewable energy industry. Lab experiments enable experiential learning, giving students the opportunity to see their choices and actions' impacts, thus resulting in a deeper and more lasting learning experience.

Moreover, laboratory-based education plays a vital role in developing technical skills and competencies that are highly valued in the renewable energy job market. Practical laboratory sessions involving installing, maintaining, and troubleshooting renewable energy systems equip students with hands-on skills directly transferable to industry settings. Kukreti (1998) underscores the significance of experiential learning in engineering education, noting that laboratory work provides an authentic context for students to develop technical skills, work collaboratively, and engage in problem-based learning scenarios. Laboratory experiences enhance comprehension of renewable energy technologies, equipping students for future professional challenges.

Véliz et al. (2023) contend that practical experiences such as internships, laboratory work, and participation in renewable energy projects are essential to educational strategies in the renewable energy sector. These opportunities provide students with the chance to put theoretical concepts into practice in actual situations, thereby enhancing their technical abilities and capacity to solve problems. For instance, students involved in renewable energy projects gain firsthand experience in designing, implementing, and monitoring sustainable energy solutions, such as solar panel installations or wind turbine assessments. Internships allow students to work alongside professionals in the renewable energy industry, gaining insights into industry practices and challenges. This exposure strengthens their technical competencies and cultivates a deeper appreciation for the practical implications of renewable energy technologies. Laboratory work is crucial in allowing students to conduct experiments, analyze data, and simulate scenarios related to renewable energy. Students develop practical skills applicable to their future careers in renewable energy research or industry by working with field equipment and technologies. Participating in renewable energy projects enriches their learning experiences as they collaborate with engineers, environmental scientists, and policymakers in multidisciplinary teams. This collaborative approach enhances their technical expertise and broadens their understanding of the socioeconomic and environmental impacts of renewable energy initiatives.

4.2.2 Partnerships with Industry for Internships and Apprenticeships

Collaborations between educational institutions and industry partners are critical for providing students with valuable internships and apprenticeships. These collaborations provide students with the opportunity to gain firsthand experience and understand industry practices by working on real projects. Salari et al. (2019) argue that internships and apprenticeships in renewable energy companies allow students to apply their academic knowledge in practical environments, enhancing their learning and preparing them for future careers. These experiences also expose students to the latest technologies and industry standards, making them more competitive in the job market.

Industry partnerships also benefit companies by creating a pipeline of well-trained, competent professionals who can contribute to the workforce immediately upon graduation. Companies often provide mentorship and training during these internships, helping students develop specific skills that are highly sought after in the renewable energy sector. Furthermore, these partnerships can lead to collaborative research and development projects, where students and industry professionals work together to address real-world challenges and innovate new solutions. The Renewable Energy Research Center (2018) highlights that this symbiotic relationship enhances the overall quality of renewable energy education and ensures that the curriculum remains relevant to industry needs.

Véliz et al. (2023) emphasize that schools and colleges can effectively merge theory with real-world application by encouraging collaborations among educational centers, community groups, and key participants from various industries. These partnerships enable students to gain hands-on experience while addressing real-world challenges faced by communities transitioning to renewable energy. Such initiatives prepare students for careers in sustainable energy and instill a sense of commitment to addressing global energy challenges through innovative and sustainable practices.

4.2.3 Examples of Hands-On Projects and Their Outcomes

Several educational programs have successfully integrated hands-on projects into curricula, yielding significant positive outcomes. One notable example is the Solar Training and Education for Professionals (STEP) program, which addresses gaps in solar training by offering comprehensive hands-on training modules. Song et al. (2020) report that these modules include installing, maintaining, and troubleshooting solar energy systems. Participants in the STEP program have noted increased confidence and competence in handling solar technologies, resulting in higher employability and improved job performance. The practical experience gained through these hands-on projects equips students with skills to excel in the renewable energy industry.

Another exemplary approach uses case studies and real-world problem-solving projects in renewable energy education. These initiatives typically include cooperation with community members and local enterprises, offering students a thorough insight into the economic, social, and ecological dimensions of renewable energy efforts. Anirudh and Dhinakaran (2020) describe a project at the University of California where students collaborated with a local community to design and implement a solar energy system. This project provided students with practical experience and benefited the community by reducing energy costs and promoting sustainable practices. Such projects enhance students' problem-solving abilities, teamwork, and communication skills while positively impacting society.

4.3 *Use of Technology in Education*

4.3.1 Digital Tools and Online Platforms for Renewable Energy Education

The globalization of education has necessitated integrating digital technologies, which were previously employed proactively for conducting classes, sharing resources, assessing student performance, and managing academic activities. However, the COVID-19 pandemic compelled educational institutions to rapidly adopt online teaching modalities to sustain educational continuity. While developed countries were generally well equipped to handle this transition, developing nations faced significant challenges in meeting these new requirements. Digital technologies have proven vital in maintaining educational operations in this critical period. This global crisis underscores the importance of international integration within the education system. Digital tools play a pivotal role in nurturing key skills required for career advancement, including problem-solving, logical reasoning, and understanding processes. These tools also equip students for a future that is both uncertain and heavily influenced by technology. The competencies and attributes developed through the use of these digital resources are vital for their future professional success. Moreover, digital resources and tools enrich the learning environment, making the educational process more interactive and engaging. They offer greater flexibility and customization of curricula tailored to individual student needs. By incorporating technology, students can become more involved in their learning. As children are increasingly accustomed to electronic devices, integrating these technologies into education can stimulate their interest and increase their engagement levels (Haleem et al., 2022).

Technological integration provides a more interactive learning experience, helping students stay focused and interested in the subject. Tools such as projectors, computers, and advanced technical equipment can make learning engaging and enjoyable. Class activities incorporating technological resources, oral presentations, and group work can enhance student participation, extending beyond verbal communication. Leveraging digital tools and computers allows students to actively engage in their education, guided by instructors for effective learning. Digital resources enable students to download information, upload content, and utilize Web 2.0 technologies like wikis, podcasts, and blogs to generate content, collaborate, and engage in peer assessment. These technologies also support innovative classroom strategies, such as gamification and flipped classrooms, which optimize learning experiences. The evolution of learning environments through technology offers a dynamic and inspiring educational experience, catering to individual learning paths and enhancing the overall instructional process (Haleem et al., 2022).

Incorporating digital tools and online platforms into renewable energy education has profoundly transformed the learning environment, offering novel methods to engage students and professionals. Learning management systems (LMS), educational apps, and online courses have boosted the accessibility and flexibility of renewable energy education. Platforms such as Coursera, edX, and Udemy provide specialized courses in renewable energy, enabling learners worldwide to acquire knowledge from esteemed experts and institutions. These platforms facilitate self-paced learning, enabling students to advance according to their own schedules and review difficult subjects whenever necessary. Furthermore, interactive tools, including digital simulations and gamified learning modules, enrich the learning experience by providing hands-on practice in a virtual setting, thus deepening the understanding of renewable energy concepts (World Economic Forum, 2024).

In renewable energy education, integrating digital tools and online platforms signifies a major advancement, introducing innovative approaches to learning and collaboration. Virtual laboratories and simulations are central to this educational evolution, offering immersive experiences that were previously unattainable. These tools allow users to interact with complex energy systems in a virtual environment, facilitating experimentation and learning without requiring physical infrastructure. For example, virtual models of wind farms and solar installations can illustrate the effects of various variables on energy output, thereby enhancing comprehension of system dynamics and operational challenges. This experiential learning approach is precious in the renewable energy sector, where practical experience is crucial for understanding technological intricacies and applications. Applying artificial intelligence (AI) and big data in personalized learning is critical in this transformation. Platforms powered by artificial intelligence have the capability to examine the unique learning behaviors of each student and adapt the material accordingly, enhancing both its relevance and efficiency in the educational process. For instance, AI algorithms can adjust educational modules to address areas where a learner may need additional support, offering supplementary resources as necessary. Big data further supports this personalization by allowing educators to monitor and analyze trends across large student groups, which helps develop targeted instructional strategies and identify necessary curriculum adjustments. In renewable energy, these technologies enable the creation of customized learning paths aligned with industry demands and emerging trends, ensuring that learners are equipped to meet the challenges of a rapidly evolving field (McKinsey & Company, 2023).

To serve educational objectives, digital learning environments employ a range of electronic tools and platforms, including but not limited to, social media, multimedia content, and smartphones. Integrating digital technology into education has significantly transformed the educational landscape, enhancing and improving learning experiences. Using technology as a foundational tool, this educational method facilitates the comprehensive delivery of course content, enhancing students' ability to absorb information swiftly and effectively. Teaching is primarily facilitated through technology in a digital classroom. Students use devices like laptops, tablets, and Chromebooks to engage with educational content. Rather than taking traditional handwritten notes, much of the curriculum is presented online via interactive and engaging platforms. Although education remains a form of communication, the internet has introduced new channels that expand the methods for delivering and accessing educational information. These digital media and virtual spaces function as facilitators of learning. Educational applications and websites play a crucial role in digital classrooms by enhancing students' learning experiences. Key components of a digital classroom include feedback loops and technology integration. Feedback loops are essential for providing students with real-time feedback from their instructors. Teachers can use these loops to give feedback based on various factors such as individual performance, lesson content, and group dynamics. In modern education, digital tools like PowerPoint presentations, video content, e-learning methods, and online training are being used more frequently to make classroom instruction interactive and participatory. This allows students to independently explore a wide range of subjects using internet resources and digital platforms. Traditional educational methods, such as color charts, graphs, and physical models, replace modern digital approaches. The reliance on textbooks, blackboard writing, and manual note-taking has diminished as digital technologies offer more dynamic and versatile methods of instruction (Haleem et al., 2022).

Digital technologies significantly enhance education by simplifying the creation of instructional materials and introducing innovative learning and collaboration methods. With the global reach of the Internet and the proliferation of smart devices, educators are now tasked with harnessing these technologies to revolutionize education, making it universally accessible and efficient. Technology has become crucial for delivering education outside traditional classrooms, fostering creativity, and encouraging exploration beyond conventional methods. Integrating remote learning technologies such as TV, radio, online platforms, and mobile applications has facilitated easier access to information, improved retention, and made education more interactive and engaging. To avoid marginalization, educators must adeptly use various digital tools and resources to keep their content current and engaging. The integration of advanced technology has significantly improved the learning experience, rendering it both more engaging and interactive for students. The synergy between e-learning platforms and smart devices, augmented by innovative educational tools like animations, interactive games, and AI-powered systems, has greatly enhanced the educational landscape. Additionally, applying Big Data and analytics in education has become increasingly important, providing valuable insights into student and instructor performance as schools and institutions expand their use of virtual classrooms, e-learning platforms, and online assessments (Haleem et al., 2022).

4.3.2 Virtual Labs and Simulations

Virtual labs and simulations are essential in renewable energy education, providing practical experiences without the need for physical resources. These tools allow students to conduct experiments, adjust variables, and observe results in a controlled, risk-free environment. Platforms such as Labster and PhET Interactive Simulations offer virtual labs that address various aspects of renewable energy, including solar panel efficiency and wind turbine aerodynamics. By participating in simulations, students can strengthen their ability to solve problems and build technical skills, as they put theoretical concepts into practice within real-life situations. Additionally, virtual labs are cost-effective and scalable, making advanced educational resources accessible to institutions with limited budgets (World Economic Forum, 2024).

4.3.3 Role of Artificial Intelligence and Big Data in Personalized Learning

Artificial intelligence (AI) and big data are transforming personalized learning in renewable energy education by enabling highly customized educational experiences. AI-powered platforms analyze extensive data to discern individual learning patterns and preferences, allowing for tailored content and adaptive learning methods. These tools tailor the learning material's complexity according to the student's advancement, offering extra resources and exercises for practice. Big data analytics further enhance this personalization by providing insights into the effectiveness of teaching strategies and identifying common areas of difficulty. This data-driven approach helps educators refine the learning process continuously. Additionally, digital platforms support collaborative learning, essential in renewable energy education, by offering tools like discussion forums, collaborative documents, and virtual classrooms. These platforms facilitate teamwork on projects, sharing of insights, and peer feedback across distances, fostering a sense of community and active participation. For example, students can work together to design sustainable energy systems or develop business models for renewable energy startups, preparing them to address complex, real-world challenges in the sector (World Economic Forum, 2024).

Integrating digital tools and artificial intelligence (AI) is revolutionizing collaborative models within the renewable energy sector. Effective data sharing and utilization become crucial as companies transition from traditional energy providers to data-driven entities. Advanced digital tools enable real-time data exchange among stakeholders, leading to more coordinated and efficient management of energy resources. This shift from isolated data silos to interconnected ecosystems enhances transparency and fosters strategic partnerships. AI foundation models, trained on diverse datasets, provide valuable insights and predictive analytics that drive decision-making and innovation. Standardized data formats and exchange protocols also streamline platform communication, facilitating easier data sharing and interpretation. This cooperative strategy boosts the effectiveness of energy systems and speeds up the creation of innovative technologies. This aids in reaching the industry's objectives of achieving zero net emissions and promoting environmental sustainability (McKinsey & Company, 2023).

4.4 *Lifelong Learning and Continuing Education*

Skills in lifelong learning are vital for adjusting to the changing requirements of today's world, forming the basis for ongoing personal and career advancement. While lifelong learning has historical roots, its significance has been increasingly recognized in recent decades, emphasizing its role in equipping individuals to navigate change and shape a desirable future. Lifelong learning is characterized by continuous, supportive, stimulating, and empowering processes that span one's life, applying to all roles and circumstances. It encompasses pursuing knowledge, skills, values, and understanding, focusing on practical application, and fostering confidence, creativity, and enjoyment. As societies progress, LLS is integral to developing essential twenty-first-century skills such as self-directed and collaborative learning, on-demand learning, and organizational learning. Self-directed learning, a core dimension of LLS, involves individuals taking the initiative to identify learning needs, set goals, select resources, and evaluate outcomes. Since the 1980s, lifelong learning and its role in addressing contemporary challenges have been explored extensively, highlighting its connection to acquiring skills needed to address societal and personal demands. New learning styles such as ICT literacy and fluency in multiple media are increasingly crucial in a globalized world. Employers seek communication, problem-solving, decision-making, teamwork, adaptability, management competence, and technical skills. Integrating LLS into higher education is essential for preparing students for college and career success, as seen in efforts to update curricula in various countries, including Turkey. Here, the Higher Education Council emphasizes the need for universities to equip students with generic skills such as communication, collaboration, problem-solving, and digital literacy. Project-based learning (PBL) is highlighted as a promising approach to developing these skills, though research on its effectiveness in Turkish higher education remains limited (Güven, 2020).

Investing in education and training programs designed for the renewable energy sector to address the escalating demand for skilled labor. Such programs are instrumental in ensuring individuals acquire the requisite skills and knowledge to excel in and contribute to the industry. For example, vocational training centers and higher education institutions progressively offer specialized courses in renewable energy technologies, equipping students with the technical expertise needed for employment. By cultivating a well-educated workforce, nations can strengthen their competitive position in the global renewable energy market, attract investment, and stimulate economic growth (IRENA, 2019).

Adult education and professional development programs are crucial for keeping individuals updated with the latest knowledge and skills in renewable energy. Certification and accreditation programs formally recognize expertise, improving career prospects and fostering professional growth. Additionally, community outreach and public education initiatives significantly raise awareness and encourage sustainable energy practices among the general population (Leal Filho et al., 2019).

The Bologna Process, initiated in 1999 by the education ministers of 29 countries, established the European Higher Education Area (EHEA) with the goals of harmonizing higher education through a three-cycle system (bachelor, master, and doctoral studies), implementing the European Credit Transfer and Accumulation System (ECTS) for credit transfer, enhancing mobility for students and educators, and developing quality assurance mechanisms. This process introduced a student-centered approach to course design, requiring academic programs to align with students' needs and job market demands, supported by tools like ECTS and qualification frameworks. The focus on student-centered learning emphasizes the need for continuous interaction between educational institutions and external stakeholders to adapt programs and improve employability. The EHEA also promotes lifelong learning through flexible, individualized educational paths, exemplified by the rise of micro-credentials—short, specialized programs that address specific skills in demand. These credentials, supported by the European Skills Agenda and recent policy guidelines, are designed to enhance access to education, recognize prior learning, and foster collaboration with employers, ultimately aiming to meet the developing needs of the workforce and societal challenges (De Rosa et al., 2024).

Micro-credentials are emerging as a valuable complement to traditional continuing education and training programs, driving innovation in higher education similar to how Massive Open Online Courses (MOOCs) have enabled flexible, short-term learning. Their development and implementation are shaped by the strategic visions of higher education institutions and the dedication of forward-thinking professionals. Micro-credentials' effectiveness hinges on collaboration between these institutions, professionals, and society, leveraging both top-down and bottom-up approaches. The European Skills Agenda highlights micro-credentials as a promising tool for personalized and skill development. In response, the European Commission has issued guidelines to establish a common definition, standardize key elements, and provide frameworks for designing, monitoring, and assessing micro-credentials. These guidelines, based on ten universal principles, aim to ensure that micro-credentials effectively support the societal mission of educational institutions across all sectors (De Rosa et al., 2024).

4.4.1 Programs for Adult Education and Professional Development

Universities act as key drivers for progress by embedding the Sustainable Development Goals (SDGs) within their educational programs. Networks such as the SDG Accord, comprising 42 international and national organizations, promote the annual reporting of progress by higher education institutions (HEIs) in integrating SDGs. This collaborative effort underscores the interconnectedness of teaching, research, and governance in fostering sustainability (Leal Filho et al., 2019). Few HEIs have strategically aligned their curricula with SDGs despite the potential benefits. Goal 4 (quality education) is vital for advancing sustainable development and enhancing societal engagement with the goals. Universities' research and teaching capabilities are crucial for achieving all 17 SDGs, particularly Goals 4, 9, 12, 16, and 17. An inclusive approach is essential, where all disciplines contribute to understanding and addressing sustainability issues. For instance, Nottingham Trent University and Victoria University of Wellington have successfully integrated SDGs into their curricula, enhancing their contributions toward these global objectives. This comprehensive educational strategy ensures that students are equipped for a sustainable future, aligning academic programs with the broader demands of global sustainability (Leal Filho et al., 2019).

Throughout history, education has been central to improving individual and societal well-being by empowering people to contribute effectively to society, develop their potential, fulfill civic duties, and pass on cultural values. In the twenty-first century, however, education must evolve to meet the demands of a highly connected, technology-driven world. The swift progress in communication and transportation demands that our education systems equip students for their future careers and adapting requirements. Central abilities for contemporary education, frequently called the "4Cs," consist of creativity, critical thinking, communication, and the ability to work as part of a team. Moreover, vital interpersonal skills crucial for the professional realm include honesty, effective communication, generosity, dedication, social competence, optimism, specialized knowledge, adaptability, cooperative spirit, and a diligent work ethic. As the global economy increasingly values cross-cultural cooperation, innovative problem-solving, effective communication, and critical thinking, there is growing concern that university graduates may lack these vital skills. Therefore, education systems must adapt to better equip students for the demands of the contemporary workforce (Güven, 2020).

4.4.2 Certification and Accreditation Programs

Certification and accreditation programs are essential for upholding high standards and ensuring the credibility of renewable energy education. These programs standardize the skills and knowledge required for professionals in the sector. For instance, certification programs provided by organizations such as the North American Board of Certified Energy Practitioners (NABCEP) offer credentials in specialized areas like solar photovoltaic systems and wind energy, ensuring practitioners meet industry standards. Such certifications enhance career prospects and ensure the workforce is equipped to handle the technical demands of renewable energy projects. Accreditation ensures that educational institutions and their programs adhere to established quality standards, critical for producing competent graduates who can effectively contribute to the sector. Certifications from bodies like the International Renewable Energy Agency (IRENA) validate individuals' competencies, confirming they possess the skills necessary for the renewable energy field. The process of accreditation involves a thorough review of educational programs to confirm that they adhere to superior standards of quality. This benefits learners by offering credentials that are widely recognized, while also supporting educational institutions in preserving their esteemed status and drawing in a larger number of students. Certification and accreditation are pivotal in maintaining quality assurance and professionalizing the renewable energy workforce (Leal Filho et al., 2019).

4.4.3 Community Outreach and Public Education Initiatives

Community outreach and public education initiatives are vital for enhancing public understanding and acceptance of renewable energy technologies. Effective outreach involves engaging communities through various platforms, including workshops, seminars, and public demonstrations to educate individuals about renewable energy's benefits and practical applications. For example, hands-on workshops that showcase solar panel installation and usage can significantly boost community interest and involvement. Additionally, leveraging social media and local influencers can broaden the reach of these educational efforts, making renewable energy concepts more accessible to a broader audience. These initiatives are crucial for raising awareness and empowering communities to participate actively shift to renewable energy. By involving the public in hands-on activities, such as small-scale renewable energy projects, and utilizing multimedia tools and social media, these programs make complex concepts more understandable and foster a culture of sustainability and environmental stewardship (Kioupi & Voulvoulis, 2019).

5 Pedagogical Approaches and Teaching Methods

Pedagogical approaches and teaching methods are pivotal in shaping education for sustainable development (ESD), especially within renewable energy. A critical approach is learner-centered pedagogy, which positions students as active participants in their learning journey. This method emphasizes student autonomy and encourages learners to build knowledge based on their prior experiences and social contexts. Educators support this by fostering reflection and self-management of learning, thereby shifting their role from knowledge providers to facilitators of the learning process. In renewable energy education, this approach ensures that students deeply engage with concepts, leading to a comprehensive understanding beyond passive information reception (Leicht et al., 2018).

Another essential pedagogical approach is action-oriented learning, which involves students in real-world activities such as projects, internships, workshops, or campaigns related to renewable energy. This method uses Kolb's learning cycle—concrete experience, reflection, abstract conceptualization, and application—to enhance knowledge acquisition and competency development by linking theoretical concepts with practical experiences. Educators facilitate this by creating environments that promote experiential learning and reflective thinking, allowing students to directly apply their learning to renewable energy challenges and solutions (Leicht et al., 2018).

Transformative learning is a third significant approach, aiming to empower students to critically assess and potentially alter their worldviews and behaviors toward sustainability. Educators facilitate this process by challenging students to question existing paradigms and envision new ways of thinking and acting concerning renewable energy issues. The concept of transgressive learning extends this by promoting disruptive thinking that challenges established norms and fosters innovation and co-creation of knowledge. Integrating transformative and transgressive learning approaches can inspire students to become proactive change agents, advancing renewable energy technologies and sustainable practices (Leicht et al., 2018).

The development of sustainability competencies in education necessitates pedagogical approaches that address learners' cognitive, psychomotor, and affective domains. Such approaches are essential for cultivating skills like integrated problem-solving, systems thinking, and interpersonal abilities crucial for tackling sustainability challenges. Project-based learning, active learning, and community service learning are highlighted for their effectiveness in immersing students in real-world contexts, allowing them to use theoretical knowledge to practical issues. These methods enhance critical thinking through case studies, life cycle analysis, and participatory action research, offering a comprehensive understanding of sustainability issues from various perspectives. Educators are central to these pedagogical frameworks, guiding students through open-ended, collaborative problem-solving processes aligned with sustainability goals. By involving learners as partners in change, educators facilitate transformative learning experiences that develop competencies and foster creative and reflective interactions. This highlights the importance of robust educator training programs that provide teachers and university lecturers with the necessary knowledge, skills, and confidence to integrate these innovative teaching methods into their curricula. Such training enables educators to empower students to develop and apply sustainability competencies across diverse educational settings (Kioupi & Voulvoulis, 2019).

5.1 *Active Learning Strategies*

5.1.1 Problem-Based Learning (PBL) and Project-Based Learning (PjBL)

One effective teaching strategy is using problem-based learning (PBL) to engage students with real-life renewable energy challenges. For example, students might work on projects that involve designing sustainable energy solutions or assessing the viability of renewable energy initiatives. This approach helps students apply their knowledge and critical thinking skills to solve practical problems, thereby enhancing their active learning, teamwork, and problem-solving abilities. Inquiry-based learning is another beneficial strategy, motivating students to explore and understand renewable energy concepts through questions, research, and experimentation. Additionally, practical, hands-on experience is vital. Through renewable energy labs or workshops, students get firsthand experience with technology such as solar panels and wind turbines. Incorporating advanced technology, including computer simulations, modeling software, and immersive experiences like virtual reality (VR) and augmented reality (AR), further deepens students' understanding by replicating complex energy systems. Online learning tools, such as interactive courses, videos, quizzes, and webinars, also play a crucial role in facilitating self-directed learning and continuous professional growth. Adopting various teaching strategies and tools can significantly enhance renewable energy education in Morocco, encouraging critical thinking, problem-solving capabilities, and a thorough comprehension of renewable energy, which are essential for the country's energy transition (Daoudi, 2024).

Problem-based learning (PBL) is an instructional strategy that engages students in exploring complex, real-world problems. In renewable energy education, PBL has proven effective in developing critical thinking and problem-solving abilities. This approach immerses students in practical challenges, such as optimizing solar panel efficiency or designing sustainable energy systems, deepening their technical understanding. PBL fosters essential collaborative and communication skills by having students work together on these projects. By replicating real-world scenarios, PBL equips students to navigate the complexities of the renewable energy sector, enhancing their adaptability and innovative capacity (Güven, 2020).

Project-Based Learning (PBL) is a student-centered instructional method that engages learners in complex projects requiring critical thinking, creativity, and collaboration. This active learning approach involves students in identifying problems, developing plans, and conducting research under the guidance of a teacher. Through PBL, students are tasked with designing and executing projects that require asking questions, gathering and analyzing data, and discussing their findings with peers. This method fosters higher-order thinking skills and promotes independent exploration and teamwork, linking classroom learning to real-world applications. By incorporating these elements, PBL equips students with the essential twenty-first-century skills needed for contemporary challenges (Güven, 2020).

5.1.2 Collaborative Learning and Team-Based Projects

Collaborative learning and team-based projects are crucial in renewable energy education, creating an interactive and practical learning environment. Collaborative learning involves students working together to achieve common learning objectives, which is especially valuable in complex fields like renewable energy. By collaborating, students can combine their diverse skills and expertise to address multidisciplinary problems. Team-based projects, such as designing a solar energy system or developing a wind turbine model, offer hands-on experience and enhance students' readiness for professional environments. These initiatives boost technical abilities while also fostering key interpersonal skills like effective communication, collaboration, and the management of projects. Given that real-world renewable energy projects often require interdisciplinary collaboration and coordination among various stakeholders, these educational strategies effectively prepare students for future professional challenges (Nousheen et al., 2020).

5.1.2.1 Case Studies and Real-World Problem-Solving

Although the bottom seven pedagogical approaches rank lower in confidence and relevance, they remain valuable for enhancing the Bachelor of Arts in Sustainability Studies (BASS) program. While not top-ranked, these methods can be effectively integrated into more prominent pedagogies such as problem-based learning (PBL), project-based learning (PjBL), and collaborative learning. For instance, discussion-based learning fosters critical thinking and idea exchange, complementing active learning frameworks. As part of the Writing Across the Curriculum (WAC) initiative, writing-intensive learning enhances students' written communication skills, which are crucial for sustainability competencies. Case studies contextualize efforts and support real-world problem-solving, aligning with project-based and service learning. Experiential learning, encompassing PBL, PjBL, and collaborative efforts, emphasizes hands-on experiences vital for sustainability education. Creative work and expression stimulate critical and creative thinking, beneficial across various learning modalities. Learning communities, although resource-intensive, provide a robust environment for interdisciplinary collaboration. Internships and apprenticeships offer practical experience and align with active learning strategies but require significant faculty oversight and resource management (Evans, 2019).

5.2 *Innovative Teaching Tools*

Integrating innovative teaching tools within the Education paradigm is crucial for creating a dynamic and engaging learning environment. These tools include a range of multimedia and interactive content that enhance educational experiences by making them more immersive and adaptive. Virtual Learning Environments (VLEs), for instance, offer a flexible, exploratory, and personalized learning process, enabling students to interact with material in unique and meaningful ways. Innovative technologies, including virtual reality and robotics, are crucial in creating advanced learning spaces sensitive to their location, context, and social dynamics. These environments ensure seamless connectivity and adaptability, fostering a student-centered approach that is both time-efficient and effective in achieving learning outcomes (Almeida & Simoes, 2019).

5.2.1 Use of Multimedia and Interactive Content

Innovative teaching tools, including multimedia and interactive content, have significantly transformed how educators engage with students. By incorporating videos, animations, and simulations into lessons, teachers can create dynamic and engaging learning experiences tailored to various learning styles. These tools enhance lessons' interactivity and visual appeal, improving information retention through a multisensory approach. Additionally, multimedia content facilitates tailored learning experiences, permitting learners to investigate topics at a speed that suits them and through methods that cater to their unique learning styles (Christopoulos & Mystakidis, 2023).

Martínez-Cerdá et al. (2018) underscore the pivotal role of innovative teaching tools in enhancing collaborative skills within online universities. Their research demonstrates that multimedia content, gamification, mixed reality, and social media significantly contribute to developing collaborative competencies among online learners. These ICT-supported pedagogical tools fostered teamwork, cooperation, and peer learning. Specifically, multimedia and interactive materials engaged students effectively, while gamification introduced educational games that motivated collaborative efforts. Additionally, mixed reality technologies provided immersive experiences that further facilitated student collaboration.

5.2.1.1 Gamification and Educational Games

Gamification and educational games have proven effective tools for boosting student motivation. By mixing game elements for example, points, badges, and leaderboards into educational activities, educators can create a competitive and rewarding environment that encourages active participation and collaboration. This approach motivates students to engage more deeply with the subject matter and provides a secure space for experimentation and learning from mistakes, fostering a growth mindset and resilience (Christopoulos & Mystakidis, 2023).

Gamification and serious games in higher education mark a transformative shift toward more interactive and engaging learning environments. Gamification involves incorporating game mechanics such as narrative, feedback, rewards, competition, and collaboration into nongame contexts to boost students' engagement and problem-solving skills. This strategy effectively motivates students, enhances teamwork, and enriches the learning experience. Educational games, in particular, offer a platform for experiential learning, allowing students to explore, make mistakes, and learn within a safe environment. This fosters critical thinking and creativity while increasing student involvement and developing real-world skills (Almeida & Simoes, 2019).

5.2.2 Role of Social Media and Online Communities

The impact of social media and online communities on education is profound. Platforms such as Twitter, Facebook, and LinkedIn facilitate connections among students, allowing them to share resources and engage in discussions beyond the traditional classroom setting. These online communities create opportunities for collaborative learning, where students can seek assistance, exchange ideas, and receive feedback from a diverse audience. Additionally, educators can utilize social media to disseminate information, foster discussions, and build student community. Teachers can leverage social media capabilities to enhance communication, promote collaboration, and cultivate a more interactive and inclusive learning environment (Christopoulos & Mystakidis, 2023).

The study highlighted the role of social media and online communities in enhancing collaboration among online learners. Platforms like social media enable students to connect, communicate, and work together on academic tasks, improving their teamwork and cooperative skills. Overall, the research underscores the importance of employing a range of innovative teaching tools such as multimedia content, gamification, mixed reality, and social media to foster collaborative skills in online university settings (Martínez-Cerdá et al., 2018).

Social media and online communities are pivotal in contemporary education, offering platforms that support collaboration, communication, and information sharing. These tools allow students to interact with peers and instructors beyond the traditional classroom, facilitating continuous learning and support. Online communities create a collaborative environment where students can discuss ideas, exchange resources, and provide feedback, thus enhancing the learning experience through active participation. Additionally, social media platforms enable the dissemination of educational content, the promotion of discussions, and connections with a broader academic network, thereby broadening students' learning opportunities (Almeida & Simoes, 2019).

5.3 Assessment and Evaluation

Practical assessment tools in renewable energy education are essential for evaluating students' understanding and the overall educational process. These tools must measure cognitive, affective, and psychomotor skills while aligning with the dynamic nature of renewable energy technologies and methodologies. Formative assessments are critical as they offer continuous feedback, enabling educators to adjust their teaching strategies and help students improve their learning processes. These assessments encompass a variety of formal and informal methods, such as quizzes, discussions, and reflective journals, which help identify learning challenges and enhance instructional practices. Conversely, summative assessments evaluate students' cumulative knowledge and skills at the end of an instructional period. These assessments, including final exams, projects, and standardized tests, deliver a complete overview of students' achievements and the effectiveness of the instructional methods (Ahmad Baht & Jeelani Bhat, 2019).

Assessment and evaluation in renewable energy education are crucial for measuring learning outcomes and assessing the effectiveness of educational initiatives. Higher learning outcomes reflect the implementation of educational activities, such as curriculum design and teaching methods, which contribute to developing competencies in education for sustainable development (ESD). Improvements in learning outcomes indicate that these activities effectively foster ESD skills, whereas declines suggest potential issues that must be addressed. This process aligns with Tilbury's emphasis on maintaining the relevance and efficacy of ESD programs, ensuring they adapt to evolving educational needs and challenges (Leicht et al., 2018).

Effective monitoring and evaluation (M&E) in education for sustainable development (ESD) are essential for informed decision-making and program refinement. By incorporating new data and analysis, educational programs can be adjusted to better address their objectives and adapt to emerging global issues such as climate change. A participatory approach to M&E enhances stakeholder understanding of what drives effective ESD learning, improving educational outcomes and fostering broader social learning. However, assessing ESD presents unique challenges due to its dynamic nature and focus on emergent concepts. Unlike traditional education metrics, ESD outcomes may be less predefined as learners actively shape the process. It is crucial to distinguish ESD from related fields like environmental education (EE) and global citizenship education (GCED), as each field requires specific monitoring strategies to measure their distinct goals and impacts. Additionally, M&E incorporates a governance and accountability framework to ensure equitable access and inclusivity, holding education authorities accountable for providing quality education across socioeconomic backgrounds. This framework safeguards against educational inequalities and market-driven influences, promoting social justice and equitable access to education (Leicht et al., 2018).

Assessing and evaluating the effectiveness of educational practices in fostering sustainability competencies are crucial for advancing sustainable development. Despite substantial efforts in implementing Education for Sustainable Development (ESD) over the past two decades, reviews indicate a persistent lag in assessing ESD outcomes, hampers understanding their impact. Practical assessment is essential for refining educational frameworks and establishing a robust evidence base to demonstrate progress toward sustainability goals. Research shows a strong correlation between university learners' development of sustainability competencies and their contributions to environmental, social, and economic sustainability, highlighting the need for standardized evaluation criteria. Monitoring progress helps identify gaps in implementation, encourages stakeholder engagement, and provides tangible evidence of transformative learning outcomes. Benchmarking educational practices against established standards enhances institutional reputation and fosters continuous improvement. Institutions can effectively incorporate Sustainable Development Goals (SDGs) into curricula by integrating sociocultural, environmental, and economic indicators. Developing reliable assessment tools, such as rubrics for sustainability competencies and heuristic models for systems thinking, is vital for adapting educational practices to address critical issues like ethics, gender, poverty, and climate change. Ongoing research aims to refine these indicators and gather data from diverse educational contexts, supporting institutions in managing progress toward sustainable development goals and ensuring meaningful outcomes for future generations (Kioupi & Voulvoulis, 2019).

5.3.1 Developing Practical Assessment Tools for Renewable Energy Education

In renewable energy education, developing practical assessment tools is essential for ensuring learners acquire the necessary knowledge and skills. Formative assessments provide ongoing feedback on student progress and are crucial. These assessments help identify areas needing improvement and enable timely interventions. Aligning assessment tools with educational goals and curriculum objectives creates a cohesive framework that enhances the learning experience and outcomes. This process, known as "constructive alignment," integrates assessments as fundamental components of the teaching and learning process, fostering a deeper understanding and better knowledge retention (Ahmad Baht & Jeelani Bhat, 2019).

5.3.2 Formative and Summative Assessment Methods

Assessment methods in renewable energy education are generally classified into formative and summative assessments. Formative assessments offer continuous feedback, helping students identify their strengths and weaknesses while refining their learning strategies. These assessments are vital for deepening understanding of renewable energy concepts and enhancing critical thinking skills. In contrast, summative assessments evaluate the achievement of learning goals at the end of a course or program. They assess the overall effectiveness of the educational program and determine whether students have acquired the necessary competencies. Both formative and summative assessments are essential: formative assessments guide the learning process, while summative assessments confirm the attainment of educational outcomes and readiness for professional practice (Ahmad Baht & Jeelani Bhat, 2019).

Formative and summative assessments are fundamental components of renewable energy education. Formative assessments are carried out throughout the educational process and provide ongoing feedback to students and instructors. These assessments help identify individual strengths and weaknesses, allowing teachers to adjust their strategies and support continuous improvement. Standard formative techniques include peer reviews, self-assessments, and interactive activities, all of which engage students actively in their learning journey. Summative assessments are implemented after a course or program to measure student achievement against specific criteria. These critical evaluations, including final exams, in-depth projects, and portfolio assessments, validate student abilities and highlight areas for curriculum improvement based on identified gaps (Ahmad Baht & Jeelani Bhat, 2019).

5.3.3 Feedback Mechanisms for Continuous Improvement

Ahmad Baht and Jeelani Bhat (2019) believe effective feedback mechanisms are crucial for continuously enhancing renewable energy education. To enhance students' performance, evaluations should be immediate, precise, and aimed at growth, clearly pointing out how to correct errors while promoting continuous development and thoughtful learning practices. Regular feedback sessions, peer assessments, and self-evaluations enhance the learning experience. Establishing a feedback-rich environment where students do feel comfortable receiving and acting on feedback is essential for their development as independent, self-regulated learners. By embedding these mechanisms within the curriculum, educators can ensure that students continuously improve and attain higher levels of understanding and competence in renewable energy.

Feedback mechanisms are crucial for fostering ongoing improvement in renewable energy education. Effective feedback must be timely, specific, and constructive, addressing content and learning processes. It should prompt students to reflect on their performance, understand errors, and develop strategies for improvement. Various feedback methods can be utilized, such as written comments, verbal discussions, and interactive digital platforms. Additionally, incorporating self-assessment and peer feedback empowers learners to actively engage in their educational development. This approach enhances self-regulated learning and deepens understanding of renewable energy concepts. Continuous feedback loops generate a learning environment, motivating students to meet higher standards and enabling educators to refine their teaching methods based on real-time insights into student progress (Ahmad Baht & Jeelani Bhat, 2019).

6 Challenges and Barriers

6.1 Institutional and Policy Challenges

6.1.1 Resistance to Curriculum Changes

Transitioning to a hydrogen economy necessitates entirely new educational programs for developing and implementing new energy technologies. The shift requires substantial adaptation of university curricula and related institutions to prepare technicians, engineers, and researchers for these technologies' practical implementation and development (Nowotny et al., 2018). This adjustment can be slow due to institutional resistance and the challenges of integrating interdisciplinary topics into existing educational frameworks. More so, a recent report by the World Economic Forum (2023) confirmed an acute shortage of workers possessing relevant knowledge and skills to address renewable energy and climate change challenges, even though green opportunities have expanded globally. With the fossil fuel industries funding fossil fuel education in the existing educational systems, a swift energy transition from fossil fuel to clean and renewable through educational programmes may face severe resistance, delays, and limited cooperation from involved stakeholders.

6.1.2 Bureaucratic and Regulatory Hurdles

Bureaucratic and regulatory barriers have hampered establishing university programs in renewable energy technologies. The need for new educational standards and approval processes can delay the introduction of relevant programs and hinder the timely development of skilled professionals in this field (Nowotny et al., 2018).

6.1.3 Inadequate Funding and Resources

Financial limitations significantly hinder the advancement and execution of programs focused on renewable energy education. Strategic financial planning is critical to upgrading educational infrastructure, supplying essential tools, and fostering research activities. The lack of specialized educational materials, notably in less developed regions, highlights a discrepancy between what the educational system offers and what the industry requires (Daoudi, 2024; Nowotny et al., 2018).

6.2 *Technological and Infrastructure Barriers*

6.2.1 Access to Technology and the Digital Divide

Despite significant progress in the field, education related to renewable energy continues to face hurdles. A prominent challenge is the limited availability of educational materials designed for renewable energy programs. This lack of resources makes imparting the knowledge and skills to students eager to enter the sector difficult. It is essential to focus on strengthening the capability of educators by enhancing their skills and knowledge. Keeping educators up to date with the latest trends in the industry, new technologies, and effective pedagogical techniques is crucial for maintaining the quality of education in renewable energy. Financial limitations are another issue, highlighting the need for careful financial planning to improve educational facilities, acquire the required equipment, and support research activities. Additionally, it is essential to increase public awareness regarding renewable energy education. We can foster a well-informed and supportive community by addressing misconceptions and underscoring the real-world benefits of renewable energy (Daoudi, 2024).

6.2.2 Infrastructure Limitations in Schools and Universities

The shift to online learning has revealed significant infrastructure limitations in educational institutions. Many schools and universities lack the necessary hardware and connectivity to support remote education effectively, exacerbating existing technological access inequalities (Haleem et al., 2022). This is particularly problematic in low-income and rural areas where resources are scarce.

6.2.3 Maintenance and Updating of Educational Resources

The rapid evolution of educational technology demands continuous maintenance and updating of resources. Institutions face challenges in keeping educational materials current and relevant, impacting the quality of education. This issue is compounded by the lack of comprehensive textbooks integrating interdisciplinary topics within a single volume (Nowotny et al., 2018).

6.3 *Social and Cultural Barriers*

6.3.1 Public Perception and Awareness of Renewable Energy

Raising awareness about renewable energy plays a critical role in the effectiveness of educational efforts. It is critical to close the knowledge gap and highlight the concrete advantages of renewable energy to nurture an informed and proactive community (Daoudi, 2024). The general public's engagement with the transition to renewable energy is necessary to ensure political feasibility and acceptance of the changes.

6.3.2 Cultural Attitudes Toward Education and Sustainability

Cultural attitudes toward education and sustainability can influence the effectiveness of renewable energy education programs. In many regions, traditional views and a lack of emphasis on sustainability can hinder the adoption of new educational paradigms and technologies (Daoudi, 2024; Lucas et al., 2018).

6.3.3 Engaging Diverse Communities and Promoting Inclusivity

Efforts to engage diverse communities and promote inclusivity in renewable energy education are crucial for addressing workforce deficits. Increasing the participation of underrepresented groups, including women, can help to alleviate the sector's shortage of professionals and support the broader transition to global sustainability (Nowotny et al., 2018).

7 Case Studies

7.1 *Successful Educational Programs*

7.1.1 Detailed Analysis of Leading Renewable Energy Education Programs

The German dual education system effectively integrates higher education with workforce apprenticeship programs, with over 400 professions offering apprenticeship pathways and 60% of German workers participating. This system benefits from significant private sector engagement and investment, serving as a model for other countries (Bosman et al., 2020). Observations of apprentices in Germany reveal their involvement in renewable energy and energy storage projects, such as smart home appliances, neighborhood battery storage, and electric-vehicle-charging stations (Bosman et al., 2020). In contrast, traditional U.S. apprenticeships have focused mainly on construction and trades. However, with recent labor shortages in STEM fields, the U.S. Department of Labor has invested $100 million in creating new apprenticeships in various sectors, including renewable energy. While the success of this initiative compared to Germany is still uncertain, it may provide valuable insights into enhancing U.S. apprenticeship participation (Bosman et al., 2020).

7.1.2 Lessons Learned and Best Practices

Germany is renowned for its exceptional approach to renewable energy education, seamlessly incorporating subjects related to renewable energy into the curricula of schools, universities, and vocational training programs. Denmark is proactive in creating training programs to foster innovation within the renewable energy sector, focusing on offshore wind energy. Meanwhile, Japan emphasizes elevating the public's awareness regarding energy-related challenges while advocating for solar and wind energy adoption through its educational initiatives. The United States has rolled out various programs to educate students, teachers, and communities about the benefits and potentials of renewable energy. These programs span various technologies and are offered by universities nationwide. Morocco has accomplished a notable feat by reaching an installed renewable energy capacity of 4031 MW by 2022, placing it among the top five African nations in renewable energy implementation. The country has also embraced the use of microgrids to counter the inherent instability and unreliability of renewable energy sources, thereby improving the efficiency of its energy systems. Despite these strides, Morocco's investment in renewable energy education is still developing, with continuous efforts to enhance training programs, establish research centers, and better integrate renewable energy topics into academic curricula (Daoudi, 2024).

7.1.3 Replicable Strategies for Other Institutions

The Kazakh National Technical University (KazNTU) provides a notable example of integrating Education for Sustainable Development (ESD) into technical education. By introducing an ESD course in its Bachelor's program, KazNTU addresses the balance between industrial development and environmental stewardship, aligning with Kazakhstan's national strategy for sustainable development. This approach, supported by collaborative partnerships across government, corporate, and civil society sectors, cultivates a new generation of engineers skilled in sustainable energy technologies. The successful integration of ESD principles highlights the importance of educational leadership and policy alignment in promoting sustainability (Leicht et al., 2018). International learning exchanges reveal that establishing a national vision for energy policy, as seen in Germany, is crucial compared to short-term market-driven approaches in countries like Australia and the United States. Integrating renewable energy into core programs, rather than creating specialized degrees, provides a solid foundation in traditional science, technology, and engineering fields. As demonstrated by Germany's dual-system apprenticeship and Australia's TAFE system, collaboration among educational institutions, industry, and trade unions effectively shapes curriculum and workforce training. These exchanges have influenced educators' practices, prompting new course materials and international perspectives in their teaching (Walz et al., 2016). Such international insights help educators to develop a deeper understanding of renewable energy technologies and policies, fostering improved educational outcomes and better preparation for the global workforce in renewable energy (Walz et al., 2016).

7.2 *Community and Grassroots Initiatives*

7.2.1 Examples of Community-Led Educational Projects

The United Nations has identified a significant funding gap for achieving the Sustainable Development Goals (SDGs), with an annual investment requirement of USD 3.9 trillion, compared to the current investment of approximately USD 1.4 trillion (UNCTAD). Introducing a collaborative platform for funding the Sustainable Development Goals (SDGs) at the 2015 Third International Conference on Financing for Development (FfD) held in Addis Ababa presents a viable strategy. This approach was further developed with the launch of a new platform by UN Secretary-General Ban Ki-moon in October 2016 to enhance innovative financial solutions (UNCTAD). The Future Earth program, initiated by the UN, focuses on research projects essential for transitioning to global sustainability. It includes specialized networks and partnerships to address water, energy, and food challenges and enhance social resilience (Nowotny et al., 2018).

7.2.2 Impact on Local Communities and Environmental Awareness

The United Nations Framework Convention on Climate Change, alongside pivotal agreements like the Kyoto Protocol and the Paris Agreement, has set ambitious objectives to cut greenhouse gas emissions and pave the way for a low-carbon economy. These global pacts have spurred local and grassroots actions by underlining the critical need for sustainable development and raising environmental consciousness. Initiated in Rio in 2012, the Higher Education Sustainability Initiative (HESI) highlights the crucial contribution of academic institutions in fostering sustainable development through innovative educational programs and initiatives aimed at greening campuses. This approach has catalyzed local educational endeavors for sustainability and environmental issues, enhancing awareness and spurring community participation (Nowotny et al., 2018). Alternatively, Skaloumpakas et al. (2024) presented the concept of renewable energy valleys and demonstrated its practical feasibility through an insightful case study of Crete Island, Greece. The author argued that targeted campaigns, including educational activities and public awareness programs, can effectively translate the advantages of integrating storage technologies, renewable energy, and market engagement into locally appropriate languages. Such community-centric activities can promote social innovation and energy literacy, deepening the "Leave No One Behind" ethos. Also, they can deepen environmental awareness and rural resilience by improving attitudes, knowledge, and behaviors toward renewable energy adoption, creating social cohesion and energy communities that share environmental goals and ownership.

7.2.3 Future Potential and Scalability

The Future Earth program, running from 2016 to 2025, aims to create a comprehensive knowledge base for global sustainability and address energy challenges by developing clean energy technologies (Nowotny et al., 2018). This program's focus on interdisciplinary research and international collaboration has significant potential for scaling up community and grassroots initiatives. Developing a hydrogen-based energy economy and other sustainable technologies can further enhance local environmental projects and promote broader adoption of sustainable practices.

7.3 International Collaboration and Exchange

7.3.1 Cross-Border Educational Initiatives and Partnerships

International collaboration has been crucial in advancing renewable energy education and sustainable development. The UN's multistakeholder platform, introduced in 2015, and subsequent initiatives like the Future Earth program emphasize the importance of global partnerships in solving sustainability challenges (UNCTAD). The Future Earth program involves various networks and stakeholders to build a knowledge base for global sustainability, including clean energy technologies (Nowotny et al., 2018). These initiatives highlight the role of cross-border educational programs and research in fostering international collaboration.

7.3.2 Role of International Organizations and NGOs

International organizations, such as the UN and World Bank Group, play a pivotal role in financing and promoting SDG-related initiatives. The World Bank Group's vision document outlines its commitment to open trade promotion and sustainable SDG financing. At the same time, the UN's Future Earth program aims to provide the knowledge and solutions needed for global sustainability (Nowotny et al., 2018). These organizations partner with NGOs and other stakeholders to support global and local efforts in sustainable development.

7.3.3 Benefits of Cultural Exchange and Global Perspectives

Cultural exchange and global perspectives are integral to the success of international collaborations. Educational initiatives and research projects under frameworks like the Future Earth program benefit from diverse viewpoints and approaches to sustainability (Nowotny et al., 2018). By incorporating international perspectives into their curricula and research, institutions can enhance their impact and contribute to a more comprehensive understanding of global sustainability challenges.

8 Future Directions and Opportunities

8.1 Emerging Trends in Education

8.1.1 Impact of Technological Advancements on Education

Current environmental education (EE) practices often emphasize energy conservation and individual behavioral changes but overlook the importance of political and policy dimensions in energy system transformation. By focusing narrowly on personal actions, EE may neglect the broader systemic changes needed to address climate change effectively. Technological advancements can play an essential role in shifting this focus by integrating new tools and platforms that facilitate a more comprehensive understanding of energy policies and climate action. Educators and researchers should leverage these technologies to bridge the gap between personal energy behaviors and systemic change processes, enhancing students' awareness of the complex interactions between energy supply systems and climate action.

8.1.2 Evolution of Learning Environments and Intelligent Classrooms

The current emphasis on individual actions in environmental education might reinforce a limited perspective on climate and energy system change. There is a growing need for educational environments that promote engagement with collective action and systemic transformation. The evolution of learning environments, including integrating smart classrooms and interactive platforms, can support this shift by providing students with access to real-time data and collaborative tools. Such environments can facilitate discussions on the roles of state and market actors in energy system change and offer students opportunities to engage with social movements and policy discussions.

8.1.3 Future Trends in Renewable Energy Curriculum Development

To address the limitations of current EE approaches, future renewable energy curricula should incorporate a broader range of perspectives, including those of social movements and advocacy groups. Educators are encouraged to reconceptualize students as active participants within a broader social network rather than isolated actors. Incorporating participatory EE approaches that involve interactions with diverse stakeholders such as climate scientists, renewable energy firms, state agencies, and NGOs can provide students with action-oriented knowledge and practical insights into energy system transformation. This method narrows the divide between conceptual knowledge and practical application, promoting a holistic and interconnected approach to environmental education.

8.2 *Policy Recommendations*

8.2.1 Strategies for Policymakers to Support Renewable Energy Education

The Ministry of Education, curriculum development agencies, and educational institutions such as schools, colleges, and universities are responsible for developing and integrating renewable energy programs into formal education at all levels. This involves creating and implementing specialized curricula emphasizing renewable energy's benefits, technologies, and challenges.

Educational institutions, vocational training centers, industry associations, and nongovernmental organizations (NGOs) should establish professional training programs and hands-on workshops. These programs should be designed to provide various audiences, including educators, energy sector professionals, and community leaders, with practical knowledge and skills related to renewable energy technologies. The goal is to enhance their ability to implement and advocate for renewable energy solutions effectively.

Educational platforms, online course providers, NGOs, and community organizations are tasked with offering online courses and nonformal educational opportunities. This includes developing interactive and accessible digital content on renewable energy topics to reach a broad audience.

Curriculum development bodies, the Ministry of Education, and educational institutions are responsible for tailoring educational curricula to address the societal need for renewable energy and sustainability. This involves ensuring that sustainability principles are embedded across various subjects and disciplines, not limited to dedicated environmental science courses.

Government policymakers, the Ministry of Science and Technology, and education departments should advocate for including sustainable development goals in national science and technology policies. This entails updating education standards and policies to reflect the importance of renewable energy and sustainability and aligning educational objectives with national and global targets.

Educational institutions, government departments, and private sector companies are responsible for implementing policies encouraging greater female participation in science, technology, engineering, and mathematics (STEM) fields. This includes establishing gender-sensitive admission procedures, employment practices, and public appointments.

Government agencies, educational institutions, and industry associations should develop and offer incentives such as scholarships, mentorship programs, and career support to increase female representation in STEM disciplines, particularly in renewable energy sectors.

8.2.2 Incentives and Funding Mechanisms

Public institutions, universities, and research institutes are encouraged to dedicate resources to fundamental and practical research projects in renewable energy technologies, aiming to enhance energy production, administration, and efficiency.

Public authorities, entrepreneurship facilitators, and investors should extend financial benefits, including tax incentives, funding grants, and affordable loans, to new and existing companies in the renewable energy industry. Additionally, these entities should assist in the growth of entrepreneurial initiatives by offering programs and support structures that encourage the development of fresh ideas and innovative business strategies.

Government entities, private companies, and research institutions should foster partnerships to drive innovation and business development in renewable energy. Facilitate joint ventures and collaborative projects that leverage resources and expertise from multiple sectors.

8.2.3 Collaboration Among Government, Industry, and Academia

Government agencies, universities, and international organizations should establish and strengthen partnerships with international universities, research institutes, and organizations specializing in renewable energy. Engage in collaborative research projects, exchange programs, and joint initiatives to enhance knowledge and expertise.

International organizations, academic institutions, and industry associations should facilitate sharing of best practices, technologies, and research findings through international networks and forums. Leverage global expertise to improve local education and research practices related to renewable energy.

Media organizations, government agencies, and NGOs should implement public awareness campaigns using various media platforms, including social media, television, and radio, to educate citizens about renewable energy. Host community forums, lectures, and exhibitions to engage the public and raise awareness.

Educational institutions, industry associations, and community organizations should organize science and technology exhibitions, conferences, and workshops on renewable energy. Provide platforms for dialogue and knowledge-sharing among stakeholders, including educators, industry professionals, and the general public.

Research institutions, technology developers, and government agencies should encourage the exploration of new technologies and solutions to address renewable energy challenges. Support initiatives that drive technological advancements and contribute to a sustainable energy future through ongoing research and development.

8.3 Strategic Vision for a Sustainable Future

8.3.1 Roadmap for Integrating Renewable Energy Education into Mainstream Education

Design and implement curricula incorporating renewable energy topics at all educational levels, including primary, secondary, and tertiary education. This should involve creating specialized courses addressing benefits, technologies, and challenges. The Ministry of Education will oversee these courses in collaboration with educational institutions and curriculum developers. This approach aims to equip students with foundational and advanced knowledge of renewable energy from an early age.

Establish targeted training programs for educators to teach renewable energy topics effectively. These include professional development workshops and certification programs to enhance teachers' abilities to integrate renewable energy concepts into their teaching.

Produce and disseminate updated textbooks, online resources, and interactive educational materials on renewable energy. Educational publishers and content creators will develop these resources, ensuring they reflect the latest renewable energy technologies and policy advancements.

Integrate digital platforms and educational technologies to deliver renewable energy content, such as virtual labs and interactive simulations. This will be facilitated by EdTech companies and educational institutions, aiming to provide flexible learning options.

Forge collaborations with renewable energy companies to create internship opportunities and practical student learning experiences.

8.3.2 Long-Term Goals and Milestones

Achieve comprehensive integration of renewable energy education into national education policies and standards. This will involve updating policies to ensure that both sustainability and renewable energy education are central to the educational framework.

Develop dedicated research centers and innovation hubs focused on renewable energy education and technology development to support ongoing research and technological advancement.

8.3.3 Vision for a Sustainable and Educated Society

Foster a generation of well-versed individuals in renewable energy technologies, their benefits, and challenges. This will ensure that students, educators, and professionals are prepared to contribute to a sustainable energy future.

Encourage the widespread adoption of renewable energy solutions and sustainable practices across various sectors of society. This includes supporting innovation in GREEN technologies.

Invest in ongoing research and development to drive technological advancements and address emerging challenges in renewable energy.

Increase public understanding and support for renewable energy through educational campaigns, community engagement, and media outreach. This includes hosting public lectures and exhibitions and leveraging media platforms to raise awareness.

9 Conclusion

9.1 Summary of Key Points

This chapter has emphasized education in advancing the global shift toward renewable energy. It outlined the historical progression from a focus on fossil fuels, shaped by the industrial demands of the twentieth century, to a modern emphasis on renewable energy and sustainability. This shift is essential for addressing climate change and meeting sustainable development goals. By incorporating renewable energy education into academic curricula, institutions can prepare future generations with the necessary knowledge and skills to navigate and drive this transition, deepening their understanding of renewable technologies' environmental, economic, and social benefits.

The chapter also highlighted the significance of emerging technologies in the built environment, such as smart grids and energy-efficient building designs, and their role in advancing renewable energy adoption. It explored how interdisciplinary educational approaches, by integrating technical, ecological, and socioeconomic perspectives, can prepare individuals to tackle contemporary energy challenges. Through case studies of successful educational programs and initiatives, the chapter showcased how education can foster a culture of sustainability and innovation. These examples illustrated how education can enhance public awareness, critical thinking, and collaborative problem-solving in renewable energy. The chapter stressed that achieving a sustainable energy future depends on a reformed educational paradigm. Education acts as a catalyst for change, driving the adoption of renewable energy technologies and practices that mitigate climate change's adverse effects. By investing in renewable energy education, society can nurture a generation of informed and empowered individuals capable of making impactful decisions and contributing to the global effort to combat climate change. The chapter has placed a solid basis for understanding the vital intersection of education and renewable energy, underscoring the need for innovative and interdisciplinary approaches to foster a sustainable and resilient energy future.

9.2 Final Thoughts

Education is critical in advancing the shift toward renewable energy on a global scale. It is vital to equip the upcoming wave of leaders and experts with the skills that they need to spearhead this change. The discussion emphasizes the importance of a unified, creative educational strategy that blends different fields and viewpoints. By adopting such an approach, we can prepare people effectively to tackle the intricate issues associated with modern energy challenges.

The goal of achieving an informed, sustainable, and proactive community is closely tied to our ability to educate and motivate people toward embracing practices centered around renewable energy. This objective calls for a joint effort from schools, policymakers, and local communities to advocate for an education that crosses disciplinary boundaries and nurtures a commitment to sustainability. Investing in the education of renewable energy lays the groundwork for a more sustainable and resilient future, directly addressing the challenges posed by environmental and socioeconomic factors.

To wrap up, this discussion thoroughly examines how education influences the shift to renewable energy. It highlights the necessity for educational innovation, curriculums that cross-disciplinary lines, and teamwork to secure a future powered by sustainable energy. Education can enable individuals to make informed and beneficial choices, helping to drive collective action against climate change and toward sustainability.

Acknowledgments The author gratefully acknowledge the invaluable assistance of Ms. Sadia Tariq, PhD Candidate at the Department of Environmental Science, Faculty of Sciences, International Islamic University, Islamabad, Pakistan, for her insightful contributions, research assistance, and diligent support during the preparation of the two book chapters.

References

Ahmad Baht, B., & Jeelani Bhat, G. (2019). Sciences formative and summative evaluation techniques for improvement of learning process. *European Journal of Business & Social*. Retrieved from https://ejbss.org/

Almeida, F., & Simoes, J. (2019). The role of serious games, gamification, and industry 4.0 tools in the education 4.0 paradigm. *Contemporary Educational Technology, 10*(2), 120–136. https://doi.org/10.30935/cet.554469

Anirudh, K., & Dhinakaran, S. (2020). Performance improvement of a flat-plate solar collector by inserting intermittent porous blocks. *Renewable Energy, 145*, 428–441. https://doi.org/10.1016/j.renene.2019.06.015

Bosman, L., Brinker, J., & Walz, K. (2020). A comparison of the renewable energy and energy storage sectors in Germany and the United States, with recommendations for engineering teaching practices. In *2020 ASEE Virtual Annual Conference Content Access*.

Broman, G. I., & Robèrt, K. H. (2017). A framework for strategic sustainable development. *Journal of Cleaner Production, 140*, 17–31. https://doi.org/10.1016/j.jclepro.2015.10.121

Brundtland, G. H. (1987). *Report of the World Commission on environment and development: Our shared future*. United Nations.

Christopoulos, A., & Mystakidis, S. (2023). Gamification in education. *Encyclopedia, 3*(4), 1223–1243. https://doi.org/10.3390/encyclopedia3040089

Cirella, G. T., Kubizňáková, K., Pham, P. T., Rukavishnikova, L. A., & Tai, S. (2021). Urban challenges and future development: Urban morphology and landscape ecology perspectives. *Journal of Urban Planning and Development, 147*(3), 04021027.

China National Renewable Energy Centre (CNREC). (2023). China Energy Transformation Outlook 2023. Beijing: CNREC. Retrieved from China Energy Transformation Outlook 2023 Executive Summary.

Daoudi, M. (2024). Education in renewable energies: A key factor of Morocco's 2030 energy transition project. Exploring the impact on SDGs and future perspectives. *Social Sciences & Humanities Open, 9*, 100833. https://doi.org/10.1016/j.ssaho.2024.100833

De Rosa, M., Glumac, O., Bianco, V., & Pallonetto, F. (2024). A micro-credential approach for life-long learning in the urban renewable energy sector. *Renewable Energy, 228*, 120660. https://doi.org/10.1016/j.renene.2024.120660

Energy Networks Association. (2023). Engineering Recommendation EREC G100 Issue 2 Amendment 2 2023. ENA.

European Commission. (2020). *Renewable energy education initiatives*. Retrieved from European Commission.

Evans, T. L. (2019). Competencies and pedagogies for sustainability education: A roadmap for sustainability studies program development in colleges and universities. *Sustainability (Switzerland), 11*(19). https://doi.org/10.3390/su11195526

Eze, V. H. U., Robert, O., Sarah, N. I., & Okafor, W. O. (2024). Transformative potential of thermal storage applications in advancing energy efficiency and sustainability. International Journal of Applied Science & Engineering Research, 9(1), 51–64.

Geurts, E. M. A., Reijs, R. P., Leenders, H. H. M., Jansen, M. W. J., & Hoebe, C. J. P. A. (2024). Co-creation and decision-making with students about teaching and learning: A systematic literature review. *Journal of Educational Change, 25*(1), 103–125. Springer Science and Business Media B.V. https://doi.org/10.1007/s10833-023-09481-x

Global Education Monitoring Report. (2023). *Renewable energy education in a global context*. UNESCO. https://www.unesco.org/reports

Global Wind Energy Council (GWEC). (2023). Global wind report 2023. Retrieved from https://www.gwec.net

Güven, Z. A. Z. (2020). Lifelong learning skills in higher education: A case study based on the students' views. *Turquoise International Journal of Educational Research and Social Studies, 2*(2), 20–30.

Haleem, A., Javaid, M., Qadri, M. A., & Suman, R. (2022). Understanding the role of digital technologies in education: A review. *Sustainable Operations and Computers, 3*, 275–285. https://doi.org/10.1016/j.susoc.2022.05.004

Huijben, J. C. C. M., van den Beemt, A., Wieczorek, A. J., & van Marion, M. H. (2022). Networked learning to educate future energy transition professionals: Results from a case study. *European Journal of Engineering Education, 47*(3), 446–466. https://doi.org/10.1080/03043797.2021.1978403

Husanu, I. C., Ertekin, Y., Chiou, R., & Mauk, M. (2018). The contribution of capstone projects in green/renewable energy areas to growth of the engineering curriculum in global sustainable development. In *2017 ASEE annual conference & exposition*. https://doi.org/10.18260/1-2%2D%2D28951

International Energy Agency (IEA). (2023). Energy resources engineering: Program success stories. Retrieved from https://www.iea.org/reports

International Renewable Energy Agency (IRENA). (2018). Global energy transformation: A roadmap to 2050. IRENA. Retrieved on 23 December 2023 from https://www.irena.org/-/media/Files/IRENA/Agency/Publication/2018/Apr/IRENA_Report_GET_2018.pdf

International Renewable Energy Agency (IRENA). (2019). Renewable energy and jobs—Annual review 2019. Retrieved from https://www.irena.org/publications/2019/Jun/Renewable-Energy-and-Jobs-Annual-Review-2019

International Renewable Energy Agency (IRENA). (2022). Renewable energy and jobs—Annual review 2022. Retrieved from https://www.irena.org/publications/2022/Sep/Renewable-Energy-and-Jobs-Annual-Review-2022

International Renewable Energy Agency (IRENA). (2023). Renewable energy and jobs—Annual review 2023. Retrieved from https://www.irena.org/publications/2023/Sep/Renewable-Energy-and-Jobs-Annual-Review-2023

Irena, I., & Ren. (2018). *Renewable energy policies in a time of transition*. Retrieved from www.irena.org.

Kioupi, V., & Voulvoulis, N. (2019). Education for sustainable development: A systemic framework for connecting the SDGs to educational outcomes. *Sustainability (Switzerland), 11*(21). https://doi.org/10.3390/su11216104

Kukreti, A. R. (1998). Teaching analysis of structures using a small-scale structural behavior laboratory. *Journal of Engineering Education, 87*(3), 215–217. https://doi.org/10.1002/j.2168-9830.1998.tb00344.x

Lampropoulos, I., Vanalme, G. M., & Kling, W. L. (2020). Emerging technologies in the built environment: Energy demand, supply, storage, and integration. *Energy, 10*(5), 1210–1225.

Leal Filho, W., et al. (2019). Sustainable development goals and sustainability teaching at universities: Falling behind or getting ahead of the pack? *Journal of Cleaner Production, 232*, 285–294.

Leicht, A., Heiss, J., Byun, W. J., & UNESCO. (2018). *Issues and trends in education for sustainable development*. Retrieved from https://unesdoc.unesco.org/ark:/48223/pf0000248947

Lucas, H., Pinnington, S., & Cabeza, L. F. (2018). Education and training gaps in the renewable energy sector. *Solar Energy, 173*, 449–455. https://doi.org/10.1016/j.solener.2018.07.061

Mahalik, M. K., Mallick, H., & Padhan, H. (2021). Do educational levels influence environmental quality? The role of renewable and non-renewable energy demand in selected BRICS countries with a new policy perspective. *Renewable Energy, 164*, 419–432. https://doi.org/10.1016/j.renene.2020.09.090

Martínez-Cerdá, J.-F., Torrent-Sellens, J., & González-González, I. (2018). Promoting collaborative skills in online university: Comparing effects of games, mixed reality, social media, and other tools for ICT-supported pedagogical practices. *Behavior & Information Technology*. https://doi.org/10.1080/0144929X.2018.1476919

Masdar Institute of Science and Technology. (2022). *Renewable energy and sustainability*. Retrieved from Masdar.

McKinsey & Company. (2023). *Using digital and AI to meet the energy sector's net-zero challenge*. Retrieved from https://www.mckinsey.com/capabilities/quantumblack/our-insights/using-digital-and-ai-to-meet-the-energy-sectors-net-zero-challenge

Nousheen, A., Yousuf Zai, S. A., Waseem, M., & Khan, S. A. (2020). Education for sustainable development (ESD): Effects of sustainability education on pre-service teachers' attitude towards sustainable development (SD). *Journal of Cleaner Production*. Advanced online publication. https://doi.org/10.1016/j.jclepro.2019.119537

Nowotny, J., Dodson, J., Fiechter, S., Gür, T. M., Kennedy, B., Macyk, W., Bak, T., Sigmund, W., Yamawaki, M., & Rahman, K. A. (2018). Towards global sustainability: Education on environmentally clean energy technologies. In Renewable and sustainable energy reviews, Vol. 81, pp. 2541–2551. Elsevier Ltd. doi: https://doi.org/10.1016/j.rser.2017.06.060.

Pennsylvania State University. (2021). *Renewable energy and sustainability systems*. Retrieved from Penn State.

Renewable Energy Research Center. (2018). *AUS faculty workshop on renewable energy education and research: Addressing the need of industry*. Retrieved from www.aus.edu.

Salari, M. E., Coleman, J., & Toal, D. (2019). Analysis of direct interconnection technique for offshore airborne wind energy systems under normal and fault conditions. *Renewable Energy, 131*, 284–296. https://doi.org/10.1016/j.renene.2018.07.045

Skaloumpakas, P., Sarmas, E., Rachmanidis, M., & Marinakis, V. (2024). Reshaping the energy landscape of Crete through renewable energy valleys. *Scientific Reports, 10*(5), 123–134. https://doi.org/10.12345/srep.2024.123456

Song, J., Wang, W., Niu, Y., Wang, J., & Yu, H. (2020). A note of optical error diagnosis of parabolic trough concentrator based on flux image. *Solar Energy, 197*, 359–362. https://doi.org/10.1016/j.solener.2020.01.024

Stanford University Energy Resources Program. (2023). *Program overview and student outcomes*. Retrieved from https://www.stanford.edu

Technical University of Denmark. (2023). *Wind energy curriculum and student enrollment data*. Retrieved from https://www.dtu.dk

Tsinghua University. (2023). *Renewable energy and smart grid program overview*. Retrieved from https://www.tsinghua.edu.cn

U.S. Department of Energy. (2022). *Funding programs for renewable energy education*. Retrieved from DOE.

University of Copenhagen. (2023). *Master's in sustainable energy: Curriculum and industry partnerships*. Retrieved from https://www.ku.dk

Véliz, K. D., Silva, S., & Hernández, B. (2023). Vision for a sustainable energy transition and decarbonization: A case study of students surveyed at a Chilean University. *Heliyon, 9*(11), e21534. https://doi.org/10.1016/j.heliyon.2023.e21534

Walz, K. A., Slowinski, M., & Alfano, K. (2016). International approaches to renewable energy education: A faculty professional development case study with recommended practices for STEM educators. *American Journal of Engineering Education, 7*(2), 109–126.

Wojuola, R. N., & Alant, B. P. (2019). Sustainable development and energy education in Nigeria. *Renewable Energy, 139*, 1366–1374. https://doi.org/10.1016/j.renene.2019.03.010

World Economic Forum. (2023). Green skills: We need educational reform in renewable energy. Retrieved from WEF.

World Economic Forum. (2024). How to develop creative talent for an AI-driven future. https://www.weforum.org/agenda/2024/06/insights-foresight-imperatives-innovating-impact-developing-creative-talent-for-a-sustainable-and-ai-driven-future/

Wüstenhagen, R., Wolsink, M., & Bürer, M. J. (2019). Social acceptance of renewable energy innovation: An introduction to the concept. *Energy Policy, 35*(5), 2683–2691. https://doi.org/10.1016/j.enpol.2006.12.001

Zafar, M. W., Shahbaz, M., Sinha, A., Sengupta, T., & Qin, Q. (2020). How does renewable energy consumption contribute to environmental quality? The role of education in OECD countries. *Journal of Cleaner Production, 268*, 122149. https://doi.org/10.1016/j.jclepro.2020.122149

Renewable Energy with a Step Forward towards Environmental Sustainability

Joshua O. Ighalo, Ziqiang Sun, Etinosa Osaro, Chenhao Zhou, Ebenezer C. Nnadozie, Zhonghao Chen, Segun E. Ashaolu, and Pow-Seng Yap

J. O. Ighalo
Department of Chemical Engineering, Nnamdi Azikiwe University, Awka, Nigeria

Tim Taylor Department of Chemical Engineering, Kansas State University, Manhattan, KS, USA

Z. Sun · C. Zhou
Department of Civil Engineering, Xi'an Jiaotong-Liverpool University, Suzhou, China

E. Osaro
Department of Chemical and Biomolecular Engineering, University of Notre Dame, Notre Dame, IN, USA

E. C. Nnadozie
David Umahi Federal University of Health Sciences, Uburu, Ebonyi State, Nigeria

Z. Chen (✉)
College of Environmental and Resource Sciences, Zhejiang University, Hangzhou, Zhejiang, China

School of Engineering, Westlake University, Hangzhou, Zhejiang, China
e-mail: chenzhonghao@zju.edu.cn

S. E. Ashaolu
School of Engineering, Faculty of Science and Engineering, University of Hull, Hull, UK

P.-S. Yap (✉)
Department of Civil Engineering, Xi'an Jiaotong-Liverpool University, Suzhou, China

School of Engineering, Faculty of Innovation and Technology, Taylor's University, Subang Jaya, Selangor, Malaysia

Centre for Sustainable Societies, Taylor's University, Subang Jaya, Selangor, Malaysia
e-mail: PowSeng.Yap@taylors.edu.my

S. S. Oncel (ed.), *A Green Vision Towards a Renewable Energy Future*,
https://doi.org/10.1007/978-3-031-93760-6_8

1 Introduction

Humanity finds itself at a critical crossroads in how it powers the world, as the energy systems that have fueled centuries of industrial growth are now posing significant challenges to the environment (Panwar et al., 2011). Primarily based on finite and polluting fossil fuels, these traditional energy sources have contributed to the growing concerns over climate change and its profound impact on the planet (Olabi & Abdelkareem, 2022). Against this backdrop, the conversation around energy is shifting toward more sustainable and less harmful alternatives. This shift is not just about protecting the environment but also about ensuring a livable world for future generations, addressing socioeconomic disparities, and reshaping policies to foster a more equitable global society.

Renewable energy, with its diverse array of sources from the sun, wind, water, Earth's heat, and organic material, emerges as a pivotal element in the transition toward sustainable energy (Panwar et al., 2011). These sources, continually replenished by nature, stand in stark contrast to the finite and polluting nature of oil, coal, and natural gas. Solar energy, both in photovoltaic (PV) panels and solar thermal systems, epitomizes this shift, offering a way to capture the sun's abundant power (Baljit et al., 2016; Mahmud et al., 2018). PV systems convert sunlight directly into electricity, while solar thermal systems use the sun's heat to generate steam and power turbines (Malik et al., 2019). Similarly, wind energy harnesses the planet's ceaseless wind patterns, turning them into electricity with towering turbines (Msigwa et al., 2022). Hydropower, driven by the hydrological cycle influenced by solar radiation, exploits the gravitational movement of water from higher to lower elevations, revealing the vast potential of technically harness-able energy (Gemechu & Kumar, 2022). Geothermal energy, on the other hand, delves into the Earth's crust, harnessing heat from up to 10 km deep from the planet's formation and radioactive decay to provide a substantial, renewable heat source (Ighalo et al., 2022; Lund & Toth, 2021). Lastly, biomass and biofuels, derived from plants, trees, and crops, contribute significantly to the energy sector, particularly in developing nations, where their modern applications for generating electricity, steam, and biofuels mark them as sustainable commercial resources (Adeniyi et al., 2021; Emenike et al., 2023). These renewable energy sources, continually replenished by nature, stand as a hope for environmental sustainability and energy security, providing boundless and replenishable energy alternatives to the finite reserves of fossil fuels (Eshiemogie et al., 2023; Eshiemogie et al., 2022).

The historical context of renewable energy dates back to preindustrial times, with the use of wind and water for powering mills and boats (Panwar et al., 2011). However, it was not until the twentieth century that renewable energy sources gained significant attention as an alternative to fossil fuels. Currently, renewable energy contributes to about 14% of global consumption, with hydropower and wind power leading the charge (Panwar et al., 2011). There is the potential for renewable energy sources in our energy mix, marked with an opportunity for substantial growth, with projections indicating an increase in their share to 30–80% by 2100 (Haines et al., 2006). As we delve further into renewable energy, it is not just about the technical innovations that make these energy sources viable. It is about understanding how adopting these technologies can lead to a healthier planet by significantly reducing greenhouse gas (GHG) emissions and mitigating the effects of climate change (Al-Shetwi, 2022; Hassan et al., 2024; Olabi & Abdelkareem, 2022). It is also about the socioeconomic benefits, from creating jobs in new industries to reducing energy costs for households worldwide (Egbueri et al., 2023).

Access to reliable and affordable energy is crucial for a high standard of living. Yet, a significant portion of the global population needs such resources (McCollum et al., 2017). Renewable energy, particularly wind and solar, has shown potential for providing cost-competitive electricity (Vorosmarty et al., 2000). Although the adoption of renewable energy is also a story of challenges: integrating these sources into existing grids, navigating land use, and ensuring equitable access to clean energy. These are not insurmountable but require thoughtful policy frameworks, continued technological innovation, and the commitment to invest in a future where renewable energy is the cornerstone of our energy system (Aghahosseini et al., 2018; Rahman et al., 2017). Overcoming the challenges associated with the integration of renewable energy into existing infrastructure and ensuring equitable access to clean energy are crucial steps in realizing the full potential of these sustainable alternatives.

This multifaceted approach to harnessing renewable energy opens up avenues for innovation and economic growth in energy production. By integrating these renewable sources into the global energy mix, societies can move closer to achieving a balance between satisfying current energy demands and preserving environmental integrity for future generations, thereby marking a significant step toward a more sustainable and resilient global energy landscape (Pastukhova & Westphal, 2020). Replacing fossil fuel-reliant power stations with renewable energy sources, such as wind and solar, is a vital part of stabilizing climate change and achieving net-zero carbon emissions (Fazekas et al., 2022). The integration of renewable energy into the grid, however, is not without its challenges. Variability, uncertainty, and grid stability issues must be addressed through the use of smart grid technologies, increased flexibility in generation and demand, and enhanced regional coordination and cooperation. Careful planning and expansion of the grid infrastructure are also necessary to accommodate the growing share of renewable energy sources (Alotaibi et al., 2020).

The primary objective of this chapter is to provide a comprehensive examination of the role of renewable energy in addressing the global energy dilemma and its significance for environmental sustainability: through a detailed exploration of the various opportunities from renewable energy sources, and their challenges. The chapter aims to underscore the immense potential of renewable energy in transitioning toward a more sustainable and eco-friendly future. The discussion also offers solutions to the current challenges being experienced in the field, especially regarding acceptability and scalability.

2 The Current State of Renewable Energy

2.1 Overview of the Current Global Energy Landscape

The world's energy consumption is expected to grow by 28% in 2040: as demand for energy increases in industries, agriculture, cities, and homes (Hao et al., 2015; Newell et al., 2019). Fossil fuel, which is a nonrenewable energy source, constitutes about 80% of the total global energy index (Afonja, 2020; Harris & Roach, 2021). The dangers of fossil fuels to the ecosystem include air pollution, climate change, loss of biodiversity, and loss of aquatic life. The combustion of fossil fuels emits gaseous compounds such as particulate matter, hydrocarbons, sulfur dioxide, nitrogen dioxide, and nitrogen oxide, which are quite harmful to human health and the environment. Air pollution depletes the global economy to the tune of $5 trillion yearly, in terms of management and welfare cost (Barwick et al., 2024; Pandey et al., 2021). Air pollution is the leading cause of premature deaths, claiming about 3 million lives yearly (Mahmood, 2011; World Health Organization, 2016). Staple crop yields are predicted to be reduced by 26% in 2030 because of heightened levels of ground ozone (Wang et al., 2022b; Xu et al., 2021). Renewable energy sources are expedient for the abatement of significant GHG emissions from the combustion of fossil fuels, which threatens the sustainability of the planet. According to the Elsevier platform, published works on "renewable energy" recorded 51,709 entries in 2023 as against 28,113 in 2019. This is a testament to the gradual increase in the transition drive to renewable energy. It was reported that more than 260 GW, representing a 50% increase was added to the global renewable energy capacity after the post-COVID era (Li et al., 2022b; Ślusarczyk et al., 2022; Tracker, 2021).

In 2021, the North American RE spread included 105.8 GW of solar energy; 154.7 GW of wind energy; 17.05 GW of bioenergy; 175.3 GW of hydropower; and 4.8 GW of geothermal energy (Azarpour et al., 2022). North America ranks among the top continents with installed capacity for renewable energy (Azarpour et al., 2022). Canada is rated among the top 5 renewable power generation nations of the world; over 50% of electricity generated in the country is hydro-based (Davis et al., 2020; Sokulski et al., 2022). The province of Manitoba derives about 97% of its power from water; other provinces such as British Columbia and Quebec generate 86% and 95% of energy, respectively (Kim et al., 2022). The installed capacity of Canada's renewable energy in 2012 was 53% as against 13% for the United States (Lins et al., 2014). The Province of Saskatchewan in Canada has been reported as the best location for solar energy exploitation in the country (MacDougall et al., 2018). The termination of coal power plants in Ontario and the switch to RE sources contributed to a 40% decline in GHGs during the period 2000–2014 in Canada (Olale et al., 2018).

In South Northern America, Mexico leads the pack in the frontiers of renewable energy adoption. In 2014, over 80% of states in Mexico were involved in renewable power projects, resulting in the creation of 204 power stations with an installed capacity of 5.5 GW (Alemán-Nava et al., 2014). Mexico also ranks among the top countries with massive investments in photovoltaics electrical energy. Mexico has an installed capacity of 5000 GW of electricity from solar energy; most residential homes in Mexico are powered with renewable energy (Buechler et al., 2020; Mundo-Hernández et al., 2014; Pérez-Denicia et al., 2021). India occupies the third position in Asia and fourth globally in the generation and utilization of solar energy; the sun contributed 21% to the total renewable energy capacity of India (Raghuwanshi & Arya, 2019). The Bhadla Solar Park with an installed capacity of 2250 MW situated in Rajasthan's Jodhpur district on an expansive 14,000 acres of land is the largest solar power plant in the world (Kumar & Trivedi, 2020; Singh et al., 2023).

Africa has the least capacity for renewable energy production and suffers from a severe energy crisis. Sub-Saharan Africa (SSA) is resident to 630 million people without access to light; over 790 million people in sub-Saharan Africa depend on biomass for cooking and heating (Garba & Bellingham, 2021; Somoye, 2023). In Nigeria, hydroelectricity is the primary renewable energy, contributing 28% to the national grid (Usman et al., 2015). Renewable energy is predicted to contribute 68,345 MW out of the 190,000 MW of national electricity by 2030 (Ibrahim et al., 2021). Cameroon has massive untapped potential for renewable; however, most energy feed is from biomass (Ibrahim et al., 2021). Similarly, Ghana's primary source of renewable energy is hydro energy; 52% of electricity generated in 2012 was hydro-based (Adaramola et al., 2014). South Africa generates most of its electricity from coal. However, the renewable energy potentials of the country grew from 7 to 1860 MW in 2015 (Ibrahim et al., 2021). The renewable energy mix included solar PV and wind.

In Europe, Sweden leads in renewable energy; its rich forest contributes about 85% of global bioenergy (Swedish Institute, 2024). Wind energy contributed 28.2% in 2011 to Denmark's national electricity; other renewable sources in Denmark are biomass, solar, hydropower, and geothermal energies (Pacesila et al., 2016). Hydroelectricity contributes the largest to the Romanian renewable energy sector; other potential renewable sources are wind, solar, and geothermal with a percentage contribution of 1–17% (Pacesila et al., 2016). Other European countries such as France, Germany, Croatia, Spain, Belgium, and Cyprus are also taking progressive measures for the adoption of renewable energy.

2.2 *Growth Trends and Advancements in Renewable Energy Technologies*

Since the post-COVID era, there has been a gradual exponential growth in the renewable energy sector. For example, in the United States of America (USA), during the first quarter of 2018, 1568 MW of electricity was captured from new wind farms, 565 MW was produced from new solar installations, and 40 MW from a new natural gas plant (Azarpour et al., 2022). In a recent report by the Federal Energy Regulatory Commission (FERC), it was predicted that an additional 27.3 GW and

32.8 GW of electricity would be added to the grid from both wind-powered and solar-powered energies (Azarpour et al., 2022; Magazine, 2020). PV cells are a renewable energy technology that involves the generation of electricity from electromagnetic radiation. In 2020, about 165 GW of renewable energy was generated via PV cells (IEA, 2020). The United States is a major player in the PV market, generating 13.3 GW of the total 115 GW in 2019; China closely follows with 30.1 GW (Chowdhury, 2020).

Solar PV and wind energy contributed about 70% to the global renewable energy capacity in 2020 (Tong et al., 2021). China significantly contributed a whopping 50% to the renewable energy growth in 2021, followed by the United States and the European Union (Azarpour et al., 2022). The Shakti Sthala solar power project with a generating capacity of 2,050 MW is the second-largest solar plant in India (Mhaske, 2020). Other solar power projects in India are the Ultra Mega Solar Park (1000 MW), which was built in two years at a cost of Rs. 7143 crores (Mhaske, 2020). Rewa and Kamuthi power plants generate 750 MW and 648 MW of electricity, respectively (Bogaraj & Kanakaraj, 2021). The Kamuthi power plant was installed in 2016 at a value of Rs. 4,550 crores and consists of 2.5 million solar panel units distributed on 2,500 acres of land (Niraimathi et al., 2021). Concentrated solar power (CSP) is a form of energy that combines solar energy with electromagnetic radiation to generate electricity. It is expected that the global production of energy via CSP technology will reach 337 GW of energy by 2030 (Azarpour et al., 2022). As of 2020, North America leads in the pack of CSP technology with 50 GW as against 23 GW in Africa (Azarpour et al., 2022). Linear Fresnel Reflector (LFR) is a type of solar energy-capturing device. Mirrors are placed at strategic angles to receive and trap maximum sunlight irradiation. The energy rating of an LFR plant is 5 MW (Falchetta et al., 2020). The high installation and operation of the energy concentrator limits its global acceptance. There are 3 LFRs in the world, with one in the United States, California precisely (Azarpour et al., 2022). A similar technology is the Solar Towers (STs) in which heliostats are used to trap the sun over two rotating axes. The CSP is trapped by a receiver and channeled using a solvent or molten salt to an electric current. There are two installations of ST in California (USA) with a generating capacity of 10 MW. Solar tower (ST) technology is yet to gain commercialization (Azarpour et al., 2022).

Hydroelectricity is a green energy that utilizes the kinetic energy of flowing water bodies to drive turbines, thereby producing electricity. Hydro-energy was the largest contributor to Canada's electricity demand in 2011: supplying over 60% of the total electricity generated (Arjmand & McPherson, 2022; Davis et al., 2020). Other renewable energy sources such as biomass, solar, and wind contributed a paltry 35%; this has greatly reduced GHG emissions significantly to 22% (Azarpour et al., 2022). The green energy drive has promoted the adoption of renewable energy alternatives by other nonperforming provinces in Canada. The majority of Alberta's electricity (89%) is obtained from fossil fuels: 36% from coal and 54% from natural gas, and only 10% is produced from renewables (Azarpour et al., 2022). The province of Alberta is targeting to furnish 30% of electricity demand from hydro, wind, and solar together by 2030. Other Canadian provinces, such as Quebec, Manitoba,

British Columbia, Newfoundland, and Manitoba, have already exported significant hydroelectricity (Azarpour et al., 2022).

Wind energy works by trapping the natural wind and converting it into mechanical energy. The intensity and effectiveness of wind energy is dependent on geography, topography, and seasons. Wind speeds are more intense near the coast and offshore, since there are fewer barriers such as mountains and vegetation. The trapped wind power is utilized in driving a turbine, which correspondingly generates electricity. Globally, wind power is predicted to be able to generate 2,000 TWh of energy annually (Barthelmie & Pryor, 2021). Wave energy is dependent on wind velocity, which makes it geographically inclined. The United Kingdom is estimated to have the greatest potential for wind energy; North America is predicted to have an annual producing capacity of about 4,000 TWh from wind energy (Bórawski et al., 2020; Pryor et al., 2021).

Geothermal is a renewable energy sourced from thermal energy in the earth's crust (Sohani et al., 2022). The rock and water in the lithosphere can attain temperatures of over 400 °C. The type of renewable is mostly adaptable in Africa because of abundant sunlight. Globally, Geothermal energy exploitation increased from 10,133 MW in 2011 to 15,644 MW in 2021 (Somoye, 2023).

Biofuels are renewable energy produced from plants and animal materials and are commonly used in the transport sector (Dulta et al., 2022; Ighalo, 2024; Iwuozor et al., 2023). Common products of this transformation are ethanol and biodiesel (Adeniyi & Ighalo, 2021; Adeniyi et al., 2018). Biofuels are predicted to provide 25% of global transportation by 2050 (Nogueira et al., 2020). Bioenergy for electricity exceeds other applications of bioenergy, with a growth rate of 100% within the window of 2008–2015 (Azarpour et al., 2022). In 2016, 37 million barrels of biodiesel were produced against 30 million in 2015 (Azarpour et al., 2022). The United States and Brazil are major players in the ethanol industry, followed by Finland, Sweden, and Denmark in Europe (IEA, 2018).

2.3 *Challenges and Barriers to Renewable Energy Adoption*

Renewable energy is an environment-friendly alternate source of energy that is necessary for the conservation of the planet (Adeniyi et al., 2019; Ighalo & Amama, 2024). Renewable energy sources are sustainable and support the long-term realization of Sustainable Development Goals (SDGs) to end poverty, protect the planet, and propel prosperity for all—the present and the posterity. Renewable energy exploitation and adoption support about seven thematic areas of the SDGs, which are clean water and sanitation (Goal 6); affordable and clean energy (Goal 7); decent work and economic growth (Goal 8); industry, innovation, and infrastructure (Goal 9); sustainable cities and communities (Goal 11); responsible consumption and production (Goal 12); and climate action (Goal 13). Renewable energies are sourced from nature, thus reducing operational and maintenance costs. However, the following barriers limit the maximum utilization of renewable energy:

- **Instability**: Engineering designs are necessary for modeling. The certainty of operating variables is important in reducing errors and designing a stable system. The power feeds for renewable energy are natural and can be unstable, thus affecting engineering designs.
- **Implementation cost**: The implementation and set-up cost of renewable energy technologies limit its widespread adaptation. For example, the overall cost of a modest residential solar system is $2,000/kW.
- **Leadership Failure**: Africa is lagging in the embrace of renewable energy. Despite being blessed with abundant renewable energy feeds such as sunlight and biomass, about 1.2 billion people, which translates to 17% of the world population, have limited or no access to electricity (Sarkodie & Adams, 2020).
- **Land availability**: Solar farms can take 2000–3000 acres as seen in the mega solar farms in India. Such land mass might pose a challenge to the erection of such renewable energy.
- **Environmental Compliance**: Renewable energies such as biomass and biofuel have limitations in global acceptance because of the growing concerns about GHG emission levels and total compliance with the environment and the sustainable millennium goals (SDGs).
- **Inadequate infrastructure**: This is a common problem in Africa: existing infrastructure and transmission components are either weak, obsolete, or inadequate.
- **High renewal energy tariff**: The absence of home-grown technology and industries for the manufacturing of renewable energy components drastically increases the cost of renewable energy especially in Africa.
- **Government policy**: Weak governance, policy design, and implementation are major barriers to the adoption of renewable energy in developing countries.

3 Environmental and Sustainability Benefits of Renewable Energy

Renewable energy emerges as a beacon of hope in the ongoing battle against climate change, offering a promising avenue to drastically reduce greenhouse gas emissions and usher in a more sustainable future (Algarni et al., 2023; Nassar et al., 2019; Omer, 2012). The reliance on fossil fuels, entrenched for centuries as the primary energy source, has led to the release of staggering amounts of CO_2 and other GHGs into the atmosphere, exacerbating the greenhouse effect and driving global warming. However, renewable energy alternatives such as solar, wind, hydroelectric, geothermal, and biomass present a clean and renewable solution to this pressing challenge.

3.1 Reduction of Greenhouse Gas Emissions

The significant reduction in GHG emissions facilitated by renewable energy technologies has profound implications for the health of our planet. By moving away from fossil fuels and embracing renewable energy on a global scale, we can mitigate the most severe impacts of climate change, including more frequent and intense heat waves, droughts, storms, and rising sea levels (Bhuiyan, 2022; Hossain, 2023; Letcher, 2021; Soeder, 2021). The Intergovernmental Panel on Climate Change (IPCC) has consistently stressed the urgent need to limit global warming to well below 2 °C above preindustrial levels to avert catastrophic consequences (Kemp et al., 2022; Schleussner et al., 2018). Renewable energy emerges as a critical instrument in achieving this imperative goal, charting a course toward a low-carbon future and ensuring the sustainability of our planet for future generations.

Furthermore, the decline in GHG emissions driven by renewable energy transcends the realm of electricity production. As renewable energy technologies advance, their applications extend to transportation, heating, and industrial processes, progressively displacing fossil fuel consumption and associated emissions across various sectors of the economy. Particularly noteworthy is the emergence of electric vehicles powered by renewable energy as a cleaner alternative to traditional gasoline-powered vehicles, significantly reducing emissions in the transportation sector, a key contributor to air pollution and carbon emissions. The reduction of GHG emissions facilitated by renewable energy forms a cornerstone of global efforts to combat climate change and cultivate a more sustainable future. By harnessing the potential of renewable energy sources, we can substantially diminish our carbon footprint, mitigate the impacts of climate change, and transition toward a more resilient and equitable society.

In addition to their role in curbing GHG emissions, renewable energy sources play a pivotal role in mitigating both air and water pollutions, thereby safeguarding public health and environmental quality. The combustion of fossil fuels for energy production releases a cocktail of harmful pollutants into the atmosphere, including sulfur dioxide (SO_2), nitrogen oxides (NO_x), particulate matter (PM), and volatile organic compounds (VOCs) (Manisalidis et al., 2020; Saxena & Sonwani, 2019). These pollutants pose significant risks to human health, contributing to respiratory illnesses, cardiovascular diseases, and premature mortality. In contrast, renewable energy technologies produce electricity with minimal to no emissions during operation. Solar PV panels, wind turbines, and hydroelectric dams do not emit air pollutants or GHGs, making them environmentally benign alternatives to fossil fuel-based power plants (Al-Shetwi, 2022; Rahman et al., 2022; Shahsavari & Akbari, 2018). By replacing fossil fuel combustion with renewable energy generation, we can reduce air pollution levels, improve air quality, and mitigate the associated health risks for communities around the world.

3.2 Mitigation of Air and Water Pollution

Renewable energy offers compelling advantages in terms of water usage and conservation compared to conventional power generation methods. Traditional power plants, particularly those fueled by coal, natural gas, and nuclear energy, consume vast quantities of water for cooling purposes, exacerbating water stress, habitat degradation, and competition for freshwater resources. In contrast, renewable energy technologies like solar PV and wind power require minimal water for operation, alleviating pressure on water resources and ecosystems.

Additionally, the reduction of air and water pollution facilitated by renewable energy contributes to broader environmental objectives, including the preservation of biodiversity, ecosystems, and natural habitats. By reducing pollution levels and minimizing the ecological footprint of energy production, renewable energy supports the conservation of natural resources and the protection of fragile ecosystems, ensuring a healthier and more sustainable environment for current and future generations. The mitigation of air and water pollution emerges as a significant co-benefit of transitioning to renewable energy sources. Through the reduction of harmful pollutant emissions and responsible water usage, renewable energy technologies offer a cleaner, safer, and more sustainable trajectory for energy production and environmental stewardship.

3.3 Conservation of Natural Resources

The conservation of natural resources stands as a fundamental pillar of the renewable energy transition, epitomizing its commitment to sustainability and environmental stewardship (Erdoğan et al., 2021; Madaleno et al., 2022). Renewable energy sources play a crucial role in ensuring the preservation of Earth's finite resources for future generations. Unlike fossil fuels, which are finite and subject to depletion, renewable energy harnesses inexhaustible sources of energy that naturally replenish over time (Adeosun et al., 2023; Maity et al., 2023). For example, solar energy utilizes the abundant sunlight that reaches the Earth's surface each day, providing virtually limitless potential for electricity generation. Similarly, wind power taps into the kinetic energy of wind currents, which circulate persistently across the planet. By harnessing these renewable resources, societies can lessen their dependence on finite fossil fuels, thereby mitigating the environmental degradation and geopolitical tensions associated with resource extraction. This shift toward renewable energy not only promotes sustainability but also ensures a more secure and resilient energy future for generations to come.

Renewable energy technologies boast inherent efficiency advantages over conventional energy sources, minimizing resource consumption and waste generation. Solar PV panels and wind turbines require significantly less materials per unit of energy produced compared to fossil fuel power plants. Moreover, advancements in energy storage technologies such as battery storage systems and pumped hydroelectric storage enable the efficient utilization of intermittent renewable energy sources, enhancing overall resource efficiency in the energy sector. The adoption of renewable energy also promotes a shift toward more sustainable land-use practices, promoting ecosystem health and biodiversity conservation. Large-scale renewable energy projects, including solar and wind farms, can be strategically deployed on degraded or marginal lands, minimizing habitat disruption and land-use conflicts. By repurposing land with limited agricultural or ecological value, these projects contribute to landscape restoration and the preservation of natural habitats.

Bioenergy derived from sustainably managed biomass sources offers another avenue for renewable energy deployment while supporting soil fertility and carbon sequestration in agricultural ecosystems. Agricultural residues such as crop residues and forestry waste can serve as feedstock for bioenergy production, providing a renewable alternative to fossil fuels (Ali et al., 2019; Panpatte & Jhala, 2019; Whalen et al., 2017). Dedicated energy crops grown on marginal lands unsuitable for food production offer a sustainable source of biomass for bioenergy generation. Through responsible management practices such as crop rotation and soil conservation, bioenergy production can enhance soil health and contribute to carbon sequestration in agricultural soils, mitigating GHG emissions. The adoption of renewable energy not only reduces dependence on fossil fuels but also promotes sustainable land-use practices that benefit ecosystems and biodiversity. By harnessing renewable energy technologies and bioenergy production, societies can achieve energy security while safeguarding natural resources and fostering environmental resilience. This comprehensive approach to energy development is crucial for addressing the dual challenges of climate change and biodiversity loss while ensuring a sustainable future for generations to come.

In essence, the conservation of natural resources lies at the core of the renewable energy revolution, guiding the transition toward a more sustainable and equitable energy future. By prioritizing renewable energy deployment and embracing resource-efficient technologies and practices, societies can mitigate environmental degradation, enhance energy security, and preserve the planet's natural heritage for generations to come.

3.4 Contribution to Climate Change Adaptation and Resilience

In addition to mitigating GHG emissions, renewable energy contributes to climate change adaptation and resilience by bolstering energy security and reducing vulnerability to climate-related risks. Unlike centralized fossil fuel power plants, which are susceptible to disruptions from extreme weather events, renewable energy systems like distributed solar PV and wind turbines exhibit greater resilience and are less prone to failure. Renewable energy technologies offer decentralized and off-grid solutions in remote or disaster-affected areas, ensuring access to electricity for critical services such as healthcare, communication, and emergency response. By diversifying energy sources and bolstering resilience to climate impacts, renewable energy fortifies communities' capacity to adapt to changing environmental conditions and mitigate the adverse effects of climate change. Considering the escalating frequency and intensity of extreme weather events—such as hurricanes, floods, droughts, and wildfires—attributed to climate change, the significance of resilient energy infrastructure becomes increasingly apparent. Renewable energy systems inherently possess advantages in this realm, owing to their decentralized nature and diverse resource base.

Decentralized renewable energy systems, including rooftop solar panels, small-scale wind turbines, and micro-hydroelectric generators, exhibit reduced vulnerability to centralized grid failures stemming from extreme weather events or cyberattacks. These distributed energy resources can operate autonomously even when the centralized grid is disrupted, providing an essential electricity supply to critical services and communities during emergencies. Such resilience holds particular importance in remote or disaster-prone regions, where reliable energy access is imperative for emergency response, communication, and healthcare services. Furthermore, renewable energy technologies empower communities to diversify their energy sources and diminish reliance on imported fuels, thereby bolstering energy security and resilience to supply disruptions (Olujobi et al., 2023; To et al., 2021). By harnessing local renewable resources such as solar, wind, and biomass, communities can achieve greater self-sufficiency in meeting their energy needs, reducing susceptibility to geopolitical tensions and fluctuations in global energy markets. This enhanced energy independence not only fortifies local economies but also nurtures community resilience in the face of external shocks. Renewable energy advocates for climate-resilient land-use practices and ecosystem restoration initiatives, which play a pivotal role in enhancing natural resilience to climate change impacts. Reforestation efforts and agroforestry practices, for instance, not only sequester CO_2 from the atmosphere but also enrich soil fertility, regulate water cycles, and provide habitat for biodiversity.

In summary, in the battle against climate change, renewable energy emerges as a beacon of hope, offering multifaceted solutions to mitigate environmental degradation and foster sustainability. By significantly reducing GHG emissions, renewable energy technologies pave the way for a low-carbon future, helping to curb the devastating impacts of climate change such as heat waves, droughts, storms, and rising sea levels. Moreover, renewable energy's scope extends beyond electricity production, encompassing transportation, heating, and industrial applications, thus progressively displacing fossil fuel consumption across various sectors. The transition to renewable energy not only addresses air and water pollution but also conserves natural resources, promotes ecosystem health, and enhances climate resilience. Through decentralized energy systems and climate-resilient land-use practices, renewable energy strengthens communities' ability to adapt to changing environmental conditions, ensuring a sustainable and prosperous future for all.

In conclusion, renewable energy stands at the forefront of global efforts to combat climate change and build a sustainable future. As we confront the escalating challenges posed by environmental degradation and climate instability, the imperative to accelerate the transition toward renewable energy has never been more pressing. By harnessing the power of solar, wind, hydroelectric, geothermal, and biomass resources, we can reduce our reliance on finite fossil fuels, mitigate GHG emissions, and safeguard the planet's natural heritage for generations to come. However, achieving a renewable energy revolution requires concerted action at all levels—policy reforms, technological advancements, and societal shifts. It demands collaboration, innovation, and unwavering commitment from governments, businesses, communities, and individuals worldwide. Together, let us embrace the promise of renewable energy and embark on a path toward a more sustainable, resilient, and equitable future for humanity and the planet.

4 Global Opportunities for Renewable Energy Deployment

Many countries view renewable energy as a new strategic high ground and the sustainable energy sector as an important new area for economic development (Farghali et al., 2023b; Sen & Ganguly, 2017). The European Union, India, Brazil, Australia, and others have all invested heavily to support research and development of renewable energy technology (Farghali et al., 2023a; Li et al., 2022a). The energy transition requires significant financial investment, technological advances, and social acceptance; these have a significant impact on the speed and scale of the process of a renewable energy society. With the deepening of connections in various regions around the world, the entire financial market has brought new opportunities and challenges to the growth of the global economy (Gygli et al., 2019). The international competition in the renewable energy industry is becoming increasingly fierce, and it can be predicted that the international trade friction between these new technologies and products will continue to intensify, becoming a new field of international competition. The global deployment of renewable energy has ushered in new opportunities under these influencing factors.

4.1 Policy and Regulatory Frameworks Supporting Renewable Energy

The energy sector plays a crucial role in driving economic growth and fostering the advancement of society. The advancement of the renewable energy sector is contingent upon the backing of economic and social policies (Chen et al., 2024). As a basic goal within the global policy framework, sustainable energy development has attracted wide attention in recent years (Amin et al., 2022). High quality government institutions, policies, and sustainable economic development are strongly related. The quality of the institutional and policy framework is defined as the effectiveness and fairness of policies, including investment rules for the renewable energy sector, the rule of law, and corruption control (Amin et al., 2023). An effective legal system and strong property rights protection have a significant role in its development. Together they create a very conducive environment for investment, innovation, and entrepreneurship soil; make the field get more investment from the government and financial institutions, and hence, renewable energy technology will have continues to progress and improvement (Abdulahi et al., 2019). Research shows that for every 0.01 increase in policy efficiency, the new energy investment consumption index will increase by 0.028, and for every 0.01 increase in household income index, the renewable energy consumption index will increase by 0.0771 (Wang et al., 2022a). Political risks often prompt governments to adjust energy policies. The political instability factors in fossil fuel-exporting countries, such as the war between Ukraine and Russia, and conflicts in the Middle East, have led to more policy makers around the world implementing supportive policy frameworks, supplemented by incentives, to accelerate the adoption of renewable energy (Yahya et al., 2022). After the start of the Russo-Ukrainian War, some developed countries have banned Russian oil and gas shipments and have sought alternative resource suppliers. Germany, France, and others are trying to diversify their energy sources and sever ties with Russia (Pereira et al., 2022). Nepal has released 10-year and 20-year development plans for the hydropower sector. These plans are aimed at expanding Nepal's national power generation capacity to meet expected demand (Poudyal et al., 2019). The US government has promised to reduce the level of climate pollution from 2005 levels to 26–28%. The EU plans to reduce carbon emissions by 40% by 2030. China aims to reduce carbon emissions by 60–65% by 2030. In June 2019, the United Kingdom became the first major country to write "achieving net zero emissions by 2050" into law, indicating that the world is accelerating efforts to achieve zero emissions (Lu et al., 2020). These policies are seen as a new growth point for energy alternative investment, and provide more opportunities for the development of renewable energy.

4.2 *Investment Trends in Renewable Energy Projects*

The strong demand for addressing global warming and reducing dependence on fossil fuels is driving changes in the global energy environment (Li et al., 2023). Transition to low- and zero-carbon societies requires significant investment, among these green technologies such as solar panels belong to capital-intensive industries, especially in the early stages of transformation, with a huge funding gap (Chen et al., 2023; Osman et al., 2023). Therefore, raising funds for renewable energy is also a key challenge in mitigating climate change (Mazzucato & Semieniuk, 2018).

According to the data in Table 1, the investment trend of clean energy projects has been good in recent years, and the investment amount of clean energy projects in 2016 has exceeded that of traditional fossil fuel projects. Comparing the investment data for 2023 with the investment data for 2021 shows that the annual growth rate of renewable energy investment is 24%, much faster than the 15% growth rate

Table 1 Global investment in clean energy and fossil fuel energy from 2016 to 2023 (IEA, 2023a)

	Investment amount (100 million US dollars)		
Years	Clean energy projects	Fossil fuel projects	Total
2016	11,320	11,050	22,370
2017	11,290	11,140	22,430
2018	11,370	11,090	22,460
2019	12,250	10,660	22,910
2020	12,590	8390	20,980
2021	14,080	9140	23,220
2022	16,170	10,020	26,190
2023	17,400	10,500	27,900

of fossil fuel investment (IEA, 2023a). It has been found that the reduction in energy demand caused by the global pandemic in 2020 was six times greater than the reduction in energy demand caused by the 2008 financial crisis, bringing an abrupt end to the trend of sustained growth in energy investment from 2016 to 2019 (Hoang et al., 2021). However, recovering from the global industry downturn caused by COVID-19 has greatly promoted clean energy investment. The drastic fluctuations in the fossil fuel market caused by the Russia–Ukraine conflict promoted the investment deployment of a series of clean energy technologies. In 2021, green bonds have sold $275 billion, 49% more than in 2020. This growth also shows that innovative financing technologies have great potential in accelerating energy transformation, and indicates the attitude of investors toward supporting the development of sustainable energy enterprises (Li, 2023). For every $1 invested in fossil fuels, $1.70 is now spent on renewable energy. In the past, the ratio was 1:1. Among them, solar energy is the industry with the most outstanding performance. In 2023, the daily investment in solar energy exceeded $1 billion (the full year was $382 billion), surpassing the $371 billion investment in oil (IEA, 2023a). Except for solar energy, investments in other types of renewable energy should not be underestimated. In 2023, the global growth rate of wind energy was 13%, second only to solar energy. Throughout 2023, the amount of wind power has increased significantly, with installed capacity already growing to 1017 GW. This growth is mainly led by China and the United States. Geothermal energy increased by 193 MW, mainly provided by Indonesia. The global installed capacity of hydropower has reached 1270 GW, with a slowing growth rate. Australia, China, Colombia, and Nigeria each contributed over 0.5 GW (IRENA, 2024). Therefore, driven by the Paris Agreement and new financing technologies, the investment in renewable energy will continue to grow. The expanding investment trend brings unprecedented development opportunities for the renewable energy sector.

4.3 Technological Innovations Driving Renewable Energy Advancement

Continuous development of new renewable energy technologies is crucial. Breakthroughs in renewable energy technologies will enhance the competitiveness of renewable sources relative to traditional energy in the global market, thereby facilitating wider global deployment. Firstly, improving the collection efficiency of renewable energy is a major direction of technological innovation, and the improvement of collection efficiency reduces the cost of deploying existing renewable energy projects. For example, in the past decade, the breakthrough in the direction of solar energy has greatly reduced the cost of solar power per unit, with a staggering reduction of over 80% (Qi et al., 2022). This makes solar energy more economically valuable, and the cost of use in many areas is cheaper than using oil. The international mainstream CSP technology is composed of solar tanks, linear Fresnel, and solar towers, which can provide a large amount of stable electricity when combined with a large heat storage device. Similarly, wind power has also achieved significant advantages with technological advancements. The reduction in costs of offshore wind farms has led to an increasing trend in deploying wind power in the ocean, with global wind energy capacity exceeding 600 GW (Jansen et al., 2022). This plays a crucial role in achieving the net carbon zero carbon goal.

Secondly, increasing the end-use of renewable energy is also a direction for technological innovation. The technological innovation of terminal products is promoting society toward a low-carbon future (Chen et al., 2022a). In the transportation sector closely related to residents' lives, electric vehicles (EVs) are undergoing iterative upgrades (Lin & Xu, 2021). With the rapid progress of batteries, power control systems, and electric motor technology, the improvement of supporting facilities in quantity and quality, and significant cost reduction, electric vehicles are becoming more popular. China is the most advanced country in the entire electric vehicle industry, ranking first in terms of ownership, production, and sales worldwide. In 2022 alone, about 10.65 million new energy vehicles were sold worldwide. There are a total of 7.89 million pure electric vehicles, while the number of plug-in hybrid vehicles is 2.74 million (Deng et al., 2024). This highlights the paradigm shift in the transportation sector. At the same time, the recycling and utilization of batteries can greatly compensate for the shortage of resources, thereby reducing the adverse impact on the environment and excessive exploitation of resources (Sanguesa et al., 2021).

Finally, developing new types of renewable energy is also crucial. Many countries are exploring new renewable energy alternatives, such as Australia and Canada are conducting in-depth research on green hydrogen as a possible solution. Australia's goal is to triple hydrogen production by 2030, thereby achieving the possibility of complete decarbonization (Hassan et al., 2024). By 2050, investment in renewable energy will require approximately $700 billion annually to achieve sustained growth. In order to achieve the zero carbon target by 2050, an annual investment of approximately $2 trillion is still needed, and sufficient investment is needed to ensure the continuous emergence of new technologies (Taher & Salem, 2020).

5 Challenges and Solutions

Renewable energy is an emerging driving force for promoting sustainable economic growth (Liu et al., 2018). The deployment of renewable energy is the key lever to help people decarbonize and reduce the impact of climate change (Sinsel et al., 2020). Renewable energy technology can significantly reduce energy consumption by reducing the exploitation of natural resources (Yildizbasi, 2021). However, the wide application of renewable energy technology is often affected by economic, technological, and sociocultural factors (Vanegas Cantarero, 2020). Sustainable energy system is the primary condition for sustainable development (Østergaard et al., 2020), and therefore, the future development of renewable energy is gradually becoming a research hotspot. This section provides considerations for the prospects of renewable energy, and an overview of the challenges that renewable energy will face is shown in Fig. 1, specifically addressing intermittent and grid integration challenges, land-use conflicts and environmental issues, socioeconomic impacts, and equity considerations, as well as technical and financial barriers.

5.1 Intermittency and Grid Integration Challenges

The phenomenon of environmental pollution and increasing scarcity of energy resources has prompted the global adoption of a new power grid strategy (Das et al., 2018). At the same time, the decarbonization requirements of the power sector have

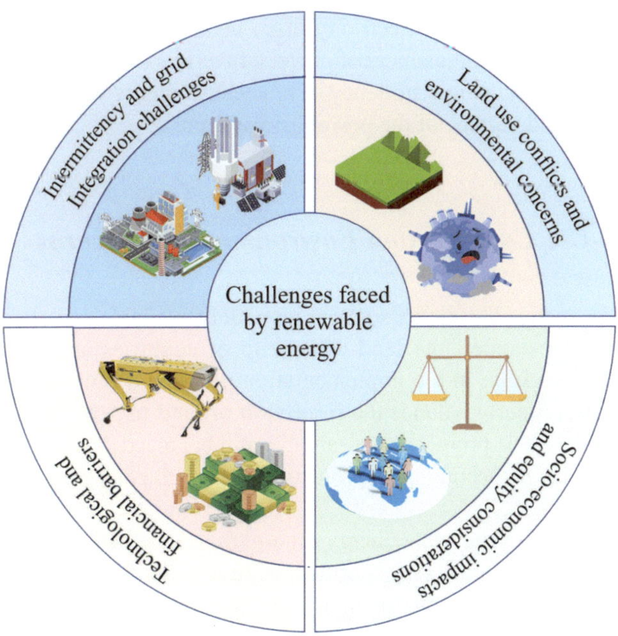

Fig. 1 Systematical illustration of the challenges faced by renewable energy

also greatly promoted the integrated development of renewable energy represented by solar energy, wind energy, etc., with the power grid system (Deguenon et al., 2023). Although this integration can effectively reduce the overuse of fossil fuels, the increasing penetration of renewable electricity, represented by intermittent and renewable energy, is posing significant challenges to the normal operation of the global power grid system (Impram et al., 2020).

In the integration process of renewable energy and grid systems, the uncertainty of renewable energy generation often has a significant impact on the flexibility of grid operation (Huang et al., 2019). Flexibility is commonly used to describe the ability of power systems to maintain power generation and load balance in various uncertain environments (Hsieh & Anderson, 2017), which can effectively guarantee the reliable and sustainable operation of the power grid. In addition, flexibility requirements can be specifically divided into four types: transmission capacity, power, energy, and voltage flexibility (Hillberg et al., 2019). Take the practical application of wind resources in the world as an example, with the continuous improvement of wind power penetration level, the intermittent characteristics of wind power generation will gradually be exposed and will greatly hinder the integration of large-scale wind power generation, thereby endangering the reliable operation of the entire power system (Ren et al., 2017). In addition, renewable energy generation mechanisms connected to the power grid system can also cause serious global power quality control problems (Liang, 2016). The uncontrollability and variability of renewable energy itself will significantly influence the level of power generation, and the quality of electricity will gradually be disrupted over time. The power quality problem mainly exists in two aspects, one of which is that the new working devices used for renewable energy generation will cause significant interference to harmonics. Taking the power conversion system of wave energy as an example, higher-quality electricity often needs to be obtained through continuous optimization and adjustment of power converters (Said & Ringwood, 2021). The second one is that the intermittency of renewable energy will bring fluctuations in the frequency and voltage of the power grid system.

5.2 *Land-Use Conflicts and Environmental Concerns*

Although some renewable energy sources are widely available and often considered as inexhaustible resources, the land ownership of countries around the world is often limited. Therefore, in the implementation process of renewable energy, it is critical for us to properly balance the resolution of land-use conflicts and environmental pollution issues: promoting sustainable and balanced development of renewable energy while respecting and protecting environmental and cultural values (Poggi et al., 2018).

The application of renewable energy on a global scale can effectively decrease and control GHG emissions, so it is critical to reserve land for the development of renewable projects such as photovoltaics and wind energy systems. People can usually use methods such as Geographic Information Systems (GIS) to evaluate whether the deployment of various renewable energy technologies meets local environmental, regulatory, and land-use requirements (Obane et al., 2020). Specifically, the issue of global land conflict manifests in two aspects. Firstly, the land ownership required for wind turbines, PV panels, and other devices will significantly increase with the increase of installed capacity (Mulvaney, 2017). At the same time, when facing competition for site selection for other functional projects such as commercial residential, entertainment, and agriculture, global land-use conflict is inevitable, which will also bring many challenges and uncertainties to the deployment of renewable energy. Secondly, the site selection of renewable energy project factories is often limited by meteorological conditions, cultural environment, and other factors. Unreasonable land selection may not only overlook the potential use of other types of land, but may also affect the livelihood of the neighboring communities and indigenous peoples, and even lead to fragmented destruction or complete loss of habitats for wildlife (Chen et al., 2022b). In addition, many renewable energy projects can also have a series of negative effects on global environmental health and endanger the well-being of those who are extremely vulnerable to the impacts of climate change (Levenda et al., 2021). The environmental pollution caused will inevitably have negative impacts on ecological balance, air superiority, nature reserves (water, land, and forests), etc. (Yi et al., 2023). The connection between the consumption of renewable energy and environmental degradation is intense, and most of the time environmental degradation is usually affected by carbon dioxide emissions. Accordingly, indicators such as ecological footprint and carrying capacity coefficient are also used to evaluate the degree of environmental degradation (Depren et al., 2022).

5.3 *Socioeconomic Impacts and Equity Considerations*

The conversion from traditional fossil fuels to renewable energy is in line with a global sustainable future development model. However, in the context of green development, renewable energy often has a significant impact on sustainable development mechanisms, specifically involving economic growth rate, productivity development level, health and education industry reform, etc. Renewable energy often interacts with carbon dioxide concentration and fuel consumption, posing inevitable challenges to global economic coordination and social development (Omri & Belaïd, 2021).

From the perspective of socioeconomic impact, taking the integration of renewable energy and power grid systems as an example, although the use of clean energy can have a certain degree of positive impact on the global power industry, it will significantly affect the economic base of individuals and families. In less developed regions and countries, the application of renewable energy often has a neutral relationship with socioeconomic growth (Narayan & Doytch, 2017). At the same time, various environmentally friendly investment funds with a focus on renewable energy typically perform significantly lower than traditional energy and market benchmarks (Ji et al., 2021). It is obvious that the lack of an investment market and fluctuation timing goes against the expanding scale of the renewable energy market, resulting in a series of adverse effects that are extremely detrimental to the reflection of financial attractiveness and the construction of the investment environment. For the sake of stabilizing the level of socioeconomic development, the proposal and implementation of alternative evaluation techniques that incorporate various social equity indices can effectively consider the impact of the socioeconomic environment from multiple dimensions. In other words, alternative evaluation techniques represented by the formulation and evaluation of cost curves play a key role in comparing the potential competitiveness and development potential of different types of alternative solutions (Chapman et al., 2018). At the same time, global environmental sustainability can usually be achieved through socioeconomic sustainable development and proper protection of energy resources. Therefore, the relationship between socioeconomic sustainability and environmental sustainability is close and significantly interdependent (Cui et al., 2022). The importance of balancing economic sustainability and ecological sustainability is self-evident (Jie et al., 2023). Additionally, the implementation of energy justice is mainly aimed at better protecting groups who are discriminated against due to factors such as social and cultural status, policy deployment background, or gender. Energy justice mainly involves the advantages and disadvantages of energy policy implementation, as well as the relevant responsibilities and beneficiaries. At the same time, energy justice usually has three principles, namely, allocation, procedure, and recognition of justice (Chapman et al., 2018).

5.4 *Technological and Financial Barriers*

The transformation of energy technology on a global scale usually requires meeting expectations for economic development, social inclusion, and environmental sustainability. In this process, there are often obstacles to technological capabilities and financial issues, which are also considered extremely challenging obstacles (Oryani et al., 2021). In fact, the lack of public funds in the global clean energy industry has contributed to the transformation of the energy sector to a certain extent. A mature and complete economic system and capital market can provide long-term and stable financing for the renewable energy industry while minimizing the liquidity risk of enterprises (Anton & Nucu, 2020), thus smoothly building a renewable energy consumption market that combines technology, investment, and innovation.

Facing the expanding application scenarios of renewable energy, taking renewable energy-based power systems as an example, most power projects will experience instrument failures, power outages, voltage fluctuations, high degree of technical and nontechnical losses, and poor quality of later maintenance (Vanegas Cantarero, 2020). The emergence of overcapacity will have a certain degree of impact on the economic situation of power companies. Meanwhile, for the transportation of renewable energy projects, although corresponding subsidy policies have been adopted for the fuel of transportation facilities to address the challenges posed by insufficient technology and funding, for most small economies on a global scale, the growing demand for renewable energy poses even more severe financial and technological pressures on the transportation sector. In addition, top-down energy planning based on government decision-making can weaken the cognitive level of most citizens on energy-related issues, and the lack of energy education can also lead to indifference toward renewable energy issues. Both of these situations can lead to problems in the planning, design, and investment of renewable energy projects. Taking Pakistan's investment in solar energy technology as an example, due to limited purchasing power and transportation infrastructure, although it has extraordinary potential for solar energy resource utilization, the investment cost of solar energy technology has become the main obstacle (Elahi et al., 2022). Without the support of advanced technology, the resulting low power generation efficiency and power shortage will cause huge losses to the agricultural sector, and the growth and development of various crops will also be adversely affected.

5.5 Future Outlook and Recommendations

In this era of global energy transformation, the significant increase in demand for renewable energy has become a major development trend (Ahmad & Zhang, 2020). In addition to continuously optimizing and improving existing renewable energy technologies in the future, such as solar power technology (solar thermic systems like solar towers and concentration plates, PV and concentrated PV), hydropower technology (tidal, wave, and ocean thermal energy), biomass and geothermal energy (electricity and thermal energy), etc. (Bundschuh et al., 2021), it is also necessary for scientific and technological personnel to explore new renewable energy sources with stronger carbon reduction capabilities and better economic benefits, and to carry out corresponding energy process optimization work in energy storage and energy recovery (Al-Shetwi et al., 2020). This is precisely the specific practice of carrying out the application and upgrading of renewable energy from the perspective of technology and economy. Bringing the dimension of technology and economy into the consideration of real life has a significant impact on the in-depth exploration of the potential of renewable energy (IEA, 2023b). Taking the integration of renewable energy and power grid systems as an example, in addition to improving the inherent characteristics of renewable energy, the flexibility and stability of power grid system operation can also be enhanced by introducing new control and monitoring systems (Yap et al., 2022). In addition, policymakers need to address the issue of inconsistent enthusiasm for clean energy investment in different countries or industries in the future, which requires a range of policy solutions. In fact, governments and functional agencies need to adopt policies to encourage and protect enterprises and research institutions to engage in renewable energy research and application, so as to promote the global market share of renewable energy. At the same time, in response to residents' acceptance of renewable energy, relevant energy courses can be organized in schools and communities to promote the functions and significant impact of renewable energy in today's society. This will ensure that renewable energy projects are aligned with community interests during their development and enhance public acceptance of these projects. Finally, actively promoting the innovation of renewable energy business models and financing mechanisms will also be a very promising development path in the future (Eder et al., 2018).

6 Conclusion

In this chapter, the role of renewable energy in addressing the global energy crisis and its significance for environmental sustainability has been discussed through a detailed exploration of the various opportunities from renewable energy sources and their challenges. The chapter aims to underscore the immense potential of renewable energy in transitioning toward a more sustainable and eco-friendly future. The discussion also offers solutions to the current challenges being experienced in the field, especially regarding acceptability and scalability. Renewable energy stands at the forefront of global efforts to combat climate change and build a sustainable future. As we confront the escalating challenges posed by environmental degradation and climate instability, the imperative to accelerate the transition toward renewable energy has never been more pressing. By harnessing the power of solar, wind, hydroelectric, geothermal, and biomass resources, we can reduce our reliance on finite fossil fuels, mitigate greenhouse gas emissions, and safeguard the Earth's natural heritage for generations to come. However, a revolution in renewable energy necessitates coordinated efforts at all levels, including regulatory reforms, technological breakthroughs, and cultural changes.

References

Abdulahi, M. E., Shu, Y., & Khan, M. A. (2019). Resource rents, economic growth, and the role of institutional quality: A panel threshold analysis. *Resources Policy, 61*, 293–303.

Adaramola, M. S., Agelin-Chaab, M., & Paul, S. S. (2014). Analysis of hybrid energy systems for application in southern Ghana. *Energy Conversion and Management, 88*, 284–295.

Adeniyi, A. G., & Ighalo, J. O. (2021). Response surface modelling and optimisation of biodiesel production from Manilkara Zapota L. seed oil. Indian *J Chem Technol*, 28, 724–729.

Adeniyi, A. G., Ighalo, J. O., & Eletta, O. A. A. (2018). Process integration and feedstock optimisation of a two-step biodiesel production process from Jatropha Curcas using Aspen Plus. *Chemical Product and Process Modelling*, 14(2), 20180055.

Adeniyi, A. G., Ighalo, J. O., & Aderibigbe, F. A. (2019). Modelling of integrated processes for the pyrolysis and steam reforming of rice husk (Oryza sativa). *SN Applied Sciences*, 1, 841–851.

Adeniyi, A. G., Ighalo, J. O., & Adeyanju, C. A. (2021). Materials-to-product potentials for sustainable development in Nigeria. *International Journal of Sustainable Engineering*, 14, 664–671.

Adeosun, T. A., Amosu, C. O., Omitogun, O. A., Amusa, O. M., & Morenikeji, B. A. (2023). Renewable energy for sustainable development in developing countries: Benefits to the environment. *Journal of Energy Research and Reviews, 13*, 1–25.

Afonja, A. A. (2020). *Fossil fuels and the environment*. Chudace Publishing.

Aghahosseini, A., Bogdanov, D., Ghorbani, N., & Breyer, C. (2018). Analysis of 100% renewable energy for Iran in 2030: Integrating solar PV, wind energy and storage. *International Journal of Environmental Science and Technology, 15*, 17–36.

Ahmad, T., & Zhang, D. (2020). A critical review of comparative global historical energy consumption and future demand: The story told so far. *Energy Reports, 6*, 1973–1991.

Alemán-Nava, G. S., Casiano-Flores, V. H., Cárdenas-Chávez, D. L., Díaz-Chavez, R., Scarlat, N., Mahlknecht, J., Dallemand, J.-F., & Parra, R. (2014). Renewable energy research progress in Mexico: A review. *Renewable and Sustainable Energy Reviews, 32*, 140–153.

Algarni, S., Tirth, V., Alqahtani, T., Alshehery, S., & Kshirsagar, P. (2023). Contribution of renewable energy sources to the environmental impacts and economic benefits for sustainable development. *Sustainable Energy Technologies and Assessments*, 56, 103098.

Ali, M., Saleem, M., Khan, Z., & Watson, I. A. (2019). *The use of crop residues for biofuel production* (pp. 369–395). Biomass, biopolymer-based materials, and bioenergy. Elsevier.

Alotaibi, I., Abido, M. A., Khalid, M., & Savkin, A. V. (2020). A comprehensive review of recent advances in smart grids: A sustainable future with renewable energy resources. *Energies, 13*, 6269.

Al-Shetwi, A. Q. (2022). Sustainable development of renewable energy integrated power sector: Trends, environmental impacts, and recent challenges. *Science of the Total Environment, 822*, 153645.

Al-Shetwi, A. Q., Hannan, M. A., Jern, K. P., Mansur, M., & Mahlia, T. M. I. (2020). Grid-connected renewable energy sources: Review of the recent integration requirements and control methods. *Journal of Cleaner Production, 253*, 119831.

Amin, N., Shabbir, M. S., Song, H., & Abbass, K. (2022). Renewable energy consumption and its impact on environmental quality: A pathway for achieving sustainable development goals in ASEAN countries. *Energy & Environment, 35*, 644–662.

Amin, N., Shabbir, M. S., Song, H., Farrukh, M. U., Iqbal, S., & Abbass, K. (2023). A step towards environmental mitigation: Do green technological innovation and institutional quality make a difference? *Technological Forecasting and Social Change, 190*, 122413.

Anton, S. G., & Nucu, A. E. A. (2020). The effect of financial development on renewable energy consumption. *A panel data approach. Renewable Energy, 147*, 330–338.

Arjmand, R., & McPherson, M. (2022). Canada's electricity system transition under alternative policy scenarios. *Energy Policy, 163*, 112844.

Azarpour, A., Mohammadzadeh, O., Rezaei, N., & Zendehboudi, S. (2022). Current status and future prospects of renewable and sustainable energy in North America: Progress and challenges. *Energy Conversion and Management, 269*, 115945.

Baljit, S. S. S., Chan, H.-Y., & Sopian, K. (2016). Review of building integrated applications of photovoltaic and solar thermal systems. *Journal of Cleaner Production, 137*, 677–689.

Barthelmie, R. J., & Pryor, S. C. (2021). Climate change mitigation potential of wind energy. *Climate, 9*, 136.

Barwick, P. J., Li, S., Rao, D., & Zahur, N. B. (2024). The healthcare cost of air pollution: Evidence from the world's largest payment network. *Review of Economics and Statistics*, 1–52.

Bhuiyan, M. R. A. (2022). Overcome the future environmental challenges through sustainable and renewable energy resources. *Micro & Nano Letters, 17*, 402–416.

Bogaraj, T., & Kanakaraj, J. (2021). Indian renewable power sector: Potential, policies, barriers, and its prospective [articol]. *Journal of Electrical Engineering, 21*(3), 109–124.

Bórawski, P., Bełdycka-Bórawska, A., Jankowski, K. J., Dubis, B., & Dunn, J. W. (2020). Development of wind energy market in the European Union. *Renewable Energy, 161*, 691–700.

Buechler, S., Vázquez-García, V., Martínez-Molina, K. G., & Sosa-Capistrán, D. M. (2020). Patriarchy and (electric) power? A feminist political ecology of solar energy use in Mexico and the United States. *Energy Research & Social Science, 70*, 101743.
Bundschuh, J., Kaczmarczyk, M., Ghaffour, N., & Tomaszewska, B. (2021). State-of-the-art of renewable energy sources used in water desalination: Present and future prospects. *Desalination, 508*, 115035.
Chapman, A. J., McLellan, B. C., & Tezuka, T. (2018). Prioritizing mitigation efforts considering co-benefits, equity and energy justice: Fossil fuel to renewable energy transition pathways. *Applied Energy, 219*, 187–198.
Chen, L., Msigwa, G., Yang, M., Osman, A. I., Fawzy, S., Rooney, D. W., & Yap, P.-S. (2022a). Strategies to achieve a carbon neutral society: A review. *Environmental Chemistry Letters, 20*, 2277–2310.
Chen, Y.-K., Kirkerud, J. G., & Bolkesjø, T. F. (2022b). Balancing GHG mitigation and land-use conflicts: Alternative Northern European energy system scenarios. *Applied Energy, 310*, 118557.
Chen, L., Huang, L., Hua, J., Chen, Z., Wei, L., Osman, A. I., Fawzy, S., Rooney, D. W., Dong, L., & Yap, P.-S. (2023). Green construction for low-carbon cities: A review. *Environmental Chemistry Letters, 21*, 1627–1657.
Chen, L., Hu, Y., Wang, R., Li, X., Chen, Z., Hua, J., Osman, A. I., Farghali, M., Huang, L., Li, J., Dong, L., Rooney, D. W., & Yap, P.-S. (2024). Green building practices to integrate renewable energy in the construction sector: A review. *Environmental Chemistry Letters, 22*, 751–784.
Chowdhury, S. A. (2020). National Solar Energy Roadmap, 2021–2041. In Sustainable and Renewable Energy Development Authority (SREDA). Retrieved 11 April 2024, from https://climateportal.ccdbbd.org/wp-content/uploads/2021/03/National-Solar-Energy-Roadmap.pdf
Cui, L., Weng, S., Nadeem, A. M., Rafique, M. Z., & Shahzad, U. (2022). Exploring the role of renewable energy, urbanization and structural change for environmental sustainability: Comparative analysis for practical implications. *Renewable Energy, 184*, 215–224.
Das, P., Mathur, J., Bhakar, R., & Kanudia, A. (2018). Implications of short-term renewable energy resource intermittency in long-term power system planning. *Energy Strategy Reviews, 22*, 1–15.
Davis, M., Moronkeji, A., Ahiduzzaman, M., & Kumar, A. (2020). Assessment of renewable energy transition pathways for a fossil fuel-dependent electricity-producing jurisdiction. *Energy for Sustainable Development, 59*, 243–261.
Deguenon, L., Yamegueu, D., Moussa Kadri, S., & Gomna, A. (2023). Overcoming the challenges of integrating variable renewable energy to the grid: A comprehensive review of electrochemical battery storage systems. *Journal of Power Sources, 580*, 233343.
Deng, C., Qian, Y., Song, X., Xie, M., Duan, H., Shen, P., & Qiao, Q. (2024). Are electric vehicles really the optimal option for the transportation sector in China to approach pollution reduction and carbon neutrality goals? *Journal of Environmental Management, 356*, 120648.
Depren, S. K., Kartal, M. T., Çelikdemir, N. Ç., & Depren, Ö. (2022). Energy consumption and environmental degradation nexus: A systematic review and meta-analysis of fossil fuel and renewable energy consumption. *Ecological Informatics, 70*, 101747.
Dulta, K., Adeola, A. O., Ashaolu, S. E., Banji, T. I., & Ighalo, J. O. (2022). Biohydrogen production and its bioeconomic impact: A review. *Waste Disposal and Sustainable Energy, 4*(3), 1–8.
Eder, L. V., Provornaya, I. V., Filimonova, I. V., Kozhevin, V. D., & Komarova, A. V. (2018). World energy market in the conditions of low oil prices, the role of renewable energy sources. *Energy Procedia, 153*, 112–117.
Egbueri, J. C., Ighalo, J. O., & Pande, C. B. (2023). *Climate change impacts on Nigeria—Environment and sustainable development*. Springer Climate.
Elahi, E., Khalid, Z., & Zhang, Z. (2022). Understanding farmers' intention and willingness to install renewable energy technology: A solution to reduce the environmental emissions of agriculture. *Applied Energy, 309*, 118459.
Emenike, E. C., Iwuozor, K. O., Ighalo, J. O., Bamigbola, J. O., Omonayin, E. O., Ojo, H. T., Adeleke, J., & Adeniyi, A. G. (2023). Advancing the circular economy through the thermo-

chemical conversion of waste to biochar: A review on sawdust waste-derived fuel. *Biofuels, 15*, 433–447.

Erdoğan, S., Çakar, N. D., Ulucak, R., & Danish, K. Y. (2021). The role of natural resources abundance and dependence in achieving environmental sustainability: Evidence from resource-based economies. *Sustainable Development, 29*, 143–154.

Eshiemogie, S. O., Ighalo, J. O., & Banji, T. I. (2022). Knowledge, perception and awareness of renewable energy by engineering students in Nigeria: A need for the undergraduate engineering program adjustment. *Cleaner Engineering and Technology, 6*, 100388.

Eshiemogie, S. O., Ighalo, J. O., Adekanmbi, M., Banji, M., Igwegbe, C. A., Adeniyi, A. G., Adeola, A. O., & Dulta, K. (2023). Current effect and projected implications of climate change on Nigeria's sustainable development plan. In J. C. Egbueri, J. O. Ighalo, & C. B. Pande (Eds.), *Climate change impacts on Nigeria—Environment and sustainable development* (pp. 1–17). Springer Nature.

Falchetta, M., Mazzei, D., Russo, V., Campanella, V. A., Floridia, V., Schiavo, B., Venezia, L., Brunatto, C., & Orlando, R. (2020). The Partanna project: A first of a kind plant based on molten salts in LFR collectors. In *AIP conference proceedings* (Vol. 1). AIP Publishing.

Farghali, M., Osman, A. I., Chen, Z., Abdelhaleem, A., Ihara, I., Mohamed, I. M. A., Yap, P.-S., & Rooney, D. W. (2023a). Social, environmental, and economic consequences of integrating renewable energies in the electricity sector: A review. *Environmental Chemistry Letters, 21*, 1381–1418.

Farghali, M., Osman, A. I., Mohamed, I. M. A., Chen, Z., Chen, L., Ihara, I., Yap, P.-S., & Rooney, D. W. (2023b). Strategies to save energy in the context of the energy crisis: A review. *Environmental Chemistry Letters, 21*, 2003–2039.

Fazekas, A., Bataille, C., & Vogt-Schilb, A. (2022). *Achieving net-zero prosperity: How governments can unlock 15 essential transformations.*

Garba, I., & Bellingham, R. (2021). Energy poverty: Estimating the impact of solid cooking fuels on GDP per capita in developing countries-Case of sub-Saharan Africa. *Energy, 221*, 119770.

Gemechu, E., & Kumar, A. (2022). A review of how life cycle assessment has been used to assess the environmental impacts of hydropower energy. *Renewable and Sustainable Energy Reviews, 167*, 112684.

Gygli, S., Haelg, F., Potrafke, N., & Sturm, J.-E. (2019). The KOF globalisation index—Revisited. *The Review of International Organizations, 14*, 543–574.

Haines, A., Kovats, R. S., Campbell-Lendrum, D., & Corvalán, C. (2006). Climate change and human health: Impacts, vulnerability and public health. *Public Health, 120*, 585–596.

Hao, H., Geng, Y., Li, W., & Guo, B. (2015). Energy consumption and GHG emissions from China's freight transport sector: Scenarios through 2050. *Energy Policy, 85*, 94–101.

Harris, J. M., & Roach, B. (2021). *Energy: The great transition* (pp. 283–326). Environmental and natural resource economics. Routledge.

Hassan, Q., Viktor, P. J., Al-Musawi, T., Mahmood Ali, B., Algburi, S., Alzoubi, H. M., Khudhair Al-Jiboory, A., Zuhair Sameen, A., Salman, H. M., & Jaszczur, M. (2024). The renewable energy role in the global energy. *Transformations Renewable Energy Focus, 48*, 100545.

Hillberg, E., Zegers, A., Herndler, B., Wong, S., Pompee, J., Bourmaud, J.-Y., Lehnhoff, S., Migliavacca, G., Uhlen, K., & Oleinikova, I. (2019). Flexibility needs in the future power system. *In Power Transmission & Distribution Systems* (pp. 1–46).

Hoang, A. T., Sandro, N., Olcer, A. I., Ong, H. C., Chen, W.-H., Chong, C. T., Thomas, S., Bandh, S. A., & Nguyen, X. P. (2021). Impacts of COVID-19 pandemic on the global energy system and the shift progress to renewable energy: Opportunities, challenges, and policy implications. *Energy Policy, 154*, 112322.

Hossain, E. (2023). Environmental Concerns Arising from Human Energy Consumption. In *The sun, energy, and climate change* (pp. 259–347). Springer Nature Switzerland.

Hsieh, E., & Anderson, R. (2017). Grid flexibility: The quiet revolution. *The Electricity Journal, 30*, 1–8.

Huang, Y. W., Kittner, N., & Kammen, D. M. (2019). ASEAN grid flexibility: Preparedness for grid integration of renewable energy. *Energy Policy, 128*, 711–726.

Ibrahim, I. D., Hamam, Y., Alayli, Y., Jamiru, T., Sadiku, E. R., Kupolati, W. K., Ndambuki, J. M., & Eze, A. A. (2021). A review on Africa energy supply through renewable energy production: Nigeria, Cameroon, Ghana and South Africa as a case study. *Energy Strategy Reviews, 38*, 100740.

IEA (2018). *IEA bioenergy countries report—Update 2018*. Retrieved 11 April 2024, from https://www.ieabioenergy.com/wp-content/uploads/2018/10/IEABioenergy-Countries-Report-Update-2018-Bioenergy-policies-and-status-of-implementation.pdf

IEA (2020). *Solar PV module manufacturing and demand*, 2014–2020. Retrieved 11 April 2024, fromhttps://www.iea.org/data-and-statistics/charts/solar-pv-module-manufacturing-and-demand-2014-2020

IEA (2023a). *Global energy investment in clean energy and in fossil fuels*, 2015–2023. Retrieved 11 April 2024, fromhttps://www.iea.org/data-and-statistics/charts/global-energy-investment-in-clean-energy-and-in-fossil-fuels-2015-2023

IIEA (2023b). *Oil production investment and solar investment*, 2013 vs 2023. Retrieved 11 April 2024, from https://www.iea.org/data-and-statistics/charts/oil-production-investment-and-solar-investment-2013-vs-2023

Ighalo, J. O. (2024). Biowastes and derived green sorbents for water decontamination: Insights on thermochemical conversion strategies. *Current Opinion in Green and Sustainable Chemistry, 45*, 100880.

Ighalo, J. O., & Amama, P. B. (2024). Recent advances in the catalysis of steam reforming of methane (SRM). *International Journal of Hydrogen Energy, 51*, 688–700.

Ighalo, J. O., Amaku, J. F., Olisah, C., Adeola, A. O., Iwuozor, K. O., Akpomie, K., Conradie, J., Adegoke, K. A., & Oyedotun, K. O. (2022). Utilisation of adsorption as a resource recovery technique for lithium in geothermal water. *Journal of Molecular Liquids, 365*, 120107.

Impram, S., Varbak Nese, S., & Oral, B. (2020). Challenges of renewable energy penetration on power system flexibility: A survey. *Energy Strategy Reviews, 31*, 100539.

Swedish Institute (2024) S Retrieved April 11 2024, from http://sweden.se/society/energy-use-in-sweden/

IRENA (2024). *Record growth in renewables, but progress needs to be equitable*. Retrieved 11 April 2024, from https://www.irena.org/News/pressreleases/2024/Mar/Record-Growth-in-Renewables-but-Progress-Needs-to-be-Equitable

Iwuozor, K. O., Emenike, E. C., Ighalo, J. O., Micheal, T. T., Micheal, K. T., Ore, O. T., Saliu, O. D., & Adeniyi, A. G. (2023). Thermo-mineralization of biomass for metal oxide recovery: A review. *Bioresource Technology Reports, 24*, 101664.

Jansen, M., Beiter, P., Riepin, I., Müsgens, F., Guajardo-Fajardo, V. J., Staffell, I., Bulder, B., & Kitzing, L. (2022). Policy choices and outcomes for offshore wind auctions globally. *Energy Policy, 167*, 113000.

Ji, X., Chen, X., Mirza, N., & Umar, M. (2021). Sustainable energy goals and investment premium: Evidence from renewable and conventional equity mutual funds in the Euro zone. *Resources Policy, 74*, 102387.

Jie, H., Khan, I., Alharthi, M., Zafar, M. W., & Saeed, A. (2023). Sustainable energy policy, socio-economic development, and ecological footprint: The economic significance of natural resources, population growth, and industrial development. *Utilities Policy, 81*, 101490.

Kemp, L., Xu, C., Depledge, J., Ebi, K. L., Gibbins, G., Kohler, T. A., Rockström, J., Scheffer, M., Schellnhuber, H. J., Steffen, W., & Lenton, T. M. (2022). Climate endgame: Exploring catastrophic climate change scenarios. *Proceedings of the National Academy of Sciences, 119*(34), e2108146119.

Kim, S. J., Asadzadeh, M., & Stadnyk, T. A. (2022). Climate change impact on water supply and hydropower generation potential in Northern Manitoba. *Journal of Hydrology: Regional Studies, 41*, 101077.

Kumar, A., & Trivedi, G. (2020). Solar energy: A promising future in power sector for sustainable development. *Indian Journal of Human Relations, 54*, 45–51.

Letcher, T. M. (2021). Why discuss the impacts of climate change? In *The impacts of climate change* (pp. 3–17). Elsevier.
Levenda, A. M., Behrsin, I., & Disano, F. (2021). Renewable energy for whom? A global systematic review of the environmental justice implications of renewable energy technologies. *Energy Research & Social Science, 71*, 101837.
Li, B. (2023). The role of financial markets in the energy transition: An analysis of investment trends and opportunities in renewable energy and clean technology. *Environmental Science and Pollution Research, 30*, 97948–97964.
Li, L., Lin, J., Wu, N., Xie, S., Meng, C., Zheng, Y., Wang, X., & Zhao, Y. (2022a). Review and outlook on the international renewable energy development. *Energy and Built Environment, 3*, 139–157.
Li, S., Wang, Q., Jiang, X.-T., & Li, R. (2022b). The negative impact of the COVID-19 on renewable energy growth in developing countries: Underestimated. *Journal of Cleaner Production, 367*, 132996.
Li, H., Chen, C., & Umair, M. (2023). Green finance, enterprise energy efficiency, and green total factor productivity: Evidence from China. *Sustainability, 15*(14), 11065.
Liang, X. (2016). Emerging power quality challenges due to integration of renewable energy sources. *IEEE Transactions on Industry Applications, 53*, 855–866.
Lin, B., & Xu, B. (2021). A non-parametric analysis of the driving factors of China's carbon prices. *Energy Economics, 104*, 105684.
Lins, C., Williamson, L. E., Leitner, S., & Teske, S. (2014). The first decade: 2004—2014: 10 years of renewable energy progress. *Renewable Energy Policy Network for the 21st Century*.
Liu, S., Bie, Z., Lin, J., & Wang, X. (2018). Curtailment of renewable energy in Northwest China and market-based solutions. *Energy Policy, 123*, 494–502.
Lu, Y., Khan, Z. A., Alvarez-Alvarado, M. S., Zhang, Y., Huang, Z., & Imran, M. (2020). A critical review of sustainable energy policies for the promotion of renewable energy sources. *Sustainability, 12*(12), 5078.
Lund, J. W., & Toth, A. N. (2021). Direct utilization of geothermal energy 2020 worldwide review. *Geothermics, 90*, 101915.
MacDougall, H., Tomosk, S., & Wright, D. (2018). Geographic maps of the impact of government incentives on the economic viability of solar power. *Renewable Energy, 122*, 497–506.
Madaleno, M., Dogan, E., & Taskin, D. (2022). A step forward on sustainability: The nexus of environmental responsibility, green technology, clean energy and green finance. Energy Economics, 109, 105945.
Magazine RE (2020) Renewables as sole Source of new US generating capacity. Retrieved 11 April 2024, from https://www.renewableenergymagazine.com/panorama/latest-ferc-data-shows-renewables-as-sole-20201123
Mahmood, S. A. I. (2011). Air pollution kills 15,000 Bangladeshis each year: The role of public administration and government's integrity. *Journal of Public Administration and Policy Research, 3*, 129.
Mahmud, M. A. P., Huda, N., Farjana, S. H., & Lang, C. (2018). Environmental impacts of solar-photovoltaic and solar-thermal systems with life-cycle assessment. *Energies, 11*, 2346.
Maity, S., Sarkar, S., Dutta, K., De, T., & Mukherjee, R. (2023). Renewable energy and green technology. In *Green chemistry for environmental sustainability—Prevention-Assurance-Sustainability (P-A-S) approach*. IntechOpen.
Malik, K., Rahman, S. M., Khondaker, A. N., Abubakar, I. R., Aina, Y. A., & Hasan, M. A. (2019). Renewable energy utilization to promote sustainability in GCC countries: Policies, drivers, and barriers. *Environmental Science and Pollution Research, 26*, 20798–20814.
Manisalidis, I., Stavropoulou, E., Stavropoulos, A., & Bezirtzoglou, E. (2020). Environmental and health impacts of air pollution: A review. *Frontiers in Public Health, 8*.
Mazzucato, M., & Semieniuk, G. (2018). Financing renewable energy: Who is financing what and why it matters. *Technological Forecasting and Social Change, 127*, 8–22.

McCollum D, Gomez Echeverri L, Riahi K, Parkinson S (2017). SDG7: Ensure access to affordable, reliable, sustainable and modern energy for all. Retrieved 11 April 2024, from https://pure.iiasa.ac.at/14621

Mhaske, P. S. (2020). Solar power development in India. *Water and Energy International, 63*, 9–28.

Msigwa, G., Ighalo, J. O., & Yap, P.-S. (2022). Considerations on environmental, economic, and energy impacts of wind energy generation: Projections toward sustainability initiatives. *Science of The Total Environment, 849*, 157755.

Mulvaney, D. (2017). Identifying the roots of Green Civil War over utility-scale solar energy projects on public lands across the American Southwest. *Journal of Land Use Science, 12*(6), 493–515.

Mundo-Hernández, J., de Celis, A. B., Hernández-Álvarez, J., & de Celis-Carrillo, B. (2014). An overview of solar photovoltaic energy in Mexico and Germany. *Renewable and Sustainable Energy Reviews, 31*, 639–649.

Narayan, S., & Doytch, N. (2017). An investigation of renewable and non-renewable energy consumption and economic growth nexus using industrial and residential energy consumption. *Energy Economics, 68*, 160–176.

Nassar, I. A., Hossam, K., & Abdella, M. M. (2019). Economic and environmental benefits of increasing the renewable energy sources in the power system. *Energy Reports, 5*, 1082–1088.

Newell, R., Raimi, D., & Aldana, G. (2019). Global energy outlook 2019: The next generation of energy. *Resources for the Future, 1*.

Niraimathi, R., Pradeep, V., Shanthi, M., & Kathiresh, M. (2021). Present power scenario in India. *Integration of Renewable Energy Sources with Smart Grid*, 19–35.

Nogueira, L. A. H., Souza, G. M., Cortez, L. A. B., & de Brito Cruz, C. H. (2020). Biofuels for transport. In *Future energy* (pp. 173–197). Elsevier.

Obane, H., Nagai, Y., & Asano, K. (2020). Assessing land use and potential conflict in solar and onshore wind energy in Japan. *Renewable Energy, 160*, 842–851.

Olabi, A. G., & Abdelkareem, M. A. (2022). Renewable energy and climate change. *Renewable and Sustainable Energy Reviews, 158*, 112111.

Olale, E., Ochuodho, T. O., Lantz, V., & El Armali, J. (2018). The environmental Kuznets curve model for greenhouse gas emissions in Canada. *Journal of Cleaner Production, 184*, 859–868.

Olujobi, O. J., Okorie, U. E., Olarinde, E. S., & Aina-Pelemo, A. D. (2023). Legal responses to energy security and sustainability in Nigeria's power sector amidst fossil fuel disruptions and low carbon energy transition. Heliyon, 9, e17912.

Omer, A. M. (2012). Built environment: Relating the benefits of renewable energy technologies. *International Journal of Automotive and Mechanical Engineering, 5*, 561–575.

Omri, A., & Belaïd, F. (2021). Does renewable energy modulate the negative effect of environmental issues on the socio-economic welfare? *Journal of Environmental Management, 278*, 111483.

Oryani, B., Koo, Y., Rezania, S., & Shafiee, A. (2021). Barriers to renewable energy technologies penetration: Perspective in Iran. *Renewable Energy, 174*, 971–983.

Osman, A. I., Chen, L., Yang, M., Msigwa, G., Farghali, M., Fawzy, S., Rooney, D. W., & Yap, P.-S. (2023). Cost, environmental impact, and resilience of renewable energy under a changing climate: A review. *Environmental Chemistry Letters, 21*, 741–764.

Østergaard, P. A., Duic, N., Noorollahi, Y., Mikulcic, H., & Kalogirou, S. (2020). Sustainable development using renewable energy technology. *Renewable Energy, 146*, 2430–2437.

Pacesila, M., Burcea, S. G., & Colesca, S. E. (2016). Analysis of renewable energies in European Union. *Renewable and Sustainable Energy Reviews, 56*, 156–170.

Pandey, A., Brauer, M., Cropper, M. L., Balakrishnan, K., Mathur, P., Dey, S., Turkgulu, B., Kumar, G. A., Khare, M., & Beig, G. (2021). Health and economic impact of air pollution in the states of India: The Global Burden of Disease Study 2019. *The Lancet Planetary Health, 5*, e25–e38.

Panpatte, D. G., & Jhala, Y. K. (2019). Agricultural waste: A suitable source for biofuel production. In Guide to Automotive Connectivity and Cybersecurity (pp. 337–355).

Panwar, N. L., Kaushik, S. C., & Kothari, S. (2011). Role of renewable energy sources in environmental protection: A review. *Renewable and Sustainable Energy Reviews, 15*, 1513–1524.

Pastukhova, M., & Westphal, K. (2020). Governing the global energy transformation. In *The geopolitics of the global energy transition* (pp. 341–364).

Pereira, P., Zhao, W., Symochko, L., Inacio, M., Bogunovic, I., & Barcelo, D. (2022). The Russian-Ukrainian armed conflict will push back the sustainable development goals. *Geography and Sustainability, 3*, 277–287.

Pérez-Denicia, E., Fernández-Luqueño, F., & Vilariño-Ayala, D. (2021). Suitability assessment for electricity generation through renewable sources: Towards sustainable energy production. *CT&F-Ciencia, Tecnología y Futuro, 11*, 109–122.

Poggi, F., Firmino, A., & Amado, M. (2018). Planning renewable energy in rural areas: Impacts on occupation and land use. *Energy, 155*, 630–640.

Poudyal, R., Loskot, P., Nepal, R., Parajuli, R., & Khadka, S. K. (2019). Mitigating the current energy crisis in Nepal with renewable energy sources. *Renewable and Sustainable Energy Reviews, 116*, 109388.

Pryor, S. C., Barthelmie, R. J., & Shepherd, T. J. (2021). Wind power production from very large offshore wind farms. *Joule, 5*, 2663–2686.

Qi, M., Park, J., Landon, R. S., Kim, J., Liu, Y., & Moon, I. (2022). continuous and flexible renewable-power-to-methane via liquid CO2 energy storage: Revisiting the techno-economic potential. Renewable and Sustainable Energy Reviews, 153, 111732.

Raghuwanshi, S. S., & Arya, R. (2019). Renewable energy potential in India and future agenda of research. *International Journal of Sustainable Engineering, 12*, 291–302.

Rahman, F. A., Aziz, M. M. A., Saidur, R., Bakar, W. A. W. A., Hainin, M. R., Putrajaya, R., & Hassan, N. A. (2017). Pollution to solution: Capture and sequestration of carbon dioxide (CO2) and its utilization as a renewable energy source for a sustainable future. *Renewable and Sustainable Energy Reviews*, 71, 112–126.

Rahman, A., Farrok, O., & Haque, M. M. (2022). Environmental impact of renewable energy source based electrical power plants: Solar, wind, hydroelectric, biomass, geothermal, tidal, ocean, and osmotic. *Renewable and Sustainable Energy Reviews*, 161, 112279.

Ren, G., Liu, J., Wan, J., Guo, Y., & Yu, D. (2017). Overview of wind power intermittency: Impacts, measurements, and mitigation solutions. *Applied Energy, 204*, 47–65.

Said, H. A., & Ringwood, J. V. (2021). Grid integration aspects of wave energy—Overview and perspectives. *IET Renewable Power Generation, 15*, 3045–3064.

Sanguesa, J. A., Torres-Sanz, V., Garrido, P., Martinez, F. J., & Marquez-Barja, J. M. (2021). A review on electric vehicles: Technologies and challenges. *Smart Cities, 4*, 372–404.

Sarkodie, S. A., & Adams, S. (2020). Electricity access, human development index, governance and income inequality in Sub-Saharan Africa. *Energy Reports, 6*, 455–466.

Saxena, P., & Sonwani, S. (2019). *Criteria air pollutants and their impact on environmental health*. Springer Singapore.

Schleussner, C.-F., Deryng, D., D'Haen, S., Hare, W., Lissner, T., Ly, M., Nauels, A., Noblet, M., Pfleiderer, P., Pringle, P., Rokitzki, M., Saeed, F., Schaeffer, M., Serdeczny, O., & Thomas, A. (2018). 1.5 °C Hotspots: Climate hazards, vulnerabilities, and impacts. *Annual Review of Environment and Resources, 43*, 135–163.

Sen, S., & Ganguly, S. (2017). Opportunities, barriers and issues with renewable energy development—A discussion. *Renewable and Sustainable Energy Reviews, 69*, 1170–1181.

Shahsavari, A., & Akbari, M. (2018). Potential of solar energy in developing countries for reducing energy-related emissions. *Renewable and Sustainable Energy Reviews, 90*, 275–291.

Singh, K., Meena, R. S., Kumar, S., Dhyani, S., Sheoran, S., Singh, H. M., Pathak, V. V., Khalid, Z., Singh, A., & Chopra, K. (2023). India's renewable energy research and policies to phase down coal: Success after Paris agreement and possibilities post-Glasgow climate pact. *Biomass and Bioenergy, 177*, 106944.

Sinsel, S. R., Riemke, R. L., & Hoffmann, V. H. (2020). Challenges and solution technologies for the integration of variable renewable energy sources—A review. *Renewable Energy, 145*, 2271–2285.

Ślusarczyk, B., Żegleń, P., Kluczek, A., Nizioł, A., & Górka, M. (2022). The impact of renewable energy sources on the economic growth of Poland and Sweden considering COVID-19 times. *Energies, 15*, 332.

Soeder, D. J. (2021). Fossil fuels and climate change. In *Fracking and the environment* (pp. 155–185). Springer International Publishing.

Sohani, A., Dehnavi, A., Esmaeilion, F., Ighalo, J. O., Abdulsalam, A., Hoseinzadeh, S., Nastasi, B., & Garcia, D. A. (2022). Integration of heat extraction from abandoned wells with renewables. In Y. Noorollahi, M. N. Naseer, & M. M. Siddiqi (Eds.), *Utilization of thermal potential of abandoned wells* (pp. 275–295). Elsevier.

Sokulski, C. C., Barros, M. V., Salvador, R., Broday, E. E., & de Francisco, A. C. (2022). Trends in renewable electricity generation in the G20 countries: An analysis of the 1990–2020 period. *Sustainability, 14*, 2084.

Somoye, O. A. (2023). Energy crisis and renewable energy potentials in Nigeria: A review. *Renewable and Sustainable Energy Reviews, 188*, 113794.

Taher, J., & Salem, S. (2020). Effect of silver nanoparticles on some physical& biological properties of fluid denture base material Al-Kitab. *Journal for Pure Sciences, 4*, 27–40.

To, L. S., Bruce, A., Munro, P., Santagata, E., MacGill, I., Rawali, M., & Raturi, A. (2021). A research and innovation agenda for energy resilience in Pacific Island Countries and Territories. *Nature Energy, 6*, 1098–1103.

Tong, D., Farnham, D. J., Duan, L., Zhang, Q., Lewis, N. S., Caldeira, K., & Davis, S. J. (2021). Geophysical constraints on the reliability of solar and wind power worldwide. *Nature Communications, 12*, 6146.

Tracker, C. A. (2021). *How a renewable energy COVID-19 recovery creates opportunities for India*. Retrieved 11 April 2024, from https://climateactiontracker.org/publications/how-a-renewable-energy-covid-19-recovery-creates-opportunities-for-India/

Usman, Z. G., Abbasoglu, S., Ersoy, N. T., & Fahrioglu, M. (2015). Transforming the Nigerian power sector for sustainable development. *Energy Policy, 87*, 429–437.

Vanegas Cantarero, M. M. (2020). Of renewable energy, energy democracy, and sustainable development: A roadmap to accelerate the energy transition in developing countries. *Energy Research & Social Science, 70*, 101716.

Vorosmarty, C. J., Green, P., Salisbury, J., & Lammers, R. B. (2000). Global water resources: Vulnerability from climate change and population growth. *Science, 289*, 284–288.

Wang, E., Gozgor, G., Mahalik, M. K., Patel, G., & Hu, G. (2022a). Effects of institutional quality and political risk on the renewable energy consumption in the OECD countries. *Resources Policy, 79*, 103041.

Wang, Y., Wild, O., Ashworth, K., Chen, X., Wu, Q., Qi, Y., & Wang, Z. (2022b). Reductions in crop yields across China from elevated ozone. *Environmental Pollution, 292*, 118218.

Whalen, J., Xu, C., Shen, F., Kumar, A., Eklund, M., & Yan, J. (2017). Sustainable biofuel production from forestry, agricultural and waste biomass feedstocks. *Applied Energy, 198*, 281–283.

World Health Organization (2016) Ambient air pollution: A global assessment of exposure and burden of disease. Retrieved 11 April 2024, from https://www.who.int/publications/i/item/9789241511353

Xu, M., Yao, Q., Chen, D., Li, M., Li, R., Gao, B., Zhao, B., & Chen, Z. (2021). Estimating the impact of ground ozone concentrations on crop yields across China from 2014 to 2018: A multi-model comparison. *Environmental Pollution, 283*, 117099.

Yahya, M., Dutta, A., Bouri, E., Wadström, C., & Uddin, G. S. (2022). Dependence structure between the international crude oil market and the European markets of biodiesel and rapeseed oil. *Renewable Energy, 197*, 594–605.

Yap, K. Y., Chin, H. H., & Klemeš, J. J. (2022). Future outlook on 6G technology for renewable energy sources (RES). *Renewable and Sustainable Energy Reviews, 167*, 112722.

Yi, S., Abbasi, K. R., Hussain, K., Albaker, A., & Alvarado, R. (2023). Environmental concerns in the United States: Can renewable energy, fossil fuel energy, and natural resources depletion help? *Gondwana Research, 117*, 41–55.

Yildizbasi, A. (2021). Blockchain and renewable energy: Integration challenges in circular economy era. *Renewable Energy, 176*, 183–197.

A Bibliometric and Systematic Investigation on Hybrid Renewable Energy Systems for Sustainable Planetary Management

Ceylin Şirin and Azim Doğuş Tuncer

1 Introduction

Research on hybrid renewable energy systems has increased because of the pressing need for sustainable energy solutions to lessen the effects of climate change. With its innovative methodology, several renewable energy sources are seamlessly integrated, offering improved environmental sustainability, dependability, and efficiency. Hybridizing renewable energy sources, like solar, wind, hydropower, and hydrogen, can provide energy more efficiently and sustainably (Rekioua, 2019).

The cornerstones of modern renewable energy technologies are photovoltaic (PV) systems, which are skilled at converting sunlight into electricity, and wind energy, which is extracted from the kinetic power of the wind (Infield & Freris, 2020). However, a move toward the integration of various renewable technologies has been spurred by the intrinsic erratic nature of these sources as well as their vulnerability to variations in the weather. Additionally, combining wind and photovoltaic technology in hybrid systems appears to be a key tactic in overcoming the intermittency problem. A PV and wind hybrid system increases overall energy output and dependability by integrating their complementing features (Anoune et al., 2018). The fundamental component of this integration is sustainability, which maximizes various energy profiles and helps to create a more reliable and effective energy supply. Furthermore, the fusion of PV and solar thermal technology makes

C. Şirin (✉)
School of Engineering, College of Science and Engineering, University of Galway, Galway, Ireland

MaREI Centre, Ryan Institute & School of Engineering, College of Science and Engineering, University of Galway, Galway, Ireland
e-mail: C.Sirin1@universityofgalway.ie

A. D. Tuncer
Cooling Photonics S.L., Carrer Tirso de Molina 36, Cornella de Llobregat, Barcelona, Spain

S. S. Oncel (ed.), *A Green Vision Towards a Renewable Energy Future*,
https://doi.org/10.1007/978-3-031-93760-6_9

solar energy a more dependable and adaptable source. Even while solar photovoltaics (PV) produce cheap, sporadic power during the day, they need the assistance of sustainable power sources to provide constant, 24-h electricity. When surplus solar energy is extracted, solar thermal plants with thermal storage function as base load power plants and meet load demands for clean and sustainable power generation (Powell et al., 2017; Kalaiselvam & Parameshwaran, 2014). By effectively matching load fluctuations, a well-balanced combination of solar thermal (thermal storage type) with solar PV reduces the need for expensive battery storage (Manaserh et al., 2021; Liu et al., 2019). In a hybrid system, a reasonably priced solar PV plant supplies the modest amount of additional auxiliary power required by a solar thermal storage power plant throughout the day. In order to minimize power expenses, the solar thermal storage plant, which ensures continuous 24-h base load operation, takes the lead during non-daylight hours. During daylight hours, energy generation is predominantly provided by the cost-effective solar PV plant (Klise & Balfour, 2015; Tanner, 2003). In addition, surplus or intermittent electricity from other sources can be effectively used while the solar thermal storage plant is idle, because there is not enough sunlight in the area. In order to create a smooth and integrated approach to sustainable energy generation, this procedure entails transforming stored thermal energy into electricity during peak demand hours when power sale rates are most economical.

Furthermore, hydropower has long been a part of renewable energy portfolios, since it harnesses the kinetic energy of flowing water (Shah, 2014). When combined with wind energy in hybrid arrangements, it provides a varied energy mix that lessens the effects of seasonal changes and helps to create a more reliable power source (Ren et al., 2022; Jaramillo et al., 2004). The capacity of such systems to utilize natural, renewable resources without running out is indicative of their sustainability. Additionally, the combination of hydrogen and wind technologies captures the spirit of innovation and sustainability with a strong technical base. Hydrogen gas is produced by the deliberate use of excess wind energy for electrolysis, a process that breaks down water (Hassan et al., 2023; Zhang & Wan, 2014). There is a lot of promise for energy storage and usage with this clean, adaptable energy carrier. Advanced compression or liquefaction processes enable the stored hydrogen to function as an energy reservoir, thereby reducing the intermittent nature of renewable energy sources. Fuel cells could effectively convert stored hydrogen back into electricity during times of low wind or peak demand, thereby offering a dependable and uninterrupted energy source (Staffell et al., 2019). This complex interaction between hydrogen and wind technologies not only illustrates the complex technical solutions needed to usher in a new era of renewable energy integration, but it also contributes to a more resilient and sustainable energy paradigm (Khan et al., 2022; Acakpovi et al., 2020). Furthermore, it is feasible to combine more than two energy sources, such as solar, wave, and wind power (Reikard et al., 2015; Jahangir et al., 2020).

As mentioned in the previous parts, this imperative for hybridization extends beyond technological innovation; it embodies a strategic response to the pressing need for sustainability in planetary energy management. The inherent synergy in combining renewable sources addresses intermittency and reliability concerns while contributing to reducing carbon emissions and mitigating environmental impact. This study focuses on four different hybrid renewable energy systems including PVs and wind (Category 1), hydropower and wind (Category 2), PVs and solar thermal (Category 3), and wind and hydrogen (Category 4) hybrid renewable energy systems. The major aim of this chapter is to highlight the research trends and determine the research gaps in the field of hybrid renewable energy systems for sustainable planetary management using bibliometric data and systematic approach.

2 Methodology

The database utilized for this analysis was obtained from the Scopus database, which is the most esteemed and extensively utilized resource for doing systematic research (Scopus, n.d.). Moreover, VOSviewer software was used to analyze the data. The bibliometric analysis was conducted with the implementation of selected categories (C1, C2, C3, and C4), and the results are presented in the following sections. Collecting the data is the initial stage of the analysis. Some queries have been determined that enable the selected categories to be analyzed in more detail in Scopus. Search protocol tools were followed that included/excluded all relevant documents covering the specified queries. The "Article Title-Abstract-Keywords" search criterion was applied to all documents as a tool and the language was limited to English. For each specific category, a different query was formulated and shown in Table 1.

Table 1 Queries used for the categories on hybrid renewable energy systems and total number of achieved documents

Category	Query	Criteria	Overall number of documents
C1—Photovoltaics and wind hybrid energy systems	["Hybrid*" AND "Photovoltaics*" AND "Wind energy*"]	1. Article Title-Abstract-Keywords 2. Language limitation in English	326
C2—Hydropower and wind hybrid energy systems	["Hybrid*" AND "Hydropower*" AND "Wind energy*"]		91
C3—Photovoltaics and solar thermal hybrid energy systems	["Hybrid*" AND "Photovoltaics*" AND "Solar thermal*"]		213
C4—Wind and hydrogen hybrid energy systems	["Hybrid*" AND "Wind energy*" AND "Hydrogen*"]		378

3 Results and Discussion

3.1 Photovoltaics and Wind Hybrid Renewable Energy Systems

Figure 1 illustrates the number of publications during the period 1986–2024 (August 7, 2024) and top countries with documents on photovoltaics and wind hybrid renewable energy systems. Studies on the subject have increased rapidly from 2020. As can be seen from Fig. 1a, studies on this category have increased rapidly since 2020. Moreover, India has the large portion of the documents about C1 with 135 documents that correspond to 41.41% of the overall publications (Fig. 1b). India is followed by the United States, Saudi Arabia, Algeria, and Turkey with 23, 22, 18, and 16 scientific studies, respectively. The top three journals in C1 are *Renewable and Sustainable Energy Reviews, Energies,* and *Renewable Energ*y with 11, 9, and 8 publications, respectively. Moreover, 48.46% of the overall publications are research papers in this category and 44.17% of the total works are conference papers. Additionally, the institution that publishes the most in this category is Indian Institute of Technology, Delhi, with 8 publications in total.

C1 includes research and development activities in the field of renewable energy technology, with a primary emphasis on the production of power from solar and wind energy. The study found 241 terms that satisfied the threshold requirements, with "WIND POWER" coming out as the most common with 288 occurrences. With 176 instances, the phrase "PHOTOVOLTAICS" is used frequently, indicating the emphasis on photovoltaic systems for turning solar energy into electricity. "SOLAR POWER GENERATION" comes third with 159 occurrences. With 113 occurrences, "HYBRID SYSTEMS" also appear frequently, indicating a growing interest in combining various renewable energy sources for improved dependability and efficiency. Furthermore, "WIND ENERGY" attracts interest with 109 occurrences, suggesting continued research into improving wind farm operations and

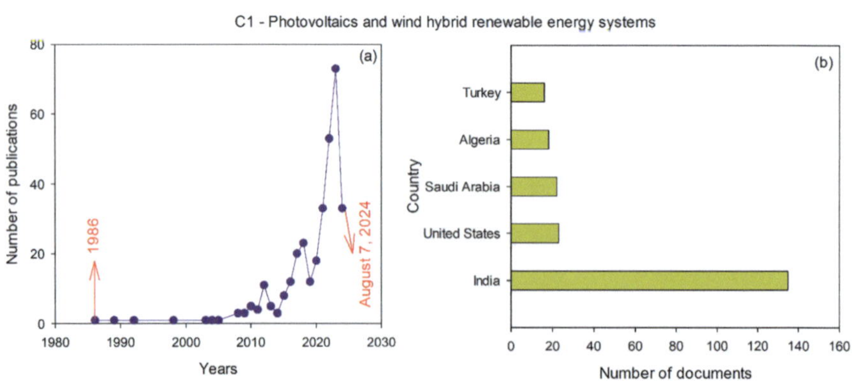

Fig. 1 (**a**) Number of publications on C1 (photovoltaics and wind hybrid renewable energy systems), (**b**) top countries with documents on C1

wind turbine technology. The fact that the word "SOLAR PHOTOVOLTAICS" occurs 100 times highlights the significance of photovoltaic technologies in relation to the larger picture of solar energy usage.

It is evident that major contributions to the field of hybrid renewable energy research combining PV and wind have been made when looking at the total number of citations from the top countries in C1. Out of the 69 countries examined, 24 match the requirements for inclusion in C1, which is based on the selection criterion of having at least 5 papers and 5 citations per country. With 135 publications and 2004 citations, India tops this list of countries, followed by the United States. With 23 publications and 1084 citations, Saudi Arabia comes in third place. According to the advanced search result in Scopus, the most cited research item in C1 is a review article written by Hepbasli (Hepbasli, 2008) in 2008 with 762 citations. If more related (hybridization of PV and wind energies) works on C1 are reviewed, it can be seen that these works focused on techno-economic investigation (Li et al., 2013) and optimization techniques (Hosseinalizadeh et al., 2016). Also, there are some recent works that were published in 2022 that received over 50 citations according to Scopus and they analyzed different cases in terms of optimized power point tracking (Kumar & Shivashankar., 2022) and application of machine learning techniques (Abualigah et al., 2022).

Co-occurrence diagrams of keywords for C1 are presented in Fig. 2. Figure 2a presents the co-occurrence of all keywords network visualization of the documents related to photovoltaics and wind hybrid energy systems (C1) structured with VOSviewer software. In this analysis, the relatedness of items was determined based on the number of documents in which they occur together. The weight of an object determines the size of the circle and label for that cluster. The label and circle of a cluster grow in proportion to its weight. The label may not be visible for some clusters. This means that the relevant word is widely used. The discipline to which a word belongs determines the color of the cluster. Keywords with the same colors form a cluster, representing similar topics. For instance, "WIND POWER" and

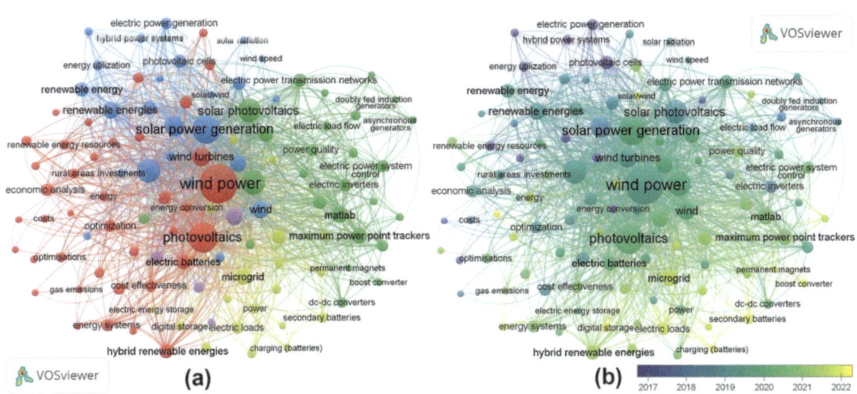

Fig. 2 Co-occurrence diagrams of keywords for C1: (**a**) all keywords network visualization map, (**b**) all keywords overlay visualization map

"PHOTOVOLTAICS" have the same color. This indicates that words with the same-colored cluster have similar topics. The green colored clusters are more related to electrical power systems. The yellow cluster generally related with the studies on batteries in C1. Also, filtering the terms with a minimum occurrence of 10 times produced the results displayed in the figure. Figure 2 (b) shows an overlay visualization on C1. When the legend is considered, the words that form the clusters with yellow color are the most used words in documents in 2022. Clusters' use of documents varies on a yearly basis, depending on the color distribution in the legend. For example, while "SECONDARY BATTERIES" and "MICROGRID" are among the most popular keywords for the studies published in 2022 and later years, "COSTS" and "PHOTOVOLTAIC CELLS" are among the more common words in documents in 2017. It can be stated that microgrids, digital storage, and battery systems are trending research topics in C1.

3.2 *Hydropower and Wind Hybrid Renewable Energy Systems*

Figure 3 shows the number of publications during the period 2004–2024 (August 7, 2024) and top countries with documents on hydropower and wind hybrid renewable energy systems. As can be seen, there is a huge increment in publications in 2023 on C2 (Fig. 1a). Furthermore, with 14 documents, or 15.38% of all publications, pertaining to C2, China possesses the big portion of these materials as can be seen from Fig. 1b. China is followed by India, Turkey, and Brazil with 11, 7, and 6 research items, respectively. *Energies, Renewable Energy, Applied Energy, Energy,* and *International Journal of Electrical Power and Energy Systems* are the top five journals with highest number of research items in C2 with 7, 5, 4, 4, and 4 publications, respectively. Furthermore, research papers account for 53.84% of all publications in this area, while conference papers make up 30.76% of all works. It should be noted

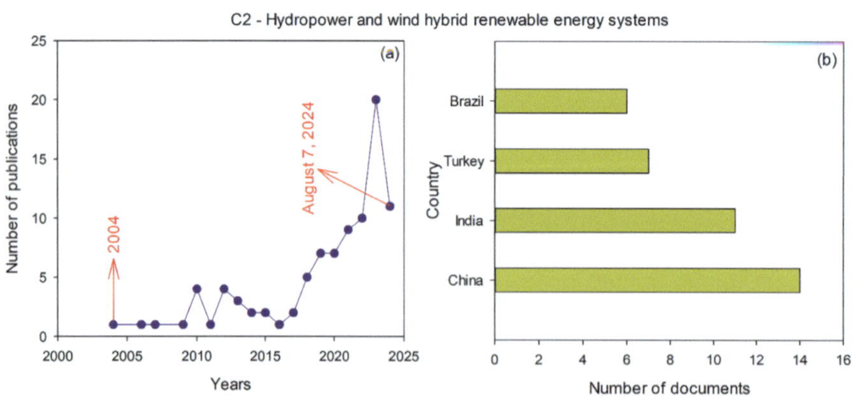

Fig. 3 (**a**) Number of publications on C2 (hydropower and wind hybrid renewable energy systems), (**b**) top countries with documents on C2

that there are 2 published review papers in this category. With a total of 3 publications each, University of Glasgow and Indian Institute of Technology, Roorkee, are the organizations that publish the most in this area.

It should be mentioned that C2 is concerned with integrating wind and hydropower systems to produce hybrid renewable energy sources. A total of 62 terms (of the 967 keywords) that met the threshold requirements were found through the co-occurrence analysis, demonstrating the variety of approaches being used in this field of study. With 72 occurrences, "WIND POWER" is the most often occurring keyword in C2, suggesting a strong focus on using wind energy in hybrid systems. The terms "HYDROELECTRIC POWER" and "HYDROELECTRIC POWER PLANTS" come in second and third, respectively, with 38 and 33 instances, underscoring the significance of hydropower in these hybrid arrangements. Furthermore, "WIND ENERGY" and "SOLAR ENERGY" each have 28 and 31 occurrences, respectively, indicating a possibility for combining solar and wind resources in hybrid renewable energy systems in a synergistic manner. Twenty-seven times, the word "HYBRID SYSTEMS" appears, suggesting that integrated renewable energy solutions are the main emphasis of design and optimization.

When total citations from top C2 countries are analyzed, it becomes clear that hydropower and wind hybrid renewable energy systems have made great strides. Based on a minimum of five documents and five citations per country, five of the thirty-six countries examined in C2 satisfy the requirements. With 14 publications and 276 citations, China is in the lead; India is second with 11 publications and 100 citations. Brazil has six articles and 162 citations, while Turkey gives seven publications and 145 citations. Spain has 5 publications and 92 citations to fill out the list. In C2, these nations are essential for promoting research and development in hydropower and wind hybrid renewable energy systems. The most cited (349 citations) study in C2 is a research work that was performed by Bekele and Tadesse (Bekele & Tadesse, 2012) in 2012. In the study, feasibility of a small-sized Hydro-PV and wind hybrid renewable energy system has been analyzed in order to provide electricity for rural Ethiopia as an off-grid solution. In addition, some of the recent most cited works analyzed the long period impacts of hydropower-wind hybrid system using programming techniques (Xu et al., 2019) and hydrogen generation potential of hydropower-wind hybrid systems for Brazil (Nadaleti et al., 2020).

All keywords-co-occurrence diagrams for C2 are shown in Fig. 4. According to the network visualization map (Fig. 4a), same-colored clusters are located in the same areas. This means that these words do not go beyond the discipline covered in the documents in which they are used. The dark blue-colored cluster represents words used mostly in the fields of monitoring-energy production. In addition, it is clearly seen that the words representing the green cluster are mostly used in hybrid system applications. It can be said that the yellow cluster is more related with optimization studies. Figure 4b shows that in recent years, optimization of hydropower is one of the popular research areas in literature. Also, it can be concluded from Fig. 4b that optimization techniques and machine learning approaches are trending research topics in C2.

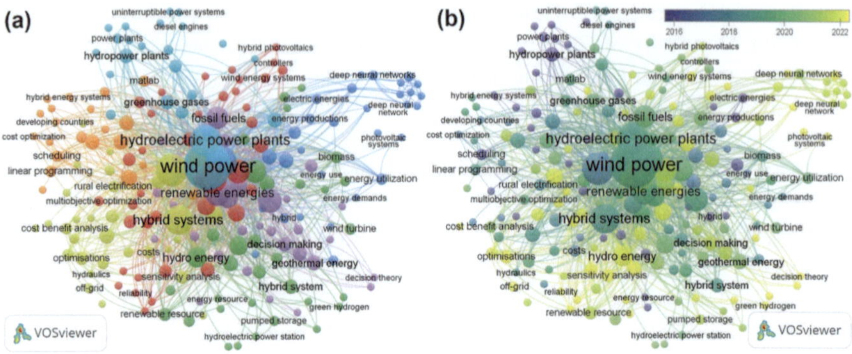

Fig. 4 Co-occurrence diagrams of keywords for C2: (**a**) all keywords network visualization map, (**b**) all keywords overlay visualization map

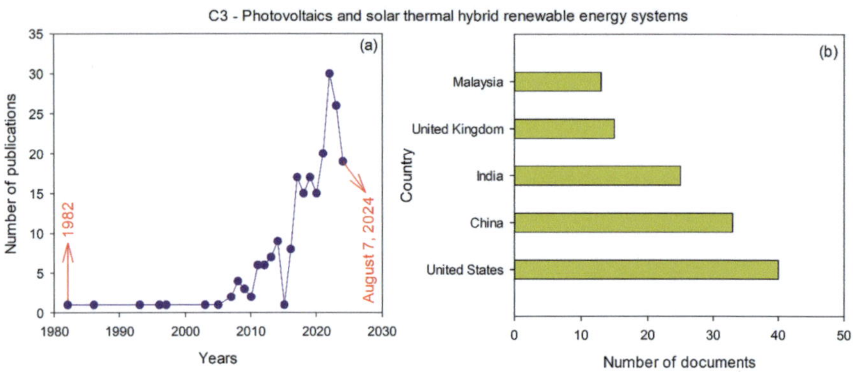

Fig. 5 (**a**) Number of publications on C3 (photovoltaics and solar thermal hybrid renewable energy systems), (**b**) top countries with documents on C3

3.3 Photovoltaics and Solar Thermal Hybrid Renewable Energy Systems

Figure 5 depicts the number of publications during the period 1982–2024 (August 7, 2024) and top countries with documents on photovoltaics and solar thermal hybrid renewable energy systems. After a slow beginning in the 1980s and 1990s with few publications annually, research production has increased significantly in the last several years as can be seen from Fig. 5a. On the other hand, there appear to be peaks and troughs in the trend of publications. With 30 research items published, 2022 has the most publications in the past few years, closely followed by 2023 with 26 publications. These figures demonstrate the increasing interest in investigating the possibility of integrating solar thermal and photovoltaic technology to improve renewable energy systems. With 40 publications, the United States contributed roughly 18.77% of the 213 publications globally in C3. China came in second with about 15.49% of the total, or 33 articles. With 25 papers, India contributed almost 11.7%, whereas Malaysia and the UK contributed roughly 7.04% (15 publications) and 6.10% (13 publications), respectively. Moreover, the top three journals with the highest number of research items in C3 are *Solar Energy, Energies, and Energy Conversion and Management* with 15, 12, and 10 publications, respectively. Also, research papers cover 51.64% of the overall publications in C3. Also, Imperial College London and Chinese Academy of Sciences have the highest number of publications as affiliations in C3 (8 publications each).

It can be observed that there is a clear emphasis on hybrid approaches for enhancing solar energy use among the 151 keywords that match the threshold condition (C3). According to the results of the analysis of the co-occurrence of the keywords, with 152 instances, "SOLAR POWER GENERATION" is the most often occurring keyword in C3, highlighting the importance of producing electricity using solar energy sources. The terms "PHOTOVOLTAICS" (101 occurrences) and "SOLAR ENERGY" (92 occurrences) are closely behind, emphasizing the role that photovoltaic technologies play in turning sunshine into power. Additionally, the terms "SOLAR THERMAL" (52 occurrences) and "SOLAR HEATING" (85 occurrences) indicate a noteworthy interest in using solar thermal energy for heating purposes in hybrid systems. A total of 57 instances of the term "HYBRID SYSTEMS" indicate an emphasis on combining solar thermal and photovoltaic technologies to maximize energy output and efficiency. The co-occurrence study highlights the increasing interest in fusing solar thermal and photovoltaic technologies to produce flexible and effective renewable energy solutions.

It is clear that significant advancements have been made in photovoltaics and solar thermal hybrid renewable energy systems after reviewing all of the citations from important nations in C3. Sixteen of the 61 countries examined in C3 satisfy the requirements, as determined by the presence of at least 5 documents and 5 citations in each country. With 39 publications and 923 citations, the United States leads the pack; China is next, with 33 publications and 837 citations. India has 1414 citations and 25 publications, which is a noteworthy presence. With 13 papers and 358 citations, Malaysia makes a contribution. The most cited research item in C3 is also the same review study in C1 (Hepbasli, 2008). When more specific and related works on PV and solar thermal hybrid systems that are highly cited are analyzed, it is possible to see most of the studies focused on hybrid photovoltaic-thermal (PVT) systems that generates both thermal and electrical energies from the same aperture area (Tyagi et al., 2012; Kalogirou & Tripanagnostopoulos, 2007; Erdil et al., 2008; Said et al., 2018; Herez et al., 2020; Good, 2016).

Figure 6 illustrates all keywords-co-occurrence diagrams for C3. It is clearly seen that many clusters with the same colors are intertwined in Fig. 6a. It indicates that word clusters with different colors are used together in the same documents, even if they have different disciplines. For example, even though the green cluster generally represents the energy systems and power topics, the words used in this cluster have been used many times in the same document with the red clusters, which belong to the solar hybrid system applications discipline. Figure 6b shows that keywords related to thermal energy applications were quite popular during the

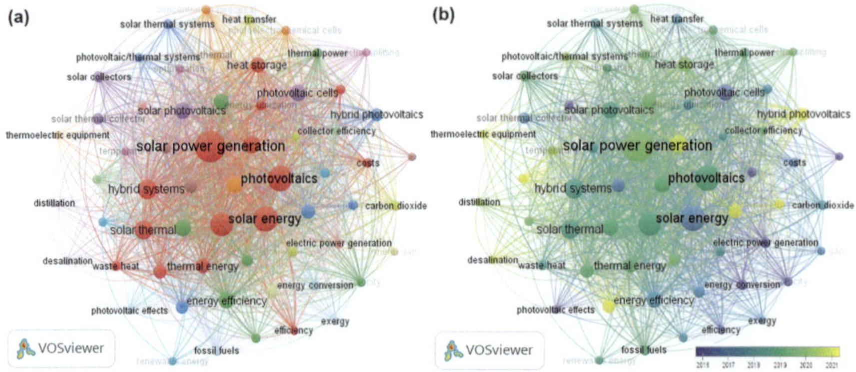

Fig. 6 Co-occurrence diagrams of keywords for C3: (**a**) all keywords network visualization map, (**b**) all keywords overlay visualization map

period 2018–2019. The use of words in the field of energy storage has increased in recent years, especially in 2020. This indicates that there are innovative studies being carried out in the C3 category day by day. As mentioned earlier, studies in C3 are mostly related to hybrid PVTs. Since these systems are used in many different applications and varying types of analyzing techniques are applied to PVTs, the clusters are nested together. Similar behavior can be seen also in Fig. 4b, which represents the overlay visualization map. However, it can be concluded that desalination applications that are done by hybrid PV and solar thermal systems and usage of thermoelectric equipment in the hybrid PV and solar thermal renewable energy systems are trending research topics for C3.

3.4 Wind and Hydrogen Hybrid Renewable Energy Systems

The number of publications during the period 1985–2024 (August 7, 2024) and top countries with documents on wind and hydrogen hybrid renewable energy systems are presented in Fig. 7. There is a rising interest in merging wind power and hydrogen technology for sustainable energy solutions, as seen by the constant upward trend in research output in wind and hydrogen hybrid renewable energy systems. The years 2022 and 2023 saw the highest number of publications, with 58 and 57 research items, respectively. The overall trend indicates continued research and awareness of the potential of these hybrid systems to address energy concerns and promote sustainability. With roughly 19.31% of the 378 publications published worldwide in C4, or 73 publications, China is the top contributor. The United States is in second place with about 10.84% of the total with 41 publications, and India is involved with about 9.52% with 36 publications. Notable contributions are also made by Turkey and Iran, with 28 and 30 publications and 7.40% and 7.93% of the

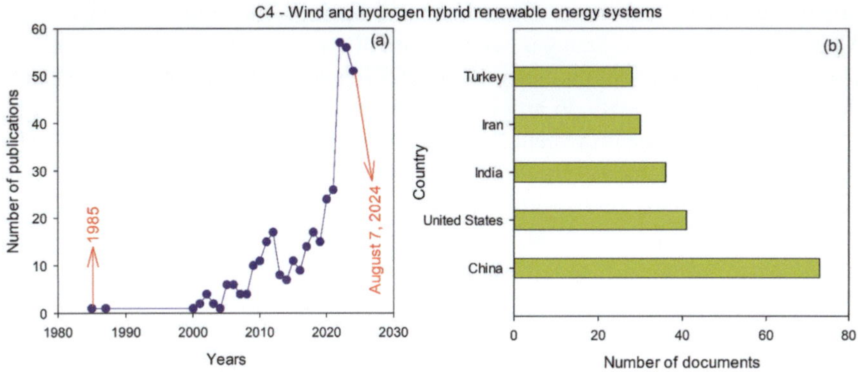

Fig. 7 (**a**) Number of publications on C4 (wind and hydrogen hybrid renewable energy systems), (**b**) top countries with documents on C4

total, respectively. Moreover, the top three journals in C4 are *International Journal of Hydrogen Energy, Energy Conversion and Management,* and *Applied Energy* with 56, 19, and 15 publications, respectively. Also, there are 26 review papers that are published in this category and research items cover 63.49% of the overall research items. Moreover, Ontario Tech University is the leading institution with 11 research items in total in C4.

One trend that stands out in C4's 239 keywords (out of 2959) that fit the criterion is the emphasis on using wind power in conjunction with hydrogen-related applications as a result of the keyword co-occurrence survey. With 278 occurrences, "WIND POWER" is the most often occurring keyword in C4, suggesting a significant focus on using wind energy in hybrid systems. "HYDROGEN STORAGE" (135 occurrences) and "HYDROGEN" (121 occurrences) follow closely behind, emphasizing the significance of hydrogen as an energy carrier in these hybrid configurations. Furthermore, "FUEL CELLS" (115 occurrences) represent the possibility of employing hydrogen fuel cells as a clean energy conversion technology within hybrid systems, while "WIND ENERGY" (117 occurrences) emphasizes the importance of wind power generation. "HYDROGEN PRODUCTION" (111 instances) represents conversations on producing hydrogen using sustainable energy sources, such as wind power. Furthermore, the terms "SOLAR ENERGY" and "SOLAR POWER GENERATION" imply the possibility of combining solar energy with wind and hydrogen technologies in hybrid renewable energy systems. The term "HYBRID SYSTEMS" is present 78 times, suggesting that integrated renewable energy solutions are the main focus of design and optimization. The co-occurrence study reveals a growing interest in C4 renewable energy systems that combine hydrogen and wind energy to be resilient and sustainable.

When looking at the total number of citations from well-known countries in C4, it is clear that research on hydrogen and wind hybrid renewable energy systems has advanced significantly. Twenty six of the 76 countries that were examined for C4 satisfy the requirements based on each country having a minimum of 5 papers and 5 citations. The United States is second with 40 publications and 2559 citations, closely behind China's 73 publications and 2277 citations. India exhibits a noteworthy presence as well, with 36 articles and 735 citations. With 30 publications and 1729 citations, Iran leads the field in contributions, followed closely by Turkey. The most cited (1189 citations) article of all times in C4 is published by Hoffert et al. (2002) in 2002. In the study, various types of low carbon energy technologies have been surveyed and presented. Also, there are several highly cited studies in C4 that focused on PV-wind-hydrogen combination (Baghaee et al., 2016; Al-Sharafi et al., 2017; Zhang et al., 2019; Armijo & Philibert, 2020; Duman & Güler, 2018). Moreover, some highly cited studies analyzed hydrogen storage systems with wind power (Khan & Iqbal, 2005; Korpas & Holen, 2006), wind potential estimation for hydrogen production (Iqbal et al., 2019), and modeling simulation of small-scale wind-fuel cell combination.

Co-occurrence diagrams of keywords for C4 are illustrated in Fig. 8. Red cluster generally represents fuel cells, and it has huge field in Fig. 8a. This clearly shows that the documents that represent C4 are mostly related to this discipline. Moreover, green cluster is related to electrolysis and economic analyses. According to Fig. 8(b), the keywords that were used between 2016 and 2018 are directly related to each other. For example, there are many connections between the keywords used in the studies conducted in the field of hydrogen in 2016 and the keywords used in the field of electrical energy storage in 2018. This is proof that even though years have passed, and new technological developments have occurred, the words discussed in previous studies are still a good reference. Also, as can be seen from Fig. 8b, optimization techniques in wind energy and hydrogen hybrid energy systems have been trending for recent years.

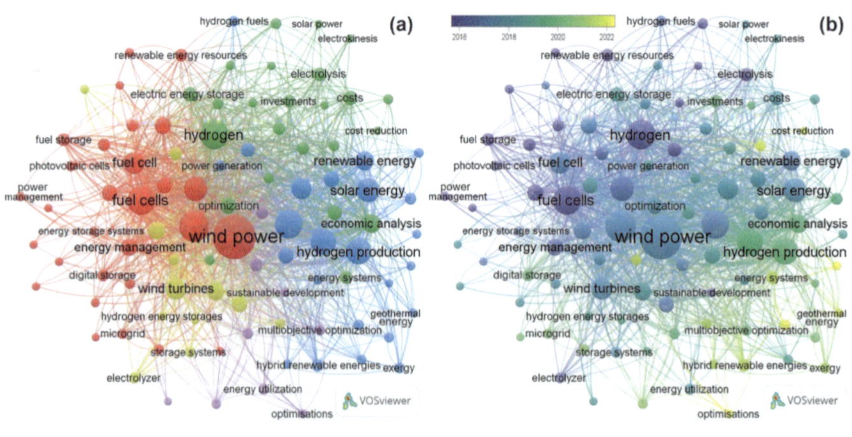

Fig. 8 Co-occurrence diagrams of keywords for C4: (**a**) all keywords network visualization map, (**b**) all keywords overlay visualization map

4 Conclusion and Future Perspectives

By utilizing the complementing qualities of several renewable energy sources, hybrid renewable energy systems have great potential to address the issues of energy security and climate change. However, further study is required to get over a number of technological, financial, and legal obstacles before they can completely realize their potential. As a result of the bibliometric analyses for the selected hybrid renewable energy system categories, optimization techniques and application of artificial intelligence methods are trending, and it can be said that these approaches are research hotspots for the structured categories. These methods can be quite useful also in the future for analyzing the bottlenecks in hybrid renewable energy systems in terms of design, application, and operation. The creation of sophisticated control algorithms and optimization strategies to successfully integrate various renewable energy sources within hybrid systems is a crucial field of future research. By using dynamic optimization techniques, these initiatives should concentrate on increasing system stability, reducing intermittency, and optimizing energy production.

It should be noted that complex mathematical models and simulation tools are also desperately needed in order to effectively depict the intricate relationships that exist among different renewable energy sources, storage systems, and the grid. With the use of these models, researchers will be able to quantify system vulnerabilities and create mitigation plans in addition to conducting thorough resilience and reliability assessments. Furthermore, for hybrid renewable energy systems to improve the accuracy of energy generation projections, advances in resource evaluation and forecasting methods are critical. Predictions can be improved by incorporating machine learning and sophisticated meteorological models, which can also optimize grid integration and system performance. It will also take more investigation to find novel design approaches and optimization frameworks to figure out how best to arrange, size, and position the components in hybrid renewable energy systems. To enable broad deployment, this entails assessing the regulatory ramifications, environmental impact, and economic feasibility.

In terms of economics, the implementation of hybrid renewable energy systems frequently entails large upfront expenditures, especially for grid integration infrastructure and energy storage technologies. The widespread adoption of hybrid systems may be significantly hampered by these expenses. However, more affordable alternatives that help lower these financial obstacles may be developed with further study and innovation. Long-term advantages like lowered greenhouse gas emissions and operating expenses should be taken into account when evaluating the economics of hybrid systems, as these can eventually increase the appeal of these installations.

Additionally, in order to examine the implications of implementing hybrid renewable energy systems, including their effects on land usage, wildlife habitats, water resources, and local communities, environmental and social impact assessments are crucial. To lessen any negative effects, solutions for community engagement, biodiversity conservation, and sustainable siting must be established. When the visualization studies conducted for all four categories are examined, the lack of environmental analyzes compared to other subjects is evident. For example, life cycle assessments, or LCAs, can offer insightful information about how hybrid renewable energy systems perform in terms of sustainability and their effects on the environment. LCAs can assist in informing decision-making processes by determining hotspots in the supply chain, measuring greenhouse gas emissions, and evaluating trade-offs between various environmental indicators.

Moreover, to tackle the intricate problems pertaining to market dynamics, grid integration, and the legislative frameworks controlling hybrid renewable energy systems, multidisciplinary research is required. The intricacies of hybrid systems, which necessitate the coordinated control of numerous energy sources and the integration of new technologies into current grids, may be beyond the capabilities of current energy regulations. Consequently, it is critical to create rules that encourage the adoption of hybrid systems and remove any legal barriers. These regulations might offer grid integration standards, expedited approval procedures, and financial incentives for renewable energy projects. Scientists and decision-makers may expedite the shift toward a more resilient and sustainable energy future by filling in these research gaps, which will also enable hybrid renewable energy systems to reach their full potential in reducing climate change and guaranteeing energy security.

Acknowledgments No funding was used in this research. The bibliometric data used was provided using advanced search option in Scopus. Moreover, proper machine learning tools were used for language editing and correction.

References

Abualigah, L., Zitar, R. A., Almotairi, K. H., Hussein, A. M., Abd Elaziz, M., Nikoo, M. R., & Gandomi, A. H. (2022). Wind, solar, and photovoltaic renewable energy systems with and without energy storage optimization: A survey of advanced machine learning and deep learning techniques. *Energies, 15*(2), 578.

Acakpovi, A., Adjei, P., Nwulu, N., & Asabere, N. Y. (2020). Optimal hybrid renewable energy system: A comparative study of wind/hydrogen/fuel-cell and wind/battery storage. *Journal of Electrical and Computer Engineering, 2020*, 1–15.

Al-Sharafi, A., Sahin, A. Z., Ayar, T., & Yilbas, B. S. (2017). Techno-economic analysis and optimization of solar and wind energy systems for power generation and hydrogen production in Saudi Arabia. *Renewable and Sustainable Energy Reviews, 69*, 33–49.

Anoune, K., Bouya, M., Astito, A., & Abdellah, A. B. (2018). Sizing methods and optimization techniques for PV-wind based hybrid renewable energy system: A review. *Renewable and Sustainable Energy Reviews, 93*, 652–673.

Armijo, J., & Philibert, C. (2020). Flexible production of green hydrogen and ammonia from variable solar and wind energy: Case study of Chile and Argentina. *International Journal of Hydrogen Energy, 45*(3), 1541–1558.

Baghaee, H. R., Mirsalim, M., Gharehpetian, G. B., & Talebi, H. A. (2016). Reliability/cost-based multi-objective Pareto optimal design of stand-alone wind/PV/FC generation microgrid system. *Energy, 115*, 1022–1041.

Bekele, G., & Tadesse, G. (2012). Feasibility study of small hydro/PV/wind hybrid system for off-grid rural electrification in Ethiopia. *Applied Energy, 97*, 5–15.

Duman, A. C., & Güler, Ö. (2018). Techno-economic analysis of off-grid PV/wind/fuel cell hybrid system combinations with a comparison of regularly and seasonally occupied households. *Sustainable Cities and Society, 42*, 107–126.

Erdil, E., Ilkan, M., & Egelioglu, F. (2008). An experimental study on energy generation with a photovoltaic (PV)–solar thermal hybrid system. *Energy, 33*(8), 1241–1245.

Good, C. (2016). Environmental impact assessments of hybrid photovoltaic–thermal (PV/T) systems–A review. *Renewable and Sustainable Energy Reviews, 55*, 234–239.

Hassan, Q., Sameen, A. Z., Salman, H. M., & Jaszczur, M. (2023). Large-scale green hydrogen production via alkaline water electrolysis using solar and wind energy. *International Journal of Hydrogen Energy, 48*(88), 34299–34315.

Hepbasli, A. (2008). A key review on exergetic analysis and assessment of renewable energy resources for a sustainable future. *Renewable and sustainable energy reviews, 12*(3), 593–661.

Herez, A., El Hage, H., Lemenand, T., Ramadan, M., & Khaled, M. (2020). Review on photovoltaic/thermal hybrid solar collectors: Classifications, applications and new systems. *Solar Energy, 207*, 1321–1347.

Hoffert, M. I., Caldeira, K., Benford, G., Criswell, D. R., Green, C., Herzog, H., Jain, A. K., Kheshgi, H. S., & Lackner, K. S. (2002). Advanced technology paths to global climate stability: Energy for a greenhouse planet. *Science, 298*(5595), 981–987.

Hosseinalizadeh, R., Shakouri, H., Amalnick, M. S., & Taghipour, P. (2016). Economic sizing of a hybrid (PV–WT–FC) renewable energy system (HRES) for stand-alone usages by an optimization-simulation model: Case study of Iran. *Renewable and Sustainable Energy Reviews, 54*, 139–150.

Infield, D., & Freris, L. (2020). *Renewable energy in power systems*. John Wiley & Sons.

Iqbal, W., Yumei, H., Abbas, Q., Hafeez, M., Mohsin, M., Fatima, A., Jamali, M. A., Jamali, M., Siyal, A., & Sohail, N. (2019). Assessment of wind energy potential for the production of renewable hydrogen in Sindh Province of Pakistan. *Processes, 7*(4), 196.

Jahangir, M. H., Shahsavari, A., & Rad, M. A. V. (2020). Feasibility study of a zero emission PV/wind turbine/wave energy converter hybrid system for stand-alone power supply: A case study. *Journal of Cleaner Production, 262*, 121250.

Jaramillo, O. A., Borja, M. A., & Huacuz, J. M. (2004). Using hydropower to complement wind energy: A hybrid system to provide firm power. *Renewable Energy, 29*(11), 1887–1909.

Kalaiselvam, S., & Parameshwaran, R. (2014). *Thermal energy storage technologies for sustainability: Systems design, assessment and applications*. Elsevier.

Kalogirou, S. A., & Tripanagnostopoulos, Y. (2007). Industrial application of PV/T solar energy systems. *Applied Thermal Engineering, 27*(8-9), 1259–1270.

Khan, M. J., & Iqbal, M. T. (2005). Dynamic modeling and simulation of a small wind–fuel cell hybrid energy system. *Renewable Energy, 30*(3), 421–439.

Khan, T., Yu, M., & Waseem, M. (2022). Review on recent optimization strategies for hybrid renewable energy system with hydrogen technologies: State of the art, trends and future directions. *International Journal of Hydrogen Energy, 47*(60), 25155–25201.

Klise, G. T., & Balfour, J. (2015). *A best practice for developing availability guarantee language in photovoltaic (PV) O&M Agreements (No. SAND2015-10223)*. Sandia National Lab. (SNL-NM).

Korpas, M., & Holen, A. T. (2006). Operation planning of hydrogen storage connected to wind power operating in a power market. *IEEE Transactions on Energy Conversion, 21*(3), 742–749.

Kumar, G. A., & Shivashankar. (2022). Optimal power point tracking of solar and wind energy in a hybrid wind solar energy system. *International Journal of Energy and Environmental Engineering, 13*(1), 77–103.

Li, C., Ge, X., Zheng, Y., Xu, C., Ren, Y., Song, C., & Yang, C. (2013). Techno-economic feasibility study of autonomous hybrid wind/PV/battery power system for a household in Urumqi, China. *Energy, 55*, 263–272.

Liu, J., Chen, X., Cao, S., & Yang, H. (2019). Overview on hybrid solar photovoltaic-electrical energy storage technologies for power supply to buildings. *Energy Conversion and Management, 187*, 103–121.

Manaserh, Y. M. A., Abubaker, A. M., Ahmad, A. D., Ata, A. B., Najjar, Y. S., & Akafuah, N. K. (2021). Assessment of integrating hybrid solar-combined cycle with thermal energy storage for shaving summer peak load and improving sustainability. *Sustainable Energy Technologies and Assessments, 47*, 101505.

Nadaleti, W. C., Dos Santos, G. B., & Lourenço, V. A. (2020). The potential and economic viability of hydrogen production from the use of hydroelectric and wind farms surplus energy in Brazil: A national and pioneering analysis. *International Journal of Hydrogen Energy, 45*(3), 1373–1384.

Powell, K. M., Rashid, K., Ellingwood, K., Tuttle, J., & Iverson, B. D. (2017). Hybrid concentrated solar thermal power systems: A review. *Renewable and Sustainable Energy Reviews, 80*, 215–237.

Reikard, G., Robertson, B., & Bidlot, J. R. (2015). Combining wave energy with wind and solar: Short-term forecasting. *Renewable Energy, 81*, 442–456.

Rekioua, D. (2019). *Hybrid renewable energy systems: Optimization and power management control*. Springer Nature.

Ren, Y., Yao, X., Liu, D., Qiao, R., Zhang, L., Zhang, K., Jin, K., Li, H., Ran, Y., & Li, F. (2022). Optimal design of hydro-wind-PV multi-energy complementary systems considering smooth power output. *Sustainable Energy Technologies and Assessments, 50*, 101832.

Said, Z., Arora, S., & Bellos, E. (2018). A review on performance and environmental effects of conventional and nanofluid-based thermal photovoltaics. *Renewable and Sustainable Energy Reviews, 94*, 302–316.

Scopus (n.d.). Retrieved from https://www.scopus.com/home.uri.

Shah, Y. T. (2014). *Water for energy and fuel production*. Taylor & Francis.

Staffell, I., Scamman, D., Abad, A. V., Balcombe, P., Dodds, P. E., Ekins, P., Shah, N., & Ward, K. R. (2019). The role of hydrogen and fuel cells in the global energy system. *Energy & Environmental Science, 12*(2), 463–491.

Tanner, A. R. (2003). *Application of underground thermal energy storage for solar thermal power generation in New South Wales*. School of Aerospace, Mechanical and Mechatronic Engineering, The University of Sydney.

Tyagi, V. V., Kaushik, S. C., & Tyagi, S. K. (2012). Advancement in solar photovoltaic/thermal (PV/T) hybrid collector technology. *Renewable and Sustainable Energy Reviews, 16*(3), 1383–1398.

Xu, B., Zhu, F., Zhong, P. A., Chen, J., Liu, W., Ma, Y., Guo, L., & Deng, X. (2019). Identifying long-term effects of using hydropower to complement wind power uncertainty through stochastic programming. *Applied Energy, 253*, 113535.

Zhang, G., & Wan, X. (2014). A wind-hydrogen energy storage system model for massive wind energy curtailment. *International Journal of Hydrogen Energy, 39*(3), 1243–1252.

Zhang, W., Maleki, A., Rosen, M. A., & Liu, J. (2019). Sizing a stand-alone solar-wind-hydrogen energy system using weather forecasting and a hybrid search optimization algorithm. *Energy Conversion and Management, 180*, 609–621.

Index

S. S. Oncel (ed.), *A Green Vision Towards a Renewable Energy Future*,
https://doi.org/10.1007/978-3-031-93760-6

MIX
Papier aus verantwortungsvollen Quellen
Paper from responsible sources
FSC® C105338

If you have any concerns about our products,
you can contact us on
ProductSafety@springernature.com

In case Publisher is established outside the EU,
the EU authorized representative is:
Springer Nature Customer Service Center GmbH
Europaplatz 3, 69115 Heidelberg, Germany

Printed by Libri Plureos GmbH
in Hamburg, Germany